Evolutionary Trends

edited by
Kenneth J. McNamara

The University of Arizona Press
Tucson

First published in Great Britain in 1990 by
Belhaven Press (a division of Pinter Publishers),
25 Floral Street, London WC2E 9DS

Published by arrangement with Belhaven Press:
THE UNIVERSITY OF ARIZONA PRESS
1230 North Park Avenue, Suite 102,
Tucson, Arizona 85719-4140

Library of Congress Cataloging-in-Publication data have been requested.

British Library Cataloguing in Publication Data are available.

ISBN: 0-8165-1233-7 (cloth)
      0-8165-1234-5 (paper)

Printed in Great Britain

CONTENTS

LIST OF FIGURES

LIST OF TABLES

LIST OF CONTRIBUTORS

Robert L. Anstey, Department of Geological Sciences, Michigan State University, East Lansing, Michigan 48824–1115, USA.

Michael J. Benton, Department of Geology, University of Bristol, Queen's Road, Bristol BS8 1RJ, UK.

John Damuth, Department of Biology, University of California, Santa Barbara, California 93106, USA.

Jean-Louis Dommergues, UA CNRS 157, Centre des Sciences de la Terre, Université de Bourgogne, 6 boulevard Gabriel, F-21100 Dijon, France.

Richard A Fortey, Department of Palaeontology, British Museum (Natural History), Cromwell Road, London SW7 5BD, UK.

Stephen J. Gould, Museum of Comparative Zoology, Harvard University, Cambridge, Massachusetts 02138, USA.

Christine M Janis, Division of Biology and Medicine, Brown University, Providence, Rhode Island 02912, USA.

John A. Long, Department of Earth and Planetary Sciences, Western Australian Museum, Francis Street, Perth, Western Australia 6000, Australia.

Michael L. McKinney, Department of Geological Sciences, University of Tennessee, Knoxville, Tennessee 37996–1410, USA.

Kenneth J. McNamara, Department of Earth and Planetary Sciences, Western Australian Museum, Francis Street, Perth, Western Australia 6000, Australia.

Arnold I. Miller, Department of Geology, University of Cincinnati, Cincinnati, Ohio 45221–0013, USA.

Robert M. Owens, Department of Geology, National Museum of Wales, Cardiff CF1 3NP, UK.

Michael J. Simms, Department of Geology, Trinity College, Dublin 2, Ireland.

'The past has revealed to me the structure of the future'

Pierre Teihard de Chardin

PREFACE

Kenneth J. McNamara

The study of the history of life on this planet encompasses the origins of species to their demise: evolution and extinction. This latter aspect has been the main focus of attention since the late 1970s, and has been treated by a host of textbooks. Since the earlier 1970s, evolutionary studies have been dominated by analyses of rates of evolution. However, there is another fundamental aspect of life's history that the fossil record is ideally placed to deal with, yet which has received relatively little attention: the directionality of evolution. Arguably this is one of the most important aspects of evolution. As Mike McKinney points out in Chapter 2, it is more important to know where evolutionary change is going, rather than how fast it goes.

Patterns of directionality revealed by the fossil record are usually described as 'evolutionary trends'. This is a rather vague term which has been used in a number of ways to describe everything from microevolutionary changes within single species lineages, to large-scale macroevolutionary trends. Many 'classic' evolutionary trends, such as the simplistic picture of horses evolving through just five forms, from *Hyracotherium* to *Equus* have, quite rightly, come under attack, largely because of the confusion between small-scale and larger-scale trends (see Chapter 1). Arguments that few gradual interspecific trends can actually be documented from the fossil record have further seen the importance of evolutionary trends assigned a relatively minor place in modern evolutionary studies. Many trends described from the fossil record have been said to be little more than the fanciful wanderings of some palaeontologists' trend-orientated imaginations, being perhaps just the outcome of narrow conceptual preconditioning.

Classically, evolutionary trends have been seen as being adaptive. The great evolutionary biologist, George Gaylord Simpson, wrote in his book *The*

Major Features of Evolution, published in 1953, that 'all long- and most short-range trends consistent in direction are adaptively orientated.' The implication is that all trends arise from extrinsically induced natural selection: only those better adapted survive to continue the lineage. However, recent ideas, such as Steve Stanley's idea of 'species selection', Elizabeth Vrba's 'effect hypothesis', Steve Gould and Richard Lewontin's non-adaptive explanation of trends, and Steve Gould's more recent suggestion that changes in variance can explain many trends, have seriously questioned the role of adaptation, particularly in evolutionary trends at higher taxonomic levels. Yet papers are still being published in the palaeontological literature which interpret trends in an adaptionist light. Who is right? Or are both schools of thought correct? Could it be that trends seen at different taxonomic levels can be generated in different ways? To just what extent are the patterns of evolutionary trends between species reflected at higher taxonomic levels? And what are the mechanisms that drive these trends?

It is the aim of this book to try and answer these questions by looking at the dynamics of evolutionary trends and presenting a synthesis of evolutionary trends in a range of fossil taxa, with suggestions of the mechanisms that led to their generation. The book is divided into three parts. Part One, called 'The Dynamics of Evolutionary Trends', contains four chapters that look at trends from a theoretical viewpoint. The introductory chapter examines the question of the operation of evolutionary trends at different hierarchical levels. A classification and quantitative analysis of evolutionary trends is then presented in Chapter 2. This is followed by an examination of the role of developmental changes, notably heterochrony, in trend generation (Chapter 3) and a chapter on trends in body-size evolution (Chapter 4). Parts Two and Three deal with invertebrates and vertebrates, respectively. Arthropods are represented by trilobites (Chapter 5); molluscs by bivalves (Chapter 6) and ammonoids (Chapter 7); echinoderms by crinoids (Chapter 8) and echinoids (Chapter 9); and colonial organisms by bryozoans (Chapter 10). The vertebrates that are dealt with in Part Three are fishes (Chapter 11), reptiles (Chapter 12) and mammals (Chapter 13). Following Part Three there is a short epilogue in which I comment on some of the ideas engendered by the chapters in order to see if any patterns emerge. Lastly, a glossary is provided of some of the conceptual terms used in the book.

I apologise to those readers who had hoped to see chapters on other groups, such as brachiopods, graptolites or corals. Hopefully some of the ideas presented in this book might engender studies of evolutionary trends in some of these other groups. The criteria that I used for selecting chapters were many and various. They included completeness of the fossil record and a worker in the field who had an active interest in evolutionary trends in the particular group, combined, of course, with that person's availability to contribute a chapter. In the case of invertebrates my aim was to cover a range of organisms showing different growth strategies. Colonial organisms are covered by bryozoans; solitary organisms that grow periodically by trilobites; solitary organisms that grow by accretion by one extinct group of molluscs, the ammonoids, and one extant group of molluscs, the bivalves. Two groups (echinoids and crinoids) within a single phylum (the Echinodermata) are examined in order to compare and contrast patterns of evolutionary trends in a phylum exhibiting more than one growth strategy—growth involves differentiation of multiple elements (e.g. plates and spines) which subsequently grow by accretion.

When I approached the various chapter authors I asked them to look at evolutionary trends from a number of points of view. Depending on the groups concerned, some of these topics are covered more extensively in some chapters than in others, depending on the amount of research that has been carried out in the particular areas. I asked the authors to look, as far as possible, at evolutionary trends at different taxonomic levels, from the interspecific to the ordinal, in order to see if patterns would emerge to indicate whether different processes operate at different levels or whether those seen at higher levels reflect those generated at lower levels. In view of the considered importance of cladogenesis in speciation, I also requested authors to document, where possible, examples of anagenesis to determine its frequency and to examine the importance of cladogenesis and changing variance as important factors in controlling evolutionary trends.

The next request was for authors to look at which characters had been selected for in trends and to see how far they could go in interpreting why trends had gone in certain directions. In particular, what were the principal targets of selection: were they morphology alone, or were size or ecological strategies also important targets? In other words, to what extent are morphological changes adaptive or non-adaptive? With the increased interest (mine at least) in the role of intrinsic factors, principally heterochrony, in evolution, I asked authors also to examine their data to see to what extent heterochrony influenced evolutionary trends. (This produced some unexpected results in inducing a new interpretation for the evolution of amphibians from fish—see Chapter 11). The authors were also asked to consider to what extent evolutionary trends combined with environmental data can be used to establish how particular trends might have been channelled along environmental gradients. What, therefore, would have been the relative importance of intrinsic and extrinsic factors in such situations? My last request to the, by now befuddled, contributors was to ask them to assess what they thought the driving forces behind evolutionary trends in the particular groups might be. For example, are palaeontological data adequate to document the potential role of predation or of competition in influencing the directionality of evolutionary trends? Or are large-scale trends merely the outcome of changes in species variance caused by other factors? So, while each chapter documents a variety of evolutionary trends, at many taxonomic levels, attempts are made to explain why evolution has progressed in the direction that it has.

Evolution is not a random phenomenon. It operates within a constrained framework of both intrinsic and extrinsic factors. The fossil record is testimony to the directionality of evolution. The chapters in this book synthesise the current evidence on the multitude of factors constraining and directing the course of evolution. I hope that both students and researchers of palaeontology and biology will find much food for thought and be stimulated to examine further these and other groups for evolutionary trends.

The idea for this book arose from a discussion that I had with Iain

Stevenson of Belhaven Press on gaps in the palaeontological literature. I would like to thank Iain for his catalytic role and for his enthusiastic support for this book. I would like to thank a number of my colleagues at the Western Australian Museum for their help, notably Alex Baynes, John Long, Joe Ghiold, Kris Brimmel and Ken Aplin. I would also like to thank the Trustees of the Western Australian Museum for providing facilities that enabled me both to carry out research for my chapters and to edit the book. Last, but by no means least, I would like to express my gratitude to the chapter authors for their excellent contributions. It is a tribute to them that, with authors spread across the globe in the USA, the UK, Ireland, France and Australia, and with an editor in Australia and a publisher in the UK, this book has been produced within just two years of its initial conception.

PART ONE

THE DYNAMICS OF
EVOLUTIONARY TRENDS

Chapter 1

SPECIATION AND SORTING AS THE SOURCE OF EVOLUTIONARY TRENDS, OR 'THINGS ARE SELDOM WHAT THEY SEEM'

Stephen Jay Gould

POTENTIALS AND DEFINITIONS

Early in the second act of Gilbert and Sullivan's *HMS Pinafore*, Little Buttercup confronts Captain Corcoran to foreshadow the coming denouement that will demote him, by correcting some accidental baby-switching during his early days, to a simple sailor. Her song is a brilliant pastiche of major English proverbs on the subjects of deception and false appearance (see appendix 1.1). Its first line—'Things are seldom what they seem'—serves as a subtitle for this chapter. I shall cite most other lines in the course of this article (as numbers in parentheses keyed to designated lines in the Appendix); for palaeontological thinking on trends, the most important of all evolutionary subjects in our domain, has been marked by its miring in bad conceptual habits. We have, in short, been deceived by false appearances— and all the classical proverbs, quoted by Little Buttercup and the Captain, apply to our failures and hopes for rectification.

The most familiar things are often the least known—and this ironic fact emerges in many guises, from the common theme of betrayal in literature to the ignorance bred by complacency in science. Evolutionary trends—most generally defined as sustained biostratigraphic character gradients[1]—fall into the maximally treacherous category of things we all think we know and understand. I shall, against this feeling of comfort, argue that most trends have not been described correctly; that if described correctly, they have usually been improperly explained; and that the most comforting features in traditional interpretations should be viewed as the most puzzling.

The primary fallacy lies at the core of our descriptive language for trends (5). The basic phenomenon has always been expressed as lineage anagenesis.

3

Think about any classic case, from complexity of ammonite sutures, to size of mammalian brains, to number of graptolite stipes, to symmetry of the crinoid calyx. A group goes from an inadequate 'before' to a better 'after'. How can this be done save by incremental improvements (13) in an entity moving somewhere through time (or in several entities moving in parallel, and thus supporting the idea of a clear and predictable betterment)?

Yet an elementary consideration of the topology of macroevolution should illustrate the enormous potential for fallacy in descriptions so unconsciously based on lineage anagenesis. Except for the smallest-scale expression of a trend, any sustained character gradient must pass through multiple episodes of speciation—that is, must be transferred from one historical entity to another, usually a multiplicity of times in any long sustained sequence. And transference just is not the same phenomenon as flow. (At the most elementary level, you win a relay race by some combination of skill in passing the baton and speed in the flat, not by the simple cumulation of speed. And remember that when punctuated equilibrium holds, nothing much happens in the flat at all!)

Grant (1989) recently complained that his concept of 'speciational trends' has been inadequately acknowledged by later writers on macroevolution, particularly by Futuyma (1987), but also by yours truly (Gould and Eldredge, 1977). Yet if we must conduct a search for patrimony of this idea, the plaudits can only go to Darwin himself. In his long and important section on the Principle of Divergence in Chapter 4 of the *Origin* (the crucial chapter entitled 'Natural Selection'), Darwin clearly recognises (and illustrates in the *Origin*'s only figure) that sustained trends must pass through multiple events of speciation. His entire discussion is framed in this light as an argument for the efficacy of natural selection in forming new species, not only in yielding change within populations (Darwin, 1859, pp. 111–26).

Yet such a venerable formulation for the speciational character of all sustained trends produces a paradox: if evolutionists have known this all along, then why are trends invariably described in the language of lineage anagenesis, with the transformation beginning in one section of morpho-space, ending in another, and just rolling on in between? Why have we not seen the problem, the missing piece (the passing of the baton), the funda-mental gap in logic of argument?

Such examples of 'cognitive dissonance' always intrigue me, for they must record a conceptual barrier that blocks the recognition of a fallacy (or at least a lacuna) at the very heart of a crucial argument—and the documentation of such conceptual barriers, and their alteration through time, forms one of the most fascinating subjects in the intellectual history of our species. I think that we have been able to speak comfortably of trends as lineage anagenesis, while knowing that they must be produced by multiple transfers across species boundaries, for two basic reasons rooted in two major Darwinian conven-tionalisms now ripe for change. (The conceptual sorting out of trends and their meaning therefore promises to make a contribution to the more general task of revising Darwinism by expansion—see Gould, 1982).

The first reason is adaptation. In a conceptual climate of panadaptationism (see Gould and Lewontin, 1979)—the caricatured version, not Darwin's own, that became so popular in evolutionary circles during the heyday of neo-Darwinian orthodoxy during the 1950s and 1960s—all major features of morphology must arise, persist, and change as a direct consequence of natural selection for individual advantage. Thus, whatever accumulates in a trend must be the causal basis of that trend as read through its adaptive advantages. In such a myopic view (not wrong but applicable to only a subset of cases, probably a minority), it does not matter much whether the adaptive advantage accrues in anagenesis of a lineage or across several species boundaries. Speciation is only an aspect of the flux, and the flux is set by natural selection acting on the item of the trend.

The second reason concerns levels. If all evolution, as the hardest-line reductionism of the modern synthesis held, operated at the single focus of shifting gene frequencies in local populations, with all higher-level properties produced by extrapolation from this causal base, then speciation would represent nothing special or different, but only a measure for the extent of continuous change. In this perspective, speciation becomes an aspect of lineage anagenesis—and even though the topologies of branching and transformation are so different, the two phenomena become conceptually indistinguishable.

This conflation must underlie the conventional iconography of a trend (Figure 1.1) as full of branching points, but carried forward almost entirely by anagenesis within a supposed main trunk. Thus, speciation becomes, at best, an iterating device that puts a good adaptation into several baskets, perhaps as a hedge against extinction, while the main work of progressive change proceeds by mainline anagenesis. Thus, for example, Ayala (1976, p. 18) wrote:

Anagenesis, or phyletic evolution, consists of changes occurring within a given phyletic lineage as time proceeds. The stupendous changes from a primitive form of life some 3 billion years ago to man, or some other modern form of life, are anagenetic evolution. Cladogenesis occurs when a phylogenetic lineage splits into two or more independently evolving lineages. The great diversity of the living world is the result of cladogenetic evolution.

I would argue, by contrast, that anagenesis in this sense is illusory, and almost always the product of accumulated cladogenesis filtered through a higher-level process of species sorting. Similarly, Julian Huxley (1942), to Ernst Mayr's (1963, p. 621) distress burdened speciation with the status of a luxury or bauble of diversification, while the real work progressed in an anagenetic main line: 'Species formation constitutes one aspect of evolution; but a large fraction of it is, in a sense, accident, a biological luxury, without bearing upon the major and continuing trends of evolutionary process.'

The subject of trends is paramount in importance among the inputs that palaeontology can make to the structure of evolutionary theory and its proper extension into deep time. Minor changes and variation within local populations are better studied by neontologists; broadest-scale faunal turnovers and patterns remain our distinctive province, and provide great insight into the workings of evolution, but are too far from the daily activity and concern of working biologists. Trends are the meeting ground of our disciplines, and in

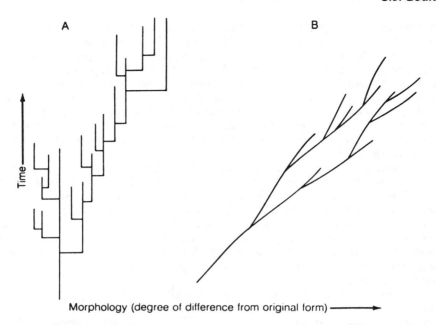

Morphology (degree of difference from original form) ⟶

Figure 1.1. Evolutionary trends under conventional anagenetic iconography (right), where directionality arises by within-species change and speciation merely increases the number of lineages bearing the trend; and under punctuated equilibrium (left), with no change in species after their origin.

this sense the most important evolutionary subject in all of palaeontology. I devote this chapter to arguing that a proper recognition of the speciational character of trends will not only build a framework for our own understanding, but also provide a large dose of unconventional insight to the reform, expansion and integration of evolutionary theory in general.

THE RESTRICTED ROLE OF LINEAGE ANAGENESIS AND THE CENTRALITY OF SPECIATION

Three fallacies

Lineage anagenesis seems so right-minded. What could be wrong, what the alternative? Add up little things of the moment through the vastness of geological time, and you eventually get what you need at any scale or magnitude. Trends as lineage anagenesis, as the summation of immediate advantage over time, are the apotheosis of the uniformitarian research programme, the fulfilment of Darwin's vision as expressed in his most striking metaphor:

Natural selection is daily and hourly scrutinising, throughout the world, every variation, even the slightest; rejecting that which is bad, preserving and adding up all that is good; silently and insensibly working, whenever and wherever opportunity offers, at the improvement of each organic being in relation to its organic and inorganic conditions of life. We see nothing of these slow changes in progress until the hand of time has marked the long lapse of ages (Darwin, 1859, p. 84).

But the fallacies become apparent as soon as we probe. First of all, a lineage rarely moves very far as a discrete population before division (often multiple) by speciation intervenes to split the patrimony among several lines. Thus, nothing much can accumulate in the purely transformational mode (11). (I can envision situations where break up by branching would not disrupt a reasonable concept of lineage anagenesis—but these must be rare or, at the very least, certainly not standard. For example, if divisions usually placed most of the original lineage—or at least the part destined for persistence—into a population predisposed to continue the anagenetic direction of a discrete ancestral population, then we might ignore the branching nodes, and trace anagenetic lines right through the evolutionary bush. But I know no possible justification for such a claim. Is it not just as likely that a discrete population might be increasing in body size in classical Cope's-Rule fashion, then split, with one lineage continuing to increase and the other being a progenetic dwarf—and that the increasing branch soon dies, leaving all genetic patrimony in the dwarf lineage? Obviously, I do not deny that single paths unite any extant twig on an evolutionary bush with the ultimate common ancestor of the monophyletic group. But this fact of genealogy permits no picture of an anagenetic highway from common source to current occupant. These paths are labyrinthine wanderings through the bush; they cross numerous speciational nodes and must often shunt from a within-species trend to a descendant species of differing direction—as in the example of progenesis above. I also do not deny that some taxa may, for reasons of unique biology, be particularly prone to sustained anagenesis without branching. Planktonic foraminifera may provide our best examples of sustained gradualism not only by virtue of excellent preservation, but largely because such vast and world-wide (or at least ocean-mass wide) populations branch infrequently or inconsequentially. But, again, this cannot be a canonical mode in a world of such riotous and mercurial diversity.)

Second, when we do find a case of sustained and gradual anagenesis within a discrete lineage, we should feel intense puzzlement rather than an urge to shout Hosanna for Darwinian vindication. We are so used to viewing geologically gradual change as a proper expression of natural selection's ordinary pace that we miss the crucial paradox. Millions of years is too generous, too much for sustained, unidirectional change of such infinitesimal magnitude. All known and empirically studied cases of gradualism at eco-logical scales would be completed in a geological twinkling of an eye. To be measurable at all against experimental error, a transformation must be too rapid for sustainability across millions of years. The runaway character and positive feedback loops of sexual selection, in particular, provide no conceiv-able model for geographical transformation. If Irish Elks evolved large antlers for sexual combat, we could not trace the increase up a geological column (8). Moreover, even if unilinear trends did not imply too impalpable an immediate pace, is it plausible that selective pressures could be sustained in one direction for so long in our world of ever-changing geological circumstances?

Gradual anagenetic change, rather than acting as the bright vindication of convention as so often portrayed in textbooks, is intensely puzzling and cannot usually represent the pure power of selective improvement in the accumulative mode. Lande (1976) calculated the per-generation rate of change from Gingerich's (1974; 1976) classical sequences for mammalian anagenesis—and found the immediate changes so impalpable that they could easily be explained by genetic drift, even in continually large populations. When translated into coefficients of selection, such a rate corresponds to one death per 100,000 individuals per generation, or a model of truncation selection that only eliminates individuals more than four standard deviations from the mean. Since most populations have no non-teratological variation so far from the norm, we must conclude that gradualism at this pace does not yield to classical selectionist explanation—that is, to natural selection 'daily and hourly scrutinising every variation, even the slightest'. Small does eventually grade into effective invisibility.

I regard this situation as akin to the insight that eventually dawned on several palaeontologists who had claimed evidence for classical ecological succession in geological sequences (Bretsky and Bretsky, 1975; Walker and Alberstadt, 1975). The changes in composition and abundances looked right, but palpable geological verticality (in most cases) represents thousands of years at least (often more)—and classical succession is over in a fraction of that time. Something interesting is occurring in these geological changes, but it cannot be succession. Likewise, something puzzling and fascinating lies behind our few examples of sustained anagenesis within fossil lineages, but it cannot be classical selection working relentlessly and unidirectionally for the adaptive advantages of characters representing the trend.

Third, and finally, when we do encounter genuine cases of lineage anagenesis (probably fairly common at appropriately small scale), we must question and abandon the old tradition of a priori interpretation as consequences of adaptative advantage for the character forming the trend. In a narrow conceptual world, the adaptationist version makes sense, indeed seems ineluctable—for a trend sustained so long and so far in anagenesis cannot be produced by drift, and what else can be envisaged in a neo-Darwinian universe? But recent revisions suggest alternatives at high relative frequency. Under concepts of higher-level causation as species sorting (see next section), trends in organismal characters must often be effects or consequences, hitch-hiking on the focal level of species treated as entities with intrinsic reasons for differential success. Under concepts of constraint (not inconsistent with neo-Darwinian principles, but underplayed to conceptual invisibility until recently), developmental and architectural side-consequences and forced correlations can be as numerous and as prominent as the selected basis of a trend itself.

In this context, I particularly admire the numerous and well-documented examples recently presented (for example, McNamara, 1982; 1988; McKinney,

1988) of trends in developmental heterochrony—the paedo- and pera-morphoclines of McNamara (1986). Of course, the documentation of extensive correlation says nothing *per se* about cause and effect; selection could be operating upon the feature brought to prominence by heterochrony, the change in developmental timing itself, or both in their happy linkage. But such cases do establish a *prima-facie* basis for strong (and testable) suspicions about non-adaptive side-consequences—and we have good reason to think (see Gould, 1977) that selection will often act primarily on the timing of development, leaving morphological consequences 'free floating' and therefore especially available for incorporation into major evolutionary shifts in the exaptive mode (Gould and Vrba, 1982). If such co-optation of neutral and non-adaptive features (14) often characterises the origin of major taxa and structural innovations, then an understanding of trends in unselected characters becomes especially important.

In sum, most cases of putative lineage anagenesis are misread examples of speciational trends; while most genuine examples of lineage anagenesis cannot bear their conventional interpretation, for neither a selective basis of the phyletic motion, nor an adaptive value for the resulting characters, can be assumed, and other conceptually interesting alternatives are available.

Three misinterpretations

The next section will examine the different explantory apparatus needed for elucidating speciational trends. But jumping the gun in quick epitome (for the logic of this section needs such an explicit statement), trends produced in the speciational mode—that is, nearly all important trends in the fossil record—occur for two basic reasons: first, because either of the two factors (speciation and extinction) develops a directional component (preferential speciation towards, or differential extinction in, a component of morphospace); and second, because the variational range of a clade alters, by change either in the number of component taxa or in disparity among them (see Gould, 1988, for details and examples).

The attempt to interpret trends so produced as the result of lineage anagenesis has given rise to traditions of error that have seriously derailed our understanding of this most fundamental evolutionary phenomenon at geological scales. Three seem most important (almost ironical, in one case at least, for the precisely backward reading thus engendered). Their exposure and correction should become a primary agenda of palaeontology (15).

Life's little joke (2)

Consider the classic examples of evolutionary sequences as portrayed in museums, textbooks and TV documentaries—the parade of horses from eohippus to Secretariat, leading to larger size, higher crowned teeth and fewer toes; and the march of human progress from apes in the trees, to hairy australopithecines, to men (for biases of gender also intrude) in business suits. In a situation so ironic that I call the phenomenon of human misperception 'life's little joke' (Gould, 1987), our propensity for misrepresenting speciational trends as lineage anagenesis guarantees that our canonical examples of progressive trends must represent unsuccessful lineages at the brink of extinction.

An unbroken sequence does connect *Hyracotherium* with *Equus*, of course, but this skein is a labyrinthine pathway across numerous nodes of speciation, not a highway of lineage anagenesis. Any current twig on an evolutionary bush crowns such a pathway to a founding member. Flourishing bushes, with numerous extant species, are replete with such pathways, and none have preferred status. Such copious bushes, in a conceptual world that misinterprets trends as lineage anagenesis, cannot serve as examples of evolutionary success, for they contain too many pathways moving in too many disparate directions. But the least successful of all extant bushes—those but a single twig from extinction—become our classical trends in an evolutionary 'progress' under the conceptual strait-jacket of lineage anagenesis. For we take the one taxon still clinging to precarious existence and misread it as the goal of a central and preferred highway, and if we acknowledge the numerous extinct pathways at all, we depict them as side branches—although nothing but the accidents of death specify designation as motorway or dirt path to oblivion. (Go back 15 000 years; would the path to *Equus* clearly be the highway, and that of *Nannippus* merely a byway? Both are equally long and intricate. And if *Nannippus* had lived and *Equus* died, then the highway would include no trend to size increase.)

Equus is the only surviving genus of a once luxurious bush in a once successful order now decimated to vestiges of horses, rhinos and tapirs as artiodactyls continue to radiate. Yet, by life's little joke of misperceiving trends as lineage anagenesis and misdrawing straggling survivors as central highways, *Equus* becomes the exemplar of mammalian success in evolution. Remember that no textbook story exists at all for the real triumphs of mammalian design—the evolution of bats, rodents and antelopes.

The mistaken quantitative apparatus (7)
The standard device for quantifying and comparing evolutionary trends, the darwin of Haldane (1949), is moot and misconceived in a world of speciational trends. The darwin is a lineage-specific rate (change as increase or decrease of one-thousandth in a character per thousand years). Its interpretation as expressing a real phenomenon of evolutionary change is clear and unexceptional in a world of lineage anagenesis with continuity in rates of normal distribution, for the darwin then becomes a central tendency with a link to physical reality.

But what can such a number mean in a world of speciational trends, especially in lineages dominated by punctuated equilibrium (that is, nearly all lineages in the view of this admitted partisan)? At best, an average rate for continuous phyletic change is a curious and indirect surrogate for numbers of speciation events, or for average amount of change per event. Consider the following fable: a family lineage of frogs moves 100 metres northward in 20 years, in the following way. The founding father jumps 10 metres in a single bound, then establishes a home site exactly where he lands. He may wander about looking for a mate, but he brings her right back to the home site and raises the next generation on the spot. Exactly two years after his jump, one of his offspring makes a single jump exactly 10 m farther north, and the same process continues for eight more generations until the lineage's home site lies 100 m north of the progenitor's original position. Would any insight be gained by saying that the frog lineage moved north at a rate of 1.4 cm per day? Would such a claim not be more misleading in implication than useful in summary value?

Consider, as an example, Kurtén's (1959) classic study of rates of evolution in mammals (written for the Darwinian centennial celebration at Cold Spring Harbor). Kurtén worked within a strict (but inarticulated and, therefore, probably unconscious) assumption that trends must be read as lineage anagenesis (he mentions new species only as originating by sufficient accumulation in unbroken lineages). He argued (see Figure 1.2) that darwin rates for Tertiary and Quaternary mammals fell into three classes, named by him A, B, and C rates. The slowest C rates averaged 0.023 darwins and characterised long-ranging Tertiary lineages. B rates (for Quaternary lineages) were 22 times as fast, averaging about 0.5 darwins. Finally, the fastest A rates (12.6 darwins on average and 25 times as fast as B rates) applied to size changes in postglacial mammals.

Kurtén provided a reasonable explanation for differences between A and B rates in the anagenetic mode—the former as short-term fluctuations in a trend, the latter as the averaged direction of the anagenetic trend itself. But the much slower C rates puzzled him greatly (Kurtén 1959, p. 213). He toyed with the idea that the standard (Simpson's horotelic) rate might have been higher in the environmentally disturbed Pleistocene than earlier in the Tertiary, but more than an order of magnitude seemed too extreme for the resulting effect. He also considered the possibility of artefact due to larger geological spacing of Tertiary samples (I believe that he was on the right track in this insight), but then came up against a conceptual wall and offered no reason why mere time should lead to such profound differences in an anagenetic rate. Yet, from a speciational perspective, a potential resolution practically jumps forth (and applies equally well to Gingerich's misinterpreted evolutionary scaling—see Gingerich, 1983, and the reply by Gould, 1984): the B rates may well represent phyletic change for speciating lineages in this tiny segment of time. But the C rates, all averaged over millions of years in lineages that pass through multiple speciation nodes, must be much lower because they amalgamate periods of speciation (perhaps at characteristic B rates) with longer periods of stasis between—thus committing the same error as the averaged rate in the frog fable above.

Haldane (1949) propagated the same error in his classic paper that defined the darwin. He calculated the annual rate of increase in paracone height at 3.6×10^{-8} for the *Hyracotherium–Mesohippus* lineage—a path through the horse bush that passes through several genera via a large number of speciation nodes, most unknown. Yet Haldane (1949, p. 53) interprets his calculated rate as meaning 'that the paracone height increased on an average by 3.6% per million years'. He then makes further calculations on generation time and variation, finally concluding (Haldane, 1949, p. 54) that about half a million generations would be needed for a full standard deviation of change.

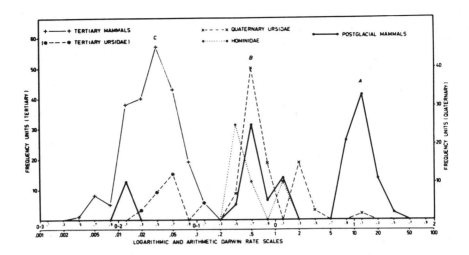

Figure 1.2.   Trimodal distribution of mammalian evolutionary rates calculated by Kurtén (1959).

Taking five standard deviations as definitional for speciation in the angenetic mode, he then calculated the duration of species when arbitrarily defined as segments of continually changing lineages. But, again, what do such figures mean if evolution operates more often as in the frog example above, or even, at the very minimum of conceptual shift, just with characteristic differences in speciation and life as a non-branching lineage (passing the baton and running in the flat)?

Trends as illusions misfocused on extremes of distributions (4)
Physical reality exists on several levels simultaneously. We may, at the same time, discuss *Homo sapiens* as the terminus of a single lineage and as one item of no preferred status within the clade Primates. Similarly, no mistake arises in singling out the lineage of *Equus* for special consideration within the perissodactyl bush. But a cardinal error occurs when we jumble categories, and describe a phenomenon at one level as the exemplar of another—in particular, and in this case, when we mistake a part for a whole.

The potential for error by this most common foible of human reasoning (see Bateson, 1979) does not arise in a world of lineage anagenesis—for only one important level exists in immediate forces of selection and their temporal extrapolation. A creature with a feature is the apex of a trend operating for the value of that character. But in a world of speciational trends, a creature with a feature is only one item in a population of taxa, one component in a spectrum of variation among species in a clade.

Most trends are fundamentally described as differential species success within clades (see next section). Key components for description are the position and range of the spectrum of variation among species in the clade through time. Our most usual sense of 'trend' might be expressed by change in position of the spectrum—that is, we may talk of a trend to increased body size within a clade if the range among species shifts from 5–10 cm in body length at time 1 to 15–25 cm at time 2. But another important sense of 'trend' acknowledges changes in variation through time (Gould, 1988)—for we may surely speak of trends in expansion if the range among species shifts from 10–15 cm at time 1 to 5–20 cm at time 2, either by increase in number of species with 'spilling out' beyond the initial tails in both directions, or by increase in the average difference between species.

We make no error in taking the 20 cm extreme species of the foregoing clade and stating that its own lineage showed an increase in body size. But we make a serious 'category mistake' if we single out that species (perhaps because its large size attracts our attention) and then argue that increasing size is a primary feature of the entire group, for the trend was built by a symmetrical increase of variance in both directions about an unchanging mean value. The error is elementary when pointed out in this simple and abstract manner; a critic might object that I am harping on something quite obvious, that anyone would see the fallacy right away. In fact, no error is more common. We make this mistake all the time, and it underlies some of our worst misconceptions about evolution (Gould, 1988). This fallacy ranks as a primary error of typological thinking, as we mistake an item in a variational range for a thing-in-itself.

We focus on outliers (perhaps because we represent one in mentality), and we easily misinterpret them as apices of particular trends, rather than extremes in spectra of variation. This tendency is exacerbated when founding lineages lie near a structural boundary and substantial evolutionary 'space' only exists in one direction. Consider two situations that could not be more fundamental for form and ecology—the size and place of origin for lineages. Important and highly corroborated arguments hold that most major lineages tend to arise at small body size for the *Bauplan* (a venerable argument based on Cope's 'law of the unspecialized'—Stanley, 1973; Gould, 1988); and that marine higher taxa tend to arise near shore (a recent argument with impressive and growing support—see Jablonski, 1986; 1988; Jablonski and Valentine, 1981; Jablonski *et al.*, 1983).

In each case, the clade can only expand in one direction if numbers of species increase—towards larger body size if the founding lineage stands at the lower limit, and offshore if physiology precludes the colonisation of land. In such expanding clades, modal values might not change at all; the most common species remains right on the ancestral spot. But means will move towards larger body size and locations further offshore because expansion has been restricted by the position of the starting point.

I showed, for example (Gould, 1988, based on data of R. Norris), that the famous threefold iterated trend of increase in size among planktonic foraminifera (Cretaceous, Paleogene and Neogene radiations) arises as a consequence of change in variation engendered by growing numbers of species, not by lineage anagenesis (Figure 1.3). Moreover, the trend is largely illusory and produced by a misfocus on extreme values as things-in-

S.J. Gould

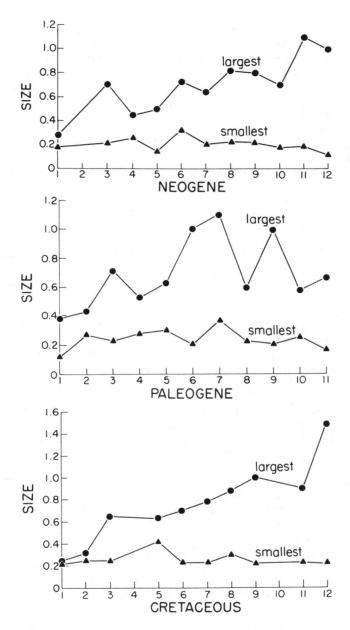

Figure 1.3   The threefold trend to larger body size in planktonic foraminifera is a consequence of asymmetrical expansion of variance around small founding lineages. From Gould (1988). Size of largest and smallest species through time is shown. Cretaceous in 5-million-year intervals; Paleogene and Neogene in 3-million-year intervals.

themselves. Founding lineages are small-sized in each radiation, and both means and extreme values do increase with time; but modal body size shows no change. In the same light, by what justification other than human hubris can we claim that evolution as a whole moves towards greater complexity through time when the earth's modal organism was at first, is now, and probably ever shall be (until the world ends) a prokaryotic cell? This much vaunted 'trend' and intrinsic-central-urge-and-tendency-of-all-life-as-a-whole is nothing but a consequence of inception (for chemical reasons, since biogenesis can hardly start with a hippopotamus) at or right near the lower boundary of preservable body size.

Increase in number of taxa (with movement of means due to asymmetries of starting points within the structural range of possible outcomes) is not lineage anagenesis, and we have made dreadful mistakes in conflating the two phenomena.

TRENDS AS SPECIES SORTING

Once we acknowledge that most trends are best described as differential success of certain kinds of discrete species within clades, implications flow towards interesting reformulations of conventional views. These centre upon the two key themes of more substantial change within traditional Darwinism —levels and adaptation.

The species level as description and causation

Vrba and Gould (1986) distinguish between 'sorting' as a descriptive term (for change by differential birth and death of entities at any level) and 'selection' as a causal claim (for sorting based upon properties functioning as interactors and emergent at the level under consideration). The shift in focus for explanation of trends—from organisms in populations (extrapolated to lineage anagenesis) to species within clades—forces a reconsideration of traditional views in both domains of description and causation.

For description, we must make a primary shift of emphasis away from transformation, and towards *number* of taxa (20)—a change embodied in Eldredge's (1979) contrast of taxic and transformational thinking. Under the speciational view, trends arise when species are differentially removed from one region of morphospace (extinction) or added to another (speciation).

A basic taxonomy of speciational trends might designate three categories, each with implications diverging from standard views rooted in lineage anagenesis.

Biases in the amount of production
Species and subclades with higher rates of speciation within a clade will, *ceteris paribus*, come to predominate, and will set a trend to whatever synapomorphic features the resulting subclade possesses. Two end-member reasons allow certain species to take over—increased persistence (with

constant rates of speciation across the clade) and enhanced rate of generating new species. The most potent trend-setting species of a clade might combine both features, but structural reasons probably enforce a negative correlation, for high speciation rates usually link with high extinction rates (Stanley, 1979). Any species that could overcome this negative correlation would be the most potent of all evolutionary agents. Their capacity to take over clades in geological time would be immense. (Is this the key to coleopteran triumph? Does this, perhaps, set the success of modern life, with its monophyletic base, against other possible experiments of the earth's infancy, now gone?)

Classically trained Darwinians often object to this category by pointing out that it seems to move a clade nowhere. So what if one in three subclades comes to dominate; it existed at the start, albeit in smaller numbers, and nothing new has happened. But newness in Darwin's world is differential success of options arising for other reasons (mutations with fortuitous adaptive benefit). Once we grasp this mode of species proliferation as the higher-level analogue of differential birth, the argument becomes clear. If an extreme subclade among three takes over, then the cladal mean shifts to this position, which becomes a new base for further exploration of previously unoccupied morphospace.

Some of the best-documented cases for macroevolution by species sorting fall into this category, including the now classic case of volutid gastropods (Hansen, 1978), where species with non-planktonic larvae take over the entire clade by virtue of their higher speciation rate (and despite a higher extinction rate as well), in the face of an extinction motor sufficiently intense, relentless and pervasive that any subclade eventually succumbs without a sufficiently high speciation rate to balance.

Biases in the direction of production
Under the speciational view, horses might become larger if all extant species within a clade produced daughter species with an equal chance of being larger or smaller than themselves, and if larger-bodied horse species either persisted longer or speciated more frequently. This scenario corresponds to the first option above. But cladal mean size might also increase if new species tended to arise at larger size than their ancestors—in other words by a bias in the direction of speciation. (Needless to say, in our real world, some combination of all processes probably operates most of the time.)

This second speciational scenario is the analogue of mutation pressure at the lower, traditional level of classical Darwinism. Mutation pressure is usually dismissed as unlikely to produce much evolutionary effect because its small power should be easily overshadowed by natural selection. But this argument rests on the fact that, at the population level, millions of ordinary births overwhelm each mutational novelty, and natural selection (working by sorting among the ordinary births) therefore has immense power over mutation pressure. In addition, we know no chemical mechanism to produce preferred, consistent, adaptive directionality in the production of new variations.

However, both these objections are cancelled and inverted at the clade level (with species as analogues of individuals), thus setting a *prima-facie* case for the relative importance of directional bias in speciational trends. First, clades produce relatively few species compared with the yield of new individuals in populations—and selection therefore has limited fuel, making production biases more potent in principle. Second, mechanisms for directional biasing are abundantly available in such themes as allometry and developmental channelling. The entire literature on heterochrony (Gould, 1977; McKinney, 1988) proclaims the plausibility and commonness, through speeding or slowing of developmental rates, of biases towards increasing juvenility or adultness of derived taxa. In addition, any such trend carries a host of sequelae in allometries and other forced correlations. Any tendency for production of species at larger or smaller than ancestral size also implies a cascade of sequelae, some by new adaptation in response to Galileo's old surface/volume principle (and therefore conventionally explainable), but others by inherited developmental correlation with this most pervasive and potent of all primary factors (these may be interpreted as legacies of past adaptations, but many must be spandrels in the sense of Gould and Lewontin, 1979). Thus, by virtue of structural differences between levels, directional biases in production are probably vitally important for speciational trends within clades, even though they may play only a minor role in ordinary natural selection within populations.

Direction by extinction (9)
Some evolutionists have trouble envisioning an extinction-driven trend in the speciational mode. Extinction makes nothing. How can the mere elimination of part of a spectrum of variation among species achieve anything new in evolution? People impressed with this claim should adopt a historical perspective to grasp its fallacy. The standard nineteenth-century argument against natural selection was framed exactly in this way—though at the lower level of organisms within populations. Natural selection could not be the cause of evolutionary change because it could only act as a negative force, a headsman or executioner for the unfit; some other 'creative' force must make the fit.

The Darwinian response correctly held that selective elimination can drive evolutionary change so long as any new modal state retains the capacity for generating a random spectrum of variation about itself. No intrinsically directed force of adaptive variation is needed, so long as the capacity for undirected variation is retained. Selective elimination can move a mode right to the extreme of a previous variational spectrum. There, a new spectrum can be constituted by the usual processes of mutation and recombination—and selective elimination can continue the trend into new morphospace.

This classical Darwinian response works just as well at the level of species elimination within clades. Suppose that patterns of speciation are entirely random with respect to the direction of a trend (no origination biases in either amount or direction by the first two arguments above). Differential extinction can move a cladal mode anywhere within the spectrum of variation among species. With a new mode at the old periphery, random speciation can reconstitute variation that moves into previously unoccupied morphospace, and directional extinction can then continue to accentuate the trend.

I suspect that the main difference between the classical anagenetic and the speciational trend by elimination must lie in the far greater possibility for random direction (not just random variation) in the speciational mode. Such a claim asserts an important departure from the key Darwinian belief that randomness operates only in the generation of raw material, while causal selection produces change.

As the long literature of genetic drift has taught us, random factors of change operate most effectively (or exclusively in some accounts) within small populations, and for obvious reasons of elementary probability theory. Random factors are often discounted at the traditional level of organisms within populations because births and deaths are usually numbered in millions per generation, and how can random death establish anything in such a context? But the level of species within clades gives a potentially great, even dominating, role to randomness for two reasons. First, the number of species within clades is relatively small in most cases, and removal of just a few can make a big difference (18). Moreover, each species is more distinctive (by possessing autapomorphic characters not subject to reconstitution) than most organisms within populations. Almost any loss in a clade, by the power of distinctive phyletic heritage, will significantly change or limit the further fortunes of the totality; but the death of almost any single individual will not derail or reset a trend within a population. Extirpate *Homo sapiens*, and what chance does the clade Primates have for language-using self-consciousness in its spectrum again? Kill almost any human and, whatever the personal tragedy, no evolutionary possibility becomes strongly altered or compromised. Second, numbers of species within clades are subject to great fluctuation and, in a world decimated by mass extinction more frequently and at greater magnitude than we once believed, particularly subject to random redirection at very small sizes. (Ammonites, for example, suffered reduction to one or two lineages at least twice in their history. Dare we say that these survivors were biomechanically the best of all ammonites then available?) Yet if population fluctuations within many species are as severe and frequent as some ecologists now believe, perhaps this style of randomness (or at least survival unrelated to reasons for developing the crucial features during normal times) is important at the traditional Darwinian level as well.

In sum, these three descriptive modes of sorting for speciational trends — all analogues to differential birth, death, and mutation pressure in populations — do their work in modes and manners different from the operation of lineage anagenesis as usually conceived. Thus, whatever the causation of trends best described by species sorting, the simple decision to narrate at the right level implies, in itself, a substantial revision and expansion of traditional views.

A note on the causation of speciational trends

The simple recognition that trends must be described at the species level has important implications for reform in evolutionary theory, as discussed in the previous section. But the theme of speciational trends also offers the prospect of more extensive reformulation, reaching to the core of evolutionary theory, if causation (and not mere description) also resides at the species level at high relative frequency.

Need for description at the species level in no way implies causation by species selection—and this separation is the key point made by Vrba and Gould (1986) in distinguishing 'sorting' (descriptive) from 'selection' (causal). In theory, all causation could reside at the traditional level of natural selection among organisms and still produce a trend requiring description as the differential success of species. For example, ceratopsian dinosaurs might show a trend to increased frilliness of horns because frillier species tended to live longer (with speciation rates equal to those of less frilly species)—while the greater survival of frillier species might result only from the Darwinian success of frilly individuals in combat.

But the need to describe at the species level also implies a large potential domain for true causation in selection at the species level—that is, on species acting as Darwinian individuals in their own right. Since classical Darwinism is a one-level theory of selection on organisms, an important role for species selection would force a major reformulation of evolutionary theory in a model of hierarchical selection, with several levels working simultaneously and often in conflict. Even so strict a Darwinian as R.A. Fisher (1958) could not deny the logical plausibility of species selection—though he discounted any possibility of a meaningful relative frequency with the false argument (see Gould, 1989) that a fully efficient natural selection, operating through millions of deaths and births per speciation event, would have to overwhelm such a 'weak' force as species selection. This argument collapses if stasis and punctuated equilibrium are norms in evolution.

We are individual organisms, and our intellectual habits only extend to items at our level (for which we have a visceral and personal appreciation) and at lower levels (thanks to centuries of practice within a reductionistic scientific tradition). We hardly know how to think about higher levels as real and active entities, rather than as abstract aggregates. We do not even have clear agreement on what should count as species selection—though we do know what to exclude under Vrba's notions of effects and upward causation (Vrba, 1980; Vrba and Eldredge, 1984). I myself continue to vacillate between a strict definition based on emergent characters (Vrba and Gould, 1986) and a more inclusive construction based on emergent fitnesses (Lloyd and Gould, in press). The first permits an identification of species-level traits as adaptations *in se*, but may restrict the domain to few cases; the second, by allowing aggregate characters to function in species selection, greatly broadens the domain by permitting (as a prime example) the important character of variation to function in species selection, but breaks the link with adaptation because aggregate characters have no emergent properties at the species level.

Thus, I am not distressed when opponents argue that 15 years of discussion on species selection have not produced a large set of uncontroversial examples (but see Jablonski, 1988, and Gilinsky, 1986). I have never encountered a more difficult or confusing conceptual domain (an impression not merely reflecting my own personal and potential stupidity, but shared by all who have seriously joined this issue), and even suspect that such hierarchical thinking may lie among the cognitive regions that the human mind is not well equipped to handle (see Tversky and Kahneman, 1974). But we must persist not only because challenges do us proud as intellectual beings, but because the rewards are great and the reformulation potentially profound.

Levels and the exaptive expansion of non-adaptive
reasons for trends

The assumption of adaptive advantage for traits defining a trend has been, at the same time, both the most pervasive assumption of our literature and the most frustrating and refractory to adequate demonstration (19). All major trends have their canonical proposals, of course—graptolite stipes for floatability, crinoid calyces for efficiency in feeding and waste disposal, ammonite sutures for strength—but we know in our heart of hearts that these are largely guesswork and that, had the trend gone in the opposite direction, we could have conjured up an adaptive explanation with equal plausibility (and lack of support).

Such a situation should pull us in two ways: we may either firm up our criteria and defend adaptationist claims more adequately, or we may, following a suspicion that should be engendered by our past failures, permit ourselves to think that our previously unquestioned allegiance to adaptation might be false, and that many of our classic trends require no adaptive explanation for the traits themselves. This second possibility is neither negative in spirit nor a shallow, opportunistic end-run around a refractory problem. Non-adaptationist alternatives are rich and positive proposals made in the light of important revisions now being proposed within evolutionary theory (Gould, 1982; 1988).

The theme of constraints, developmental and otherwise, provides powerful arguments for non-adaptationist explanations, as explored earlier in this chapter (and especially important for trends under themes of ontogenetic channelling and heterochrony). But the added theme of levels provides a second, even more potent, reason for suspecting that non-adaptation plays a great role in evolutionary trends.

Gould and Vrba (1982) introduced the term 'exaptation' to fill a terminological (and therefore conceptual) gap in the logic of evolutionary explanations. 'Adaptation' has a confusing dual use in our profession—both as a static term for an attribute now promoting fitness, and as a causal term for the process of producing such an attribute by natural selection for its current utility. Once we restrict 'adaptation' to the causal meaning (as a long tradition from Darwin himself to Williams, 1966, insists upon so strongly), we then have no term for a fundamental phenomenon in evolution—the causal formation of an attribute for a reason other than its current utility, with subsequent co-optation of the attribute by natural selection for its current function. Our unfortunate and misleading term 'preadaptation' captures part of this phenomenon—production by selection for one use followed by co-optation for a strikingly different use—but leaves undefined (and therefore unconceptualised by most scientists) the important category of attributes not arising by selection at all, but still available for co-optation.

So long as adaptationism ruled within evolutionary theory, this gap in logic posed no serious problem because the category (non-adaptive structures then co-opted for utility) was not granted a relative frequency worth dignifying with a name. But revisions imposed by themes of constraints and levels vastly expand the potential realm of this process—perhaps to a cardinal role in such key evolutionary events as the production of novel functions. The concept of exaptation has therefore become essential, and must be named.

Many exaptations occur within a level as a result of phyletic constraints, forced correlations, and developmental linkages. These are the 'spandrels' of Gould and Lewontin (1979). But the concept of levels—and, for the particular concern of this chapter, the need to describe trends as speciational —implies an even greater expansion for the role and influence of exaptation: for anything arising as the result of a causal process at one level is, by definition, an exaptation if co-opted for utility at another level. Put less abstractly, any consequence of the phenotype arising as a result of species selection in a speciational trend is either a non-aptation at the organism level or, if co-opted for utility, a product of exaptation.

When we begin to catalogue the potential reasons for such non-adaptive trends at the organism level, we begin to sense the potential power and frequency of this process. The key concept is hitch-hiking. In any subclade that begins to prevail through the differential success of its species, trends must arise for any and all apomorphic characters by simple reason of phyletic heritage, and not necessarily for any adaptive benefit of the character. Mammals prevailed and dinosaurs died, and we do not know why. Does anyone seriously argue that every single mammalian feature thus increased within the clade of vertebrates did so by adaptive superiority over its dinosaurian homologue? Surely a small minority of mammalian traits played any active role in producing the trend, especially if species selection (or drift through mass extinction) played an important role (for then, these phenotypic traits could not directly cause the trend in principle). I submit that most of the classic trends in palaeontology may owe their frustrating intransigence (to adaptive resolution) to their production at the species level, with phenotypic character gradients therefore arising as passive consequences— either non-adaptive throughout, or seized only secondarily as exaptations for utility. Is stipe reduction in graptolites better for floating? Is transition from pendant to scandant better for exposure of individual colonists to food sources? Possibly so. But is it not just as likely that fewer-stiped graptolites speciate more frequently (as a result of emergent population characters linked only by passive phylogeny to stipe number), and that the culmination of these famous trends in the monograptids occurred as a non-adaptive consequence of good fortune in the sole survivorship of this lineage through the late Ordovician extinctions?

HOW DARWINIAN IS THE WORLD OF TRENDS?

A shift to the speciational mode as our basic vantage point for trends forces a re-evaluation of all standard evolutionary questions for such broad time-scales. Consider just one example based on minimal departures from orthodoxy (the speciational component for describing directionality). Let us accept (though I would bet a considerable sum on the low relative frequency of this mode) that the characters defining a trend are adaptations, and that their causal basis lies in conventional selection among organisms. Let us then consider a classical dichotomy in the evolutionary literature (still of intense interest, as a recent symposium volume—Ross and Allman, in press— dedicated to the subject and centred on case studies, shows): does selection usually operate by biotic interactions, or the pressure of abiotic conditions upon organisms? This classic issue was central to Darwin's own world-view (see below) and underlies many evolutionary debates ever since—from Kropotkin's (1914) challenge, based on the abiotic view, against the red-in-tooth-and-claw school, to recent skirmishes (Simberloff, 1984) on the importance (and existence) of competition in ecosystems. The speciational view cannot resolve this issue in itself, but forces a rethinking of mode no matter which side prevails in any particular case. For example, if 'arms races' exist in a world of biotically driven trends—and I accept that they do for such well-documented cases as Vermeij's (1977a; 1977b) concerning the increase in shell thickness and spinosity of snails matched with growing strength in the claws of crab predators—then how do they work in the speciational mode? In this case (though the statement may sound paradoxical), the anagenetic mode is easier to visualise, but cannot possibly apply—for any continuous escalation in such positive feedback would drive the trend to its furthest point in a geological instant, while the actual events span tens of millions of years. Yet if the trend must occur more episodically by occasional frog-hops of speciation (as Vermeij, 1987, accepts), then how is the locking of biotic interaction maintained? I do not pose this question rhetorically; I find the issue personally puzzling and, mired no doubt in my own conceptual swamps, have no good solution to propose.

The importance of biotic predominance for Darwin's world-view has not been adequately appreciated (though Darwin, to say the least, scarcely hid his views!). Darwin's conceptual need arose from his ambiguous attitude to the idea of progress—a cardinal intellectual notion for any Victorian. Darwin, the philosophical radical, knew (and relished the fact) that he had constructed an evolutionary theory based on local adaptation to changing circumstances, with no reference at all to inherent or predictable progress in the bare-bones mechanics of natural selection. But Darwin, the Victorian gentleman, could not abandon this pillar of his culture. He therefore smuggled the idea of progress back into his system through an ecological argument subsidiary to natural selection.

Darwin explicitly identified 'struggle for existence' as a metaphor for any selective process, not exclusively as a statement about bloody battle. He writes that 'a plant on the edge of a desert is said to struggle for life against the drought' as much as 'two canine animals in the time of dearth, may be truly said to struggle with each other which shall get food and live' (Darwin, 1859, p. 62). But abiotic struggle (the plant at the edge of the desert) yields no vector of progress in time, for physical environments fluctuate without direction and engender no biomechanical improvement in organic response (an elephant that evolves a hairy coat in glacial times is not a better elephant in any general sense). Biotic struggle, however, can produce a plausible vector of general improvement, for success in overt battle with conspecifics should favour structural and biomechanical improvement. As Darwin says, the deer who runs faster, longer and better, escapes the wolf (16).

Darwin therefore argued, in one of the most distinctive and insistent themes of all his writing, that biotic competition must predominate in the history of life. 'The relation of organism to organism is the most important of all relations' (Darwin, 1859, p. 477). He justified this view with an ecological notion of plenitude: all ecological addresses are occupied in nature. To use Darwin's own favourite metaphor, nature is a surface covered with 10,000 wedges, each sharply driven in. New species can enter such a full world only by insinuating themselves into a crack, driving down hard, and pushing another wedge out. (In his early writing, Darwin refers to replacement by competition as 'wedging'.) If new forms succeed primarily by driving others out in overt biotic competition, then progress will accrue because generalised biomechanical improvement and increased efficiency are engendered by biotic competition (whereas abiotic struggle only produces adaptation to changing local circumstances).

The conceptual situation is no different today. A predominant relative frequency for biotic competition remains the most potent argument for a proposition still dear to our hearts—'that vague yet ill-defined sentiment . . . that organisation on the whole has progressed' (Darwin, 1859, p. 345). Thus, the frequency of biotic competition remains a vital topic, as a kind of ultimate potential vindication for larger aspects of Darwin's world-view. The most sophisticated modern version of the biotic argument, Vermeij's (1987) theory of escalation, is the best available justification for progressive and adaptive trends in evolution. (I do not doubt his well-documented cases, but I do question their high relative frequency. Somehow I do not see much of life's pattern running on the snailshell–crabclaw model). Meanwhile, the increasing importance awarded to episodes of mass extinction (a subject much feared and denigrated by Darwin, who fully appreciated its potential for wreaking havoc upon his cherished biotic mode) suggests that abiotic struggle may be more important than previously thought by palaeontologists. (At least nature's surface is not always full of wedges if mass extinctions open ecospace so often.)

In any case (and acknowledging our current inability to resolve this key issue of relative frequency), I only urge that case studies of trends, nearly all of which treat this dichotomy of biotic and abiotic, be conducted with close attention to the need for description in the speciational mode, and to the consequences of thinking at this level. For example, Lidgard and Jackson (1989) make a persuasive argument for biotically driven trends to improvement in growth forms of encrusting cheilostome bryozoans from early Cretaceous to Recent times. Yet they present all their data on changing frequencies of the various growth modes through time as percentages—a style appropriate to a concept of gradual take-over by displacement in an ana-genetic flux. But speciational trends are driven by differential changes in *number* of taxa—the precious primary datum utterly lost when all assemblages are normalised to 100 per cent. Concepts are needed to free percepts for expression, and imperfect theories can block access to the data we need while failing to deliver their own goods (17).

In *Flatland*, E.A. Abbot's (1884) classic science-fiction fable about realms of perception, a sphere from the world of three dimensions enters the plane of two-dimensional Flatland (where he is perceived as an expanding circle). In a notable scene, he lifts a Flatlander out of his own world and into the third dimension. Imagine the conceptual reorientation demanded by such an utterly new and higher-order view. I do not suggest that the move from organism to species could be nearly so radical, or so enlightening, but I do fear that we have missed much by overreliance on familiar surroundings (12).

An instructive analogy might be made, in conclusion, to our successful descent into the world of genes, with resulting insight about the importance of neutralism in evolutionary change. We are organisms and tend to see the world of selection and adaptation as expressed in the good design of wings, legs, and brains. But randomness may predominate in the world of genes—and we might interpret the universe very differently if our primary vantage point resided at this lower level. We might then see a world of largely independent items, drifting in and out by the luck of the draw—but with little islands dotted about here and there, where selection reins in tempo and embryology ties things together. What, then, is the different order of a world still larger than ourselves? If we missed the world of genic neutrality because we are too big, then what are we not seeing because we are too small? We are like genes in some larger world of change among species in the vastness of geological time. What are we missing in trying to read this world by the inappropriate scale of our small bodies and minuscule lifetimes?

In any case, we can be sure of only one thing in our intellectual struggles to understand the patterns of life, including evolutionary trends as a prominent subject. Our fondest ideas and surest feelings are subject to a law of scholarly life represented in the last proverb of Little Buttercup's song: Here to-day and gone to-morrow.

NOTE

1. Definitions and the taxonomic parsing of conceptual categories always pose problems. An apparently unitary phenomenon may have a multiplicity of causal bases if it operates at several levels. (I am not at all sure, for example, that 'extinction' should be a common category for phenomena as disparate as the death of a species in a local area and the co-ordinated wipe-outs of mass extinctions. All involve taxic death, to be sure, but different causal bases at rising levels may make the phenomena so distinct that we harm any hope of adequate explanation by our linguistic lumping.) Likewise for trends: a directional character gradient through time in a well-defined monophyletic clade may be

quite different in meaning from the more generalised, full-fauna phenomena also, if more loosely, designated as 'trends', at least in the vernacular—increase in species diversity throughout the Phanerozoic, or increase in size of the largest living creature through time (see Bonner, 1968, on levels in phyletic size increase and their disparate causal bases). This chapter concentrates on the more technical definition of character gradients in monophyletic groups. In any case, the common causal property of differential species success produces all but the most minor of trends (those attributable to true lineage anagenesis), and some explanatory union may therefore undergird all phenomena generally described as 'trends'.

BIBLIOGRAPHY

Abbot, E.A., 1884, *Flatland*, London.

Ayala, F.J., 1976, Molecular genetics and evolution. In F.J. Ayala (ed.), *Molecular evolution*. Sinauer, Sunderland, MA: 1–20.

Bateson, G., 1979, *Mind and nature*, E.P. Dutton, New York.

Bonner, J.T., 1968, Size change in development and evolution. *J. Paleont.*, **42**, supplement to no. 5 (Memoir number 2, The Paleontological Society): 1–15.

Bretsky, P.W. and Bretsky, S.S., 1975, Succession and repetition of Late Ordovician fossil assemblages from the Nicolet River Valley, Quebec, *Paleobiology*, **1**: 225–37.

Darwin, C., 1859, *On the origin of species*, John Murray, London.

Eldredge, N., 1979, Alternative approaches to evolutionary theory, *Bull. Carnegie Mus. Nat. Hist.*, **13**: 7–19.

Fisher, R.A., 1958, *The genetical theory of natural selection*, 2nd edn, Dover, New York.

Futuyma, D.J., 1987, On the role of species in anageneis, *Am. Nat.*, **130**: 465–73.

Gilinsky, N., 1986, Species selection as a causal process. *Evolutionary Biology*, **20**: 248–73.

Gingerich, P.D., 1974, Stratigraphic record of Early Eocene *Hyopsodus* and the geometry of mammalian phylogeny, *Nature*, **248**: 107–9.

Gingerich, P.D., 1976, Paleontology and phylogeny: Patterns of evolution at the species level in early Tertiary mammals, *Am. J. Sci.*, **276**: 1–28.

Gingerich, P.D., 1983, Scaling of evolutionary rates. *Science*, **222**: 159–61.

Gould, S.J., 1977, *Ontogeny and phylogeny*, Harvard University Press, Cambridge, MA.

Gould, S.J., 1982, Darwinism and the expansion of evolutionary theory, *Science*, **216**: 380–7.

Gould, S.J., 1984, Smooth curve of evolutionary rate: A psychological and mathematical artifact (Reply to Gingerich), *Science*, **226**: 994–5.

Gould, S.J., 1987, Life's Little Joke, *Nat. Hist. Magazine* (April): 16–25.

Gould, S.J., 1988, Trends as changes in variance: A new slant on progress and directionality in evolution (Presidential Address), *J. Paleont.*, **62**(3): 319–29.

Gould, S.J., 1989, Punctuated equilibrium in fact and theory, *J. Social Biol. Struct.*, **12**: 117–36.

Gould, S.J., and Eldredge, N., 1977, Punctuated equilibria: The tempo and mode of evolution reconsidered, *Paleobiology*, **3**: 115–51.

Gould, S.J., and Lewontin, R.C., 1979. The spandrels of San Marco and the Panglossian paradigm: A critique of the adaptationist programme, *Proc. R. Soc. Lond.*, Series B, **205**: 581–98.

Gould, S.J. and Vrba, E.S., 1982. Exaptation—a missing term in the science of form, *Paleobiology*, **8**: 4–15.

Grant, V., 1989, The theory of speciational trends, *Am. Nat.*, **133**: 604–12.

Haldane, J.B.S., 1949, Suggestions as to the quantitative measurement of rates of evolution, *Evolution*, **3**: 51–6.

Hansen, T.A., 1978, Larval dispersal and species longevity in Lower Tertiary neogastropods, *Science*, **199**: 885–7.

Huxley, J.S., 1942, *Evolution, the modern synthesis*, Allen & Unwin, London.

Jablonski, D., 1986, Larval ecology and macroevolution in marine invertebrates. *Bull Marine Sci.*, **39**: 565–87.

Jablonski, D., 1988, Estimates of species duration, *Science*, **240**: 969.

Jablonski, D., and Valentine, J.W., 1981, Onshore–offshore gradients in recent eastern Pacific shelf faunas and their paleobiogeographic significance. In G.G.E. Scudder and J.L. Reveal (eds), *Evolution Today*. Proc. 2nd Interntl. Congr. of Systematic and Evolutionary Biology: 441–53.

Jablonski, D., Sepkoski, J.J., Jr, Bottjer, D.J. and Sheehan, P.M., 1983, Onshore–offshore patterns in the evolution of Phanerozoic shelf communities, *Science*, **222**: 1123–5.

Kropotkin, P., 1914, *Mutual Aid*, London.

Kurtén, B., 1959, Rates of evolution in fossil mammals, *Cold Spring Harbor Symposium: Quantitative Biology*, **24**: 205–15.

Lande, R., 1976, Natural selection and random genetic drift in phenotypic evolution. *Evolution*, **30**: 314–34.

Lidgard, S. and Jackson, J.B.C., 1989, Growth in encrusting cheilostome bryozoans: Evolutionary trends, *Paleobiology*, **15**: 255–82.

Lloyd, E.A. and Gould, S.J., in press, Species selection on variability, *Am. Nat.*

Mayr, E., 1963, *Animal species and evolution*, Harvard University Press, Cambridge, MA.

McKinney, M.L., 1988, *Heterochrony in evolution*, Plenum, New York.

McNamara, K.J., 1982, Heterochrony and phylogenetic trends, *Paleobiology*, **8**: 130–42.

McNamara, K.J., 1986, A guide to the nomenclature of heterochrony, *J. Paleont.*, **60**: 4–13.

McNamara, K.J., 1988, The abundance of heterochrony in the fossil record. In M.L. McKinney (ed.), *Heterochrony in evolution*. Plenum, New York: 287–325.

Simberloff, D., 1984, The great god of competition. *The Sciences*, **24**: 16–22.

Stanley, S.M., 1973, An explanation for Cope's Rule, *Evolution*, **27**: 1–26.

Stanley, S.M., 1979, *Macroevolution: pattern and process*, Freeman, San Francisco.

Tversky, A. and Kahneman, D., 1974, Judgment under uncertainty: heuristics and biases, *Science*, **185**: 1124–31.

Vermeij, G.J., 1977a, Patterns in crab claw size. The geography of crushing, *Syst. Zool.*, **26**: 138–51.

Vermeij, G.J., 1977b, The Mesozoic marine revolution. Evidence from snails, predators and grazers, *Paleobiology*, **3**: 245–58.

Vermeij, G.J., 1987, *Evolution and escalation*. Princeton University Press, Princeton, NJ.

Vrba, E.S., 1980, Evolution, species and fossils: How does life evolve? *S. African J. Sci.*, **76**: 61–84.

Vrba, E.S. and Eldredge, N., 1984, Individuals, hierarchies and processes: Toward a more complete evolutionary theory, *Paleobiology*, **10**: 146–71.

Vrba, E.S. and Gould, S.J., 1986, The hierarchical expansion of sorting and selection: Sorting and selection cannot be equated, *Paleobiology*, **12**(2): 217–28.

Walker, K.R., and Alberstadt, L.P., 1975, Ecological succession as an aspect of structure in fossil communities, *Paleobiology*, **1**: 238–57.

Williams, G.C., 1966, *Adaptation and natural selection*, Princeton University Press, Princeton, NJ.

APPENDIX 1.1:
THE DUET OF LITTLE BUTTERCUP AND CAPTAIN CORCORAN IN W.S. GILBERT, *HMS PINAFORE*, ACT II

The proverbs are referenced in the text of this article by the numbers here provided. Most are self-explanatory. 'Catchy-catches' in line 21 are babies, the reference being to baby talk and baby games.

1. Things are seldom what they seem
2. Skim milk masquerades as cream
3. Highlows pass as patent leathers
4. Jackdaws strut in peacock's feathers
5. Black sheep dwell in every fold
6. All that glitters is not gold
7. Storks turn out to be but logs
8. Bulls are but inflated frogs
9. Drops the wind and stops the mill
10. Turbot is ambitious brill
11. Gild the farthing if you will,
    yet it is a farthing still
12. Once a cat was killed by care
13. Only brave deserve the fair
14. Wink is often good as nod
15. Spoils the child who spares the rod
16. Thirsty lambs run foxy dangers
17. Dogs are found in many mangers
18. Paw of cat the chestnut snatches
19. Worn-out garments show new patches
20. Only count the chick that hatches
21. Men are grown up catchy-catches . . .
    He will learn the truth with sorrow
22. Here to-day and gone to-morrow.

Chapter 2

CLASSIFYING AND ANALYSING EVOLUTIONARY TRENDS

Michael L. McKinney

INTRODUCTION

The concept of 'trend' is arguably the single most important in the study of evolution. As change through time, evolution, like any change, has two basic parameters, direction and rate. While the latter has received the lion's share of attention in recent years, the former would seem to be more important: it is more critical to know where change is going (and has been) than how fast it occurs. For example, Riska (1989) argues that if evolutionary theory is ever to become truly complete, macroevolutionary patterns must be interpretable in (but not necessarily reduce to) quantitative genetic terms. He lists four impediments to explaining evolutionary patterns in this way. These are difficulties in: estimating heritabilities and genetic correlations; understanding the selection regime over time; understanding of developmental processes; and characterising macroevolutionary patterns. The first and third of these are clearly best studied in living organisms, where experimental controls allow heritability estimates through breeding, and direct observation of developmental processes. However, the second and fourth are directly related to the topic of this book: trends. Understanding palaeoenvironmental trends has long been a goal of historical geology, and resolution continues to improve. Even more directly related is characterising evolutionary patterns.

In this chapter, my goal is to provide a formal outline of evolutionary patterns. Given the central role of directionality in evolutionary theory, it is astonishing how little effort has been devoted to formal definitions of what trends are, and how to study them. The approach has been largely 'common-sensical'. Hence the purpose of this chapter: to make a first step towards a formal systemisation of concepts and terms such as 'trends', 'random', 'deterministic' and 'progressive', all central to the study of evolutionary directions. In addition, I will review some of the well-developed quantitative analytical methods that may be fruitfully applied both in describing and in inferring causation of evolutionary trends. For example, correlation of palaeoenvironmental trends with evolutionary ones is the primary evidence for explaining many patterns in the record. As such, it is covered by many chapters in this book (particularly Chapters 6, 9 and 13).

CLASSIFYING EVOLUTIONARY TRENDS

Traditional trend categories

In spite of the generally non-rigorous approach towards analysing trends in the past, there has been much consistency in what evolutionists consider to be a trend. Thus, Futuyma (1986a, p. 366), Gould (1988), and Hoffman (1989) all agree that there are two basic kinds of phylogenetic pattern that have been classified as trends in the past: *anagenetic* and *cladogenetic* patterns. I will use this as a primary distinction in classifying trends. Anagenetic trends are those occurring in a single non-branching lineage, while cladogenetic trends involve changes in branching (speciation). In both cases (shown in Figure 2.1), 'trend' is defined as a persistent (rarely monotonic) change in some state variable (discussed below), resulting in a significant net gain or loss in that variable through time. In anagenesis, the net change involves only one species at any single point in time, whereas numbers of species are involved in cladogenetic trends. This distinction essentially parallels Eldredge's (1979) distinction between the 'transformational' view and the 'taxic' view of evolution. Similarly, there is a general correspondence of anagenetic trends as being microevolutionary and cladogenetic as macroevolutionary in nature. However, I would agree with Hoffman (1989) that this dichotomy need not correspond to that between 'punctuated equilibrium [and] phyletic gradualism', as was suggested by Eldredge (1979).

Anagenetic trends seem common, as shown in many fossil studies since about the mid-1970s; especially familiar are those of microfossils. Gingerich (1985) has compiled a list of such studies, showing that much of this directional change involves body size (also see Stanley and Yang, 1987; and Chapter 4 of this volume) as the 'state variable'. This is to be expected since body size is the most important single morphological trait of an organism. Charlesworth (1984) discusses anagenetic change from the quantitative genetics view.

Overall, the evidence seems to support Lande (1986), who has said that 'both gradual changes and stasis can occur together with multiple discontinuities or jumps'. That is, neither anagenesis nor rapid cladogenesis can be ruled out from current evidence. Gould (1988) associates the cladogenetic view with sorting at the species level (see Chapter 1 herein). This broaches the topic of trend causation, which I will pursue further below. However, again I see no necessary association since it has not been shown that selection is irreducible to the level of the individual (Hoffman, 1989;

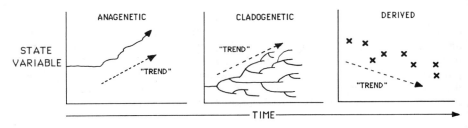

Figure 2.1.   Three kinds of trend observed in an evolutionary state variable. Morphological traits are the usual state variables analysed in non-branching (anagenetic) and branching (cladogenetic) patterns. Derived patterns are often taxonomic traits, computationally derived from (cladogenetic) morphological patterns. In all three cases, a 'trend' is a persistent but non-monotonic tendency to move in one state variable direction, either singly (anagenetic), or in a collective unit (cladogenetic, derived). As shown later, cladogenetic trends take a number of different forms.

Levinton, 1988). Trends can simultaneously occur both in a clade as a unit and within lineages constituting the clade, as illustrated below and in Chapters 4, 6, and 9 herein.

State variables

The 'state variable' referred to above, and shown in Figure 2.1, can denote virtually any characteristic that the evolutionist deems of interest. Usually, this is a morphological feature. It may be a 'specialised' organ (e.g. titanothere horn, tooth size in horses), or a more generalised feature such as body size, in which case trends will be more comparable across taxa. As discussed in Chapter 4 herein, body size is actually a highly derived composite variable and virtually impossible to measure completely, even in living organisms. Fortunately, growth of the many components constituting size is so highly covariant that any one, or a few metrics, will usually serve as a good approximation (proxy). Thus, 'body size' in both anagenetic or cladogenetic patterns is often represented by molar size, proloculus diameter, and so on. An even more problematic morphological variable is 'complexity', discussed below.

Raup (1988) has listed a number of features that can serve as state variables observed through time. Many of these will be familiar to the reader: extinction rate, origination rate, taxonomic diversity, and even ecological features such as community structure. I would place these in a second category, that of features derived from the first category of morphological traits (Figure 2.1). This is not, by any means, to relegate these features to secondary importance; indeed they provide more information at a glance when plotted and analysed. The point is that these features are ultimately *computationally* derived from changes in morphological features among many species (and are, therefore, often associated with cladogenetic patterns). Thus, extinction and origination trends, and even ecological trends, reflect morphological changes among the component organisms. (This, of course, parallels the distinction, made by Simpson, 1953, between morphological and taxonomic rates of evolution.) A number of trends in the derived category have been uncovered, and they are among the most important in evolutionary theory—for example, the apparent decline of extinction rates of many taxa through the Phanerozoic (Raup and Sepkoski, 1982) and the general increase in global diversity through the Phanerozoic (Valentine, 1985).

In terms of formal statistics, both morphological and derived state variables can be either ordinal or ratio variables. In the former case, they can be measured by some semi-quantitative scale that allows them to be ranked. For example, an organ may be more or less 'differentiated' or 'developed' than another, although it may be difficult to put a precise value on (McKinney and McNamara, 1990). The important point is that the feature must have some kind of measurable vector; thus, nominal traits (e.g., blue, green, and so on) will not do. This concept is clearly seen in heterochronoclines (Chapter 3 herein; see also McKinney and McNamara, 1990) wherein directional heterochronic (ontogenetic) changes (e.g., paedomorphosis or 'underdevelopment') form the basis of phylogenetic trends. This also illustrates another important principle: that temporal trends also often have a spatial vector. Heterochronoclines are often gradational with environmental continua (roughly speaking, a biological Walther's Law of facies change). Another spatiotemporal example, with a derived state variable, is the well-known 'nearshore–offshore' trend of cladogenetic origination and diversification (Bottjer and Jablonski, 1988; see Chapters 6 and 9 of this volume).

Finally, the metric of 'state variable' is inextricably tied to the key trend concept of 'progress'. Full discussion of this misleading concept is far outside the scope of this chapter, but suffice it to say that 'progress' as often used is highly subjective and that a universal, monotonic progression toward 'perfection' or even more mundane features clearly does not exist (see Nitecki, 1988, for a thorough review). Instead of one monotonic, universal trend in some ethereal (non-measurable) 'state variable', we see a large number of probabilistic (often reversing), non-universal (not manifested in all taxa) trends in many traits. In this latter sense, we may refer to 'progressive' trends in such traits as denoting a persistence of positive changes (increasing values) in the state variable (and 'regressive' trends as the opposite). In directional connotations, 'progress' here retains its meaning but it is stripped of the subjective baggage of 'better' (or 'worse' for regressive), meaning simply more of the state variable (clearly not always 'better', as in the case of extinction rate). Traditionally, palaeontology has focused mainly on progressive trends, perhaps because they are more common (for reasons, discussed below, dealing with initial value of the state variable) but there are many examples of regressive ones: decreasing body size in local lineages (Prothero and Sereno, 1982), loss of toes in horses (Futuyma, 1986a), decreasing tooth size in humans (Brace *et al.*, 1987), and so on.

'Complexity' is a state variable which has especially enamoured theorists with subjective ideas of 'progress' (almost invariably for philosophical reasons

—see Nitecki, 1988). Morphological complexity, for example, has obvious intuitive appeal, as shown by its long history, with roots back beyond even the Great Chain of Being, to the ancient Greeks. The problem is that even where complexity has a tangible referent in morphology, there is no consensus about just what complexity is. While the intense arguments over defining complexity in mathematics and information science are too far removed to be of direct use here (see Casti and Karlqvist, 1986), it is interesting that the most promising definitions there usually involve algorithmic approaches: complexity is the amount of information needed to specify (or compute) the system in question. This has potential application to DNA as an algorithm that specifies the ontogeny and maintenance of an organism: more 'complex' ontogenies (and therefore adults) are characterised by more coded information.

On a level more accessible to the morphologist, all formal definitions of complexity revolve around numbers and kinds of parts, and their inter-relationships. The complexity of a system is said to increase as number and kinds of parts (and relationships among those parts) increase (O'Neill *et al.*, 1986, provide a good discussion). This approach was taken by Schopf *et al.* (1975) in using the number of anatomical terms as a rough metric of complexity. More recently, Bonner's (1988) book on the subject similarly relies on number of kinds of cell in an organism as a rough metric of complexity. Another state variable that might also qualify as relating to the subjective kind of 'progress' is increasing 'efficiency' of some morphological traits. There is evidence (see Chapter 9 herein) that coevolution ('arms races') has resulted in such increases, as discussed below.

A FORMAL CLASSIFICATION OF TRENDS

In this section, I further refine the anagenetic and cladogenetic classification, using examples from particle theory to provide a frame of reference. Species, like particles, show both deterministic and random behaviour so that this section includes some discussion of trend causation, in addition to classification.

The particle view

In an unusual article, Schopf (1979) discussed the virtues of visualising species as particles. Essentially, he argued that while deterministic events govern species' evolution at fine spatiotemporal scales, at very coarse scales these deterministic factors may cancel out or otherwise show aggregate behaviour that is probabilistic and thus amenable to statistical 'laws' (analogous to the gas laws of particles). I believe this view is extremely useful in the present context, and will help considerably in visualising the kinds of species trend.

Consider a species as a particle in morphospace (Figure 2.2). It has a temporal (longitudinal) vector and moves around in morphospace as a function of that vector. Obviously, morphospace has more than the few

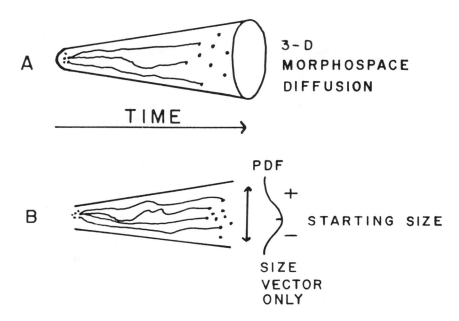

Figure 2.2. Schematic view of morphological evolution through time wherein species are particles whose direction is individually determined by environmental forces. As speciation occurs, particles replicate and these daughter particles diffuse into more distal areas of morphospace. Under random 'gas law' conditions, a multidimensional–normal particle distribution would result, with most being concentrated near the clade origin (shown in B, where size is the variable illustrated, actually representing a collapsed composite variable). For simplicity, only two and three dimensions are shown, but obviously true morphospace is *n*-dimensional, where *n* is the number of traits undergoing evolution.

dimensions shown. In fact, we may consider it to be a hypervolume with n dimensions, just as in Hutchinson's niche concept, except that instead of environmental parameters, the n dimensions are morphological ones. Any number of multivariate methods (e.g., factor analysis) are used in analysing change in this morphospace (see, for example, Bookstein *et al.*, 1985). Using these methods, it is possible to create summary (multivariate) variables to describe many state variables at once. However, most evolutionists just extract and analyse one or a few simple traits from the multidimensional morphospace (tooth area, brain volume, etc.) to maximise biological interpretability, often focusing on those that change the most. But, Cheetham (1987) has shown (in bryozoans) that it can be misleading to analyse trends only in isolated single characters, without at least some reference to other traits. This is because traits can be either coupled or decoupled as growth fields (McKinney and McNamara, 1990).

The particle view is useful not only for visual purposes but also for analytical ones because the study of the behaviour of particles is extremely well established. Berg (1983) discusses a number of quantitative methods describing particle diffusion in space through time. Some of these will be used below. Similarly, Raup's (1977) highly influential paper described morphological trends in the context of random walks. Single particles show these properties as well—they are Markovian, and Brownian motion is a random walk. Deterministic properties are modelled as 'drift'. Branching (cladogenesis) can be incorporated by allowing particles to replicate. Other analytical benefits will be seen below.

For each of the categories that follow, I subdivide the discussion into three distinct parts, for clarity. These are: description, causes and examples.

Anagenetic trends

Description
Much of the trend behaviour of one particle (species) has already been discussed. Persistent directional changes are often temporarily reversed, and can be either progressive or regressive in the long run. As noted, evidence from both the fossil record and genetic theory indicates that change can be either gradual or rapid. Berg (1983) uses the term 'drift' to denote a deterministic force externally applied to a particle that is otherwise randomly diffusing. For example, a particle in otherwise (random) Brownian motion may also be influenced by a very weak but persistent tug of gravity. Such morphospace 'diffusion with drift' would be analogous to a species' partly-random walk in morphospace: there is short-term unpredictability but in the long run there is a statistical preference. (For example, a generally cooling climate may promote larger size in the long term through Bergmann's Rule in spite of randomly fluctuating local conditions; see Figure 4.4 in Chapter 4 herein). To the evolutionist, Berg's term 'drift' is a bit confusing since, unlike genetic drift, it implies determinism, but I shall use it here since the meaning is clear when compared to diffusion.

Causes
The causes ('forces') of morphospace movement (either as deterministic drift or random movement, which essentially just means that there are multiple, often opposing, forces that produce no clear pattern) have traditionally been seen as 'external' to the species in the neo-Darwinian view. Excellent quantitative treatments of the role of selection, genetic drift and changing adaptive peaks in producing phenotypic fluctuations and trends are found in Charlesworth (1984) and Lande (1986). Futuyma (1986a) discusses the useful term *orthoselection* to denote a single dominant external force operating consistently in one direction.

An often underemphasised (mainly by palaeontologists, not by quantitative geneticists) factor in anagenetic time series has been the role of develop-mental constraints, often phenotypically expressed as correlated traits ('allometric relations'), which can limit selection response. These 'internal' forces mean that phenotypic changes (including trends) are not one-to-one reflections of environmental changes. This is because internal forces not only retard phenotypic response (create a 'lag time' to environmental changes or even dampen them out completely) but may even produce (or contribute to) trends on their own. (McNamara, Chapter 3 herein, and McKinney and McNamara, 1990, discuss ontogenetic input to trends.) Only continuing neontological work with selection response experiments (Riska, 1989, is an excellent example and discussion) can tell us directly just how great is the role of intrinsic forces.

Examples
Much of the heated debate over punctuated versus gradual patterns in the fossil record has focused on well-documented anagenetic trends in many kinds of organisms, from mammals to protozoans. These have been reviewed by Gingerich (1985, see also Table 4.1 of Chapter 4 herein). Individual anagenetic trends surveyed in this book include trilobites (Chapter 5), ammonites (Chapter 7), echinoids (Chapter 9), and fishes (Chapter 11). Even before the 'rates debate' first generated such numerous case studies in the early 1970s, anagenetic patterns were the major ones considered under the aegis of 'trends' by most workers (Gould, 1988).

Cladogenetic patterns

Description
Cladogenetic trends are directional patterns involving a number of species. This applies to any collective unit; it is not restricted to any level of clade or even phylogenetic unit. Thus, as discussed below, biosphere level trends also occur here. There are two basic kinds of cladogenetic trend: *asymmetrical* and *symmetrical* trends. In both cases, they can be further subdivided into *accretive* and *non-accretive* trends (Figure 2.3). As noted, cladogenetic trends often involve morphologically derived state variables, such as extinction rate, because a number of species are included. However, here I focus on the primary morphological state variables themselves for the sake of simplicity. Also note that the particle view still holds, with particles capable of showing drift, and also being able to replicate.

Turning first to a description of asymmetrical cladogenetic trends, Figure 2.3 illustrates that the accretive type involves origination of an ancestral species at a low value of some state variable, with subsequent expansion to higher values. Gould (1988; and Chapter 1 of this volume) calls these 'increase in variance' trends. The concept has its roots in Stanley's (1973) work which showed this in body size (see Chapter 4 herein), although Maynard Smith (1972) also discussed this 'nowhere but up' concept in other contexts (e.g., complexity trend). The asymmetry arises because expansion of the state variable is largely restricted to one direction, for one reason or another, discussed shortly. In terms of the particle view, this is a 'reflecting barrier' (Figure 2.3) which prevents particles (species) from attaining values (occupying morphospace) in that direction (Berg, 1983). The second kind of

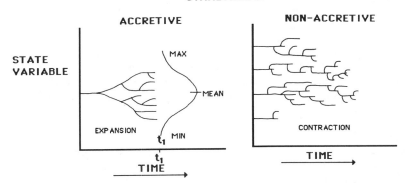

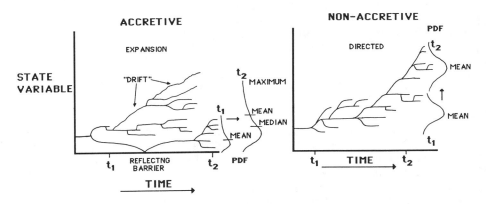

Figure 2.3.   Four basic kinds of cladogenetic trend in a state variable (usually morphology). Asymmetrical trends are the most commonly discussed in palaeontology. These can be either accretive (persistence of older values of state variable) or non-accretive. In the former case, the asymmetrical 'particle diffusion' is usually caused by a reflecting barrier that limits diffusion to one direction of morphospace. As diffusion proceeds the mean and maximum state variable values increase, as shown by the probability density function (PDF), that is, the frequency distribution of variable values. In contrast, non-accretive asymmetrical ('directed') trends tend to be motivated by biotic interactions that do not involve reflecting barriers such as minimum viable body size. Symmetrical trends show either expansion or contraction. Expansion occurs through adaptive radiation and clade proliferation in general; in this case there is no reflecting barrier to limit diffusion. Symmetrical contraction may occur from non-selective attrition, or selective elimination of extreme variable states.

asymmetrical trend is the non-accretive type, wherein state variable values at the ancestral level, in contrast to the first type, are not conserved. This produces a 'directed', as opposed to an expansionary, pattern and is due to different causes than the latter.

Symmetrical accretive trends are similar to asymmetrical accretive trends in conserving original state variable values and expanding into new ones (Figure 2.3). However, unlike the asymmetrical types, the original state variable values are not near a reflecting barrier so that diffusion to other levels is not constrained in any direction. In symmetrical non-accretive trends there is a contraction so that, after initial early expansion, the branching pattern shows a reduction in number of species (particle extinction).

Causes and examples
Asymmetrical accretive trends occur at many levels. Examples of general traits include body size in many clades and even the biosphere (Stanley, 1973; McKinney, Chapter 4 of this volume; Bonner, 1988), and complexity (especially of the biosphere, see Bonner, 1988). An example of a specialised trait showing an assymetrical accretive trend would be titanothere horn size (McKinney and Schoch, 1985), horse tooth size (MacFadden, 1986), or any organ that starts small and increases in size. Regarding causation, there are actually three aspects of asymmetrical accretive trends that need explanation: why the state variable is restricted on one side (i.e. the nature of the reflecting barrier); why the state variable starts near the reflecting barrier; and why diffusion to other values occurs.

The nature of the reflecting barrier of the state variable's size (body or trait) and complexity is easy to explain in most cases. For instance, body size in a clade is restricted because of intrinsic allometric limitations in any given body design: if one tries to 'shrink' a horse, surface/volume non-linearities will cause a lower size limit on that design, or force the design to change so much that it will cease to be a 'horse' (thus starting a new 'clade'; see Chapter 4 herein for further discussion). Of more interest is why they start there to begin with, and here the answer varies between size and complexity. In the latter, the biosphere started with simple ancestors because life is an open system and the laws of physics (especially the second law of thermodynamics) demand that ontogenies begin at minimal states of entropy and evolve complexity mainly by ontogenetic accretion, as discussed shortly. In the case of size, at the biosphere level, the same logic applies, but what about the many clade-level trends toward increasing body size ('Cope's Rule')? Extensive discussion of this is reserved for Chapter 4, but for now let us note two basic reasons: first, most clade size distributions are highly skewed to the right so that small species are much more common (and so become ancestors); and second, large species tend to go extinct more easily for a number of reasons—since many clade expansions begin after major extinctions, small species are therefore the usual ancestors. (Note that I omit Stanley's (1973) oft-cited reason that small organisms are less allometrically 'specialised' and so have more 'evolutionary potential'; I disagree with this for reasons discussed in Chapter 4.)

In at least most, if not all, major cases, the reason for the increase in upper values in a state variable is simply clade diversification into new selective regimes. However, as Maynard Smith (1972) points out, just to talk about 'going up' without being as specific as possible is not very enlightening. In the case of size, as descendant individuals migrate or otherwise find themselves in new environments, selection will sometimes favour larger size, such as in stabler environments (Chapter 4), and the upper limit of the collective unit (e.g., clade) will therefore increase. Many of McNamara's heterochronoclines (Chapter 3 herein) fall into this category, at least the ones with traits that start at values near a limit.

The evolutionary increase in complexity is a more subtle process. The environment does not as clearly 'select' for complexity so much as developmental processes 'experiment' through accretion of new ontogenetic pathways. I have discussed this at length elsewhere (McKinney and McNamara, 1990, Chapter 8), noting that while evolution by 'terminal addition' (recapitulation) was much abused in the past, there is no denying the general vector of increasing complexity via ontogenetic accretion. The reason, as argued by Katz (1987), Arthur (1984) and others, is that of the 'ratchet' principle (Levinton, 1988) whereby mutations that subtract developmental pathways from a viable system are much more likely to be detrimental than mutations that add pathways. In the latter, the pre-existing viability is retained and, by adding some, it may happen that 'new ways of doing things' are created. Since, as noted earlier, complexity consists of more parts and kinds of parts, this (generally! there are obviously reversals and many 'lateral' changes) accretive process creates a coarse vector of increasing maximum complexity while conserving ancestral simplicity.

Asymmetrical non-accretive trends (Figure 2.3) generally result from interactions (competition, predation) with other biota since the selective elimination of older groups (original, lesser-value state variables) rarely results from change in physical conditions. These trends are often co-evolutionary phenomena of the sort described by Vermeij (1987). Perhaps the best-known example is Jerison's (1973) work on brain size in carnivores and ungulates showing an evolutionary increase in both groups via an 'arms race' (Figure 2.4). Vermeij's examples focus on defensive and offensive mechanisms in marine groups. In both cases the 'race' is between general groups so that Futuyma (1986b) calls it 'diffuse coevolution': group averages shift but there is no precisely interlocking one-to-one change. Gould (1988) also discusses the brain-size example but I must disagree with his argument that it is only an increase in variance that creates the trend. In this instance, there is clearly selection against older groups (Figure 2.4) so that, unlike most body-size trends where older groups retain their place in various environments (and are thus accretive, see Chapter 4), lesser values of this state variable (brain size) are at a disadvantage and are reduced in diversity as expansion to higher values occurs. Thus, the genesis of the asymmetry is not a reflecting barrier preventing expansion to lesser values, that is, not a physical limitation, but the 'forces' of positive feedback intrinsic to much biotic interaction (O'Neill *et al.*, 1986). Some heterochronoclines fall into this category, because they are 'driven' by competition from ancestral species or predation pressure (Chapter 3). In the former case, of competition with living

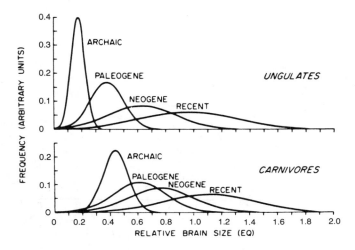

Figure 2.4.   Frequency distributions of relative brain size of ungulates and carnivores during the Caenozoic. Older forms are selected against as time progresses. Modified from Jerison (1973).

ancestors, the trend has been called 'autocatakinetic' (McKinney and McNamara, 1990) meaning it is 'self-driving' ('self' referring to the clade, 'driving' its own expansion), such as the mammalian clade driving its own brain increase.

Examples of symmetrical trends are less well known because where cladogenetic trends have been studied at all, asymmetrical ones have been the focus (since 'change' is more interesting—see Gould, 1988). Thus, good examples are rare, although one need only find a case where the initial value of a state variable is not directionally limited by physical constraint or biotic interaction. For instance, accretive trends could occur where the number of thoracic segments in a trilobite clade is both reduced and increased relative to the original ancestral species. Contractionary symmetry would occur where such early expansion was followed by selective loss of extreme values of the state variable. Gould *et al.* (1987) have shown that many clades have this pattern of early expansion and later contraction.

ANALYSING EVOLUTIONARY TRENDS

Evolution as a time series

Any sequence of observations through time is a time series. Such *longitudinal* data present certain statistical problems not found in *cross-sectional* data (the latter being values obtained from a number of measurements at a single point in time). This is because the most widely used statistical methods (*t*-tests,

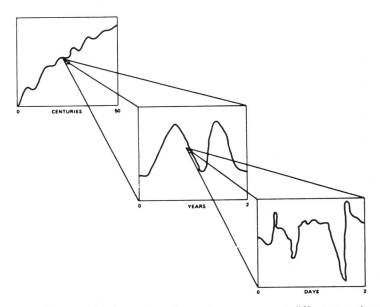

Figure 2.5.   Changes in dynamics of a system as seen at different scales of time and space. High-resolution information (time series of observations) available to the observer reveals rapid fluctuations that are partly and even completely lost at low resolution (centuries). In this case, high-resolution information does not 'see' the long-term trend that occurs across longer time-spans from persistent, large-scale 'structural' forces acting on the 'system' (e.g., species gene pool or maximum clade size), where 'system' is operationally defined as a group of components that interact more closely than those external to the 'system'. Modified from O'Neill *et al.* (1986).

correlation analysis, regression, etc.) are based on the assumption that values in the data set are drawn independently from one another. However, this is often not true for time-series data: a measurement made at a given point in time is likely to be correlated with points just before and after it. High values are often followed by high values and low by low. Such *serial correlation* (also called autocorrelation) makes intuitive sense: things can only change so fast. Temporal contingency (also a 'Markovian' property) is really the most essential aspect of evolutionary change. Jacob (1977) captures this in his well-known metaphor of evolution as 'tinkering' with pre-existing structures, and of course, comparative anatomy is based on this principle. Given this, it is surprising how little consideration has been given to time-series problems when analysing evolutionary data (Raup has been a leading exception, as discussed below).

One reason why evolutionists have been able to ignore the problem is the crucial role of scale. If measurements are taken far enough apart in time, there will be enough relaxation time for the 'memory of the system' to be erased (no serial correlation). The key is how fast the 'system' (e.g., mean morphological values, in many trends) being measured through time changes relative to the time intervals used by the investigator (Figure 2.5). It is thus uncertain just how often serial correlation has distorted evolutionary studies. McKinney and Oyen (1989) found it to be minor in two palaeontological time series. In contrast, Glass *et al.*, (1975) found that 72 per cent of time series in the social sciences showed such correlation (reflecting the much shorter, complete time series available). Obviously, a great deal depends on the resolution of the evolutionary data: large gaps may reduce serial correlation (if change is relatively fast) and the assumptions of independence are justified.

In any case, a better understanding of time series is crucial to a proper analysis of trends. On the one hand, it is needed to assess the amount of serial correlation present; on the other, it is needed to analyse data properly where such correlation is strong.

Time series: some basic statistics

Measuring serial correlation
The basic method of serial correlation is the *serial correlation coefficient*. It is calculated as follows:

1. Calculate a regression line through the time series of variable x (i.e. state variable x as a function of time).
2. Calculate the residual for each data point. (The residual is the difference between the actual data point value and that predicted by the regression line for that time.)
3. Regress the residual (e_t) of each data point versus the residual of the preceding point (e_{t-1}). The regression thus generated is:

$$e_t = r\,e_{t-1} + v_t \tag{1}$$

where v_t is white noise (unexplained flux) and r, the slope, is the serial correlation value.

In verbal terms, the above sequence involves nothing more than, first, 'detrending' the time series, which produces a stationary series (x fluctuates around a constant mean), and, second, calculating how much each preceding value, on average, affects the successive one. Thus, if r is 1, serial correlation is at a maximum since, excepting v_t (e.g., from measurement error), each x is completely determined by the preceding x. Hence, the assumption of independence is strongly violated. As r approaches 0 serial correlation diminishes accordingly, until at 0, independence is attained.

The reader may wonder at what point serial correlation becomes 'significant'. The most common procedure is to calculate the Durbin–Watson statistic from the residuals. The resulting value is then compared with a table containing various critical values, depending on sample size. An excellent, accessible discussion (and table) is found in Wonnacott and Wonnacott (1984), who have done considerable work in this area. The Durbin–Watson procedure is available in many statistical packages but where it is not, it can be computationally quite burdensome. Thus, the reader might note a quick and easy distribution-free method of testing for serial correlation described by Ostrom (1978). This 'Geary test' involves nothing more than a simple count of sign changes in the regression residuals and applying a runs test to them. (Obviously this is less powerful than tests based on theoretical distributions, such as the Durbin–Watson, but it is very useful for many purposes.)

The problem of unequally-spaced data points
The most serious problem in identifying and working with time-series data in palaeontology (and historical geology in general) is that the vagaries of the depositional process very rarely provide the equally spaced time intervals needed to carry out the above calculations (i.e. comparison of each data point with the preceding one must be 'standardised' to compare changes meaningfully). Other techniques of time-series analysis, discussed below, also need equally spaced data points. There is no panacea for resolving this dilemma; once information is lost, it can never be fully recovered. However, there are ways to estimate lost data points, so that we can at least approximately apply time-series methods. While estimating missing data may disturb some workers, it is better than ignoring the time-series problem completely, as has been the major recourse in the past.

One way around the interval problem, at least in cross-sections or cores of rocks, is to use equally spaced spatial distances among the 'data points'. That is, one is looking for trends in the morphology of fossils in the rock unit itself (see, for example, Raup and Crick, 1982). The problem, of course, is that this assumes that rates of deposition are similar between points, if this is to be meaningful in temporal terms. Nevertheless, it is a solution *vis-à-vis* computational mechanics and allows one to determine and analyse 'spatial' serial correlation.

Of more use is when one has some kind of absolute dating between data points, even if the dates are not equally spaced. If there are enough such dates, interpolation will allow the estimation of state variable values at equal time intervals. There are many ways to interpolate; a common one assumes that change between points is linear (Davis, 1986). A number of methods are discussed in Wilkes (1966). One very useful method takes serial correlation, where it occurs, and makes it work for the investigator. This is the auto-regressive (AR) model which can predict missing data on the basis of preceding values. Where serial correlation is high, this makes the predictions much more reliable than simple linear interpolation because it uses the 'memory' of the system to provide information. A simple such model is:

$$x_t = r\,x_{t-1} + v_t \tag{2}$$

where r is the serial correlation coefficient already discussed, and v is normally distributed white noise, again with zero mean and constant variance. For this to be most effective, the time series should be 'de-trended' as above (i.e., x is the residual e_t as above). Of course, this may seem circular since we have stated that the serial correlation coefficient cannot be correctly calculated without equal spacing! The solution is a 'bootstrapping' sort of technique described by Gottman (1981, p. 397): the data are fitted with a regression line and residuals computed. The most useful time interval is determined, based on that separating the original data points. The mean residual value is used as a 'dummy' at each empty point. Based on this time series, r is calculated (via equation (1)). Using this r, equation (2) is used to recalculate the missing values. The procedure is reiterated until the AR coefficient stabilises (an asymptote is reached), representing the 'best guess'. AR models that include more than one lag can be used to improve precision.

The method best used will vary with the nature of the data. The AR method works best where the original data are at least roughly equally spaced and missing values are few. Linear and other interpolations are more robust and flexible, but less precise. Obviously, all are less than ideal, but are better than nothing. Solace can be found in that, should gaps among dates be too great, serial correlation is unlikely to be a problem anyway and one can ignore these problems of estimating it and compensating for it (the latter discussed next). One 'quick and dirty' way to check for serial correlation with 'messy (non-interval) data' would be to regress each variable x against the preceding value, regardless of its temporal separation. If the resulting plot shows an uncorrelated, 'scattershot' pattern, there is probably little serial correlation in the data. Conversely, a roughly rectilinear pattern indicates high correlation (i.e. low values followed by low values for positive serial correlation).

Analysing evolutionary trends as time series

Trend-point estimates
The above is only a very basic outline of time-series principles, but it provides enough to apply to the study of evolutionary trends. Since a time series represents change in the value of a state variable through time (as, for example, Figure 2.5) it is easy to see how the above techniques apply to anagenetic time series where change in only one 'particle' is seen. However, it is less clear how to relate such a series to cladogenetic trends, where a number of particles are involved. The answer is found in the basic statistical concept of the point estimate, which provides a single value which summarises information about elements in a set. The two most familiar point estimates are the mean and standard deviation, which summarise the central tendency ('centre of gravity') and variation of a group, respectively.

For example, as shown in Figure 2.3, both central tendency and range estimates can help one characterise an evolutionary trend. In the case of asymmetrical accretion, the increasing variance in the growing 'right tail' creates a trend in the point estimate of maximum and mean values of the state variable (see Gould, 1988, Figures 4 and 5). (Other estimates of central tendency—median and mode, for instance—will be less affected by the growth of extreme values but will change nevertheless.) Range values are

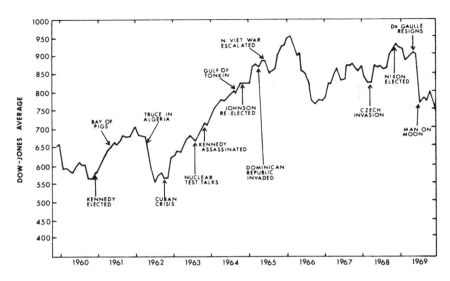

Figure 2.6. Time series of the Dow-Jones Average over a ten-year period. Some of the known proximal 'perturbations', and their effects, to the 'system' are shown (in this case a cultural 'system', the US economy, is shown; compare with hypothetical natural 'system' of anagenetic average morphology of a species in Figure 2.1). Modified from Gottman (1981).

useful in characterising symmetrical expansion, while the mean is most useful for asymmetrical directed ('coevolutionary') trends. These point estimates provide an objective way of analysing trends in collective units and reflect their objective reality. I am thus uneasy with Gould's (1988, p. 321) wording that such trends are 'byproducts or oddly-distorted perceptions of changes in variance'. They are neither purely byproducts nor perceptions. However, there is no doubt that they have a qualitatively different genesis than anagenetic trends.

Analysing single time series
The essential qualities of a single evolutionary time series are shown in Figure 2.6, the Dow-Jones Average over a ten-year period. The point estimate (mean) of this state variable is tracked, showing a large number of rises and falls. This variable represents a 'system' (a hopelessly abused term (O'Neill *et al.*, 1986) but here, as in most cases, the operational definition is clear) that is subjected to numerous 'external forces' that push either in opposite or the same directions, often coming and going, changing from in phase to out of phase with each other. These 'forces' are also operationally, and even *post hoc*, defined as any process (or 'event' if the process is very short-term) that has affected the system. Usually the (commonly small) magnitudes of such forces have a roughly log-normal distribution, which occurs when components

of the forces interact multiplicatively rather than additively (McKinney and Oyen, 1989).

The interaction of external forces on a state variable can result in a time series of virtually any conceivable pattern. 'Random' patterns do not mean that the changes were not produced by deterministic forces. It only means that there are a number of forces, fluctuating in strength and timing so that the results look chaotic and 'unpredictable' to the observer. Of course, palaeontological time series lack the resolution anywhere near that of Figure 2.6, being not only gap-ridden but having gaps of non-uniform spacing and duration. Further, fossil spatial resolution is also restricted: many spatially local events in Figure 2.6 are shown to have affected the Average (e.g., Kennedy's assassination).

I believe palaeontologists benefit from such a view as Figure 2.6 because it shows how 'non-random' evolutionary series really are. The concept of randomness has taken a strong hold as of late (discussed below) and it is important to emphasise that randomness is little more than a confession of ignorance by the observer, not an intrinsic property of evolving systems (except at the subatomic scale). Indeed, it seems surprising that palaeontologists can 'explain' trends or other time-series patterns at all given how poor the record is, relative to all that goes into determining large-scale processes (e.g., as implied by the exceptional record of Figure 2.6). Certainly a major point in our favour is the relative simplicity of natural systems (and so of their determinant 'forces') compared with cultural ones.

Of the huge number of possible patterns shown in a time series, four basic types are noted here as being useful in a discussion of evolutionary trends: interrupted, stepwise, trend, and stasis (Figure 2.7). These categories are not mutually exclusive and grade into one another; thus, a trend or stasis can be interrupted. Basically, interrupted and stepwise patterns focus on short-term

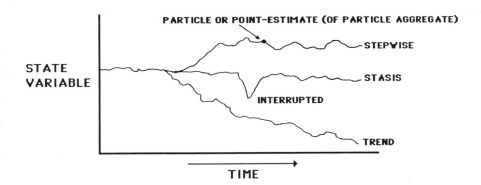

Figure 2.7.   Four basic kinds of time series in a state variable: trend, stasis, interrupted and stepwise. Causal 'forces', external to the 'system' (the state of which is summarised by the state variable) range from pulse determinism through persistent determinism, to 'random' processes (multiple determinism). Real time series may show more than one (a mixture of) these patterns.

shocks ('pulse determinism') to the state variable, while trend and stasis patterns focus on long-term behaviour ('persistent determinism'). A key point is that while analysis of each of these may tell us much about the degree and kind of identifiable determinism in a time series, they cannot prove causation of past events or specify the dynamics involved. Causation of events can never be completely proven *post hoc* and understanding of the dynamics is limited by the extreme incompleteness of the record of what happened. Nevertheless, as with a court of law, circumstantial evidence, even *ex post facto*, can be persuasive, and dynamics can be modelled and observed in modern processes.

Interrupted time series occur from 'pulse determinism': a single 'event' perturbs the system, temporarily affecting the state variable used to characterise the system (Figure 2.7). Following this the system may or may not return to its former state. For example, MacDowall *et al.* (1980) discuss the time series of the number of calls to directory enquiries at a phone company: there is a steady upward trend until a dramatic drop occurs at the time of imposition of a surcharge, whereafter the number begins to climb again. Interrupted time-series analysis is a well-developed area of study (see MacDowall *et al.*, 1980) and would seem to have much potential for the diagnosis and study of mass extinctions and other relatively short-term events. However, in a book on trends, this aspect must be glossed over as brief interruptions (literally) in longer-term processes.

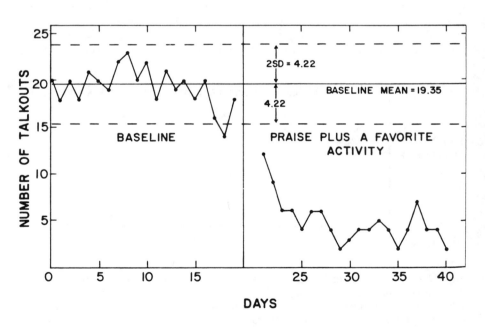

Figure 2.8 A

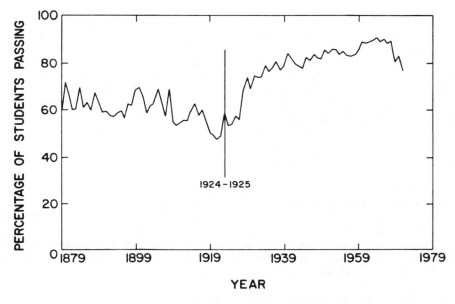

Figure 2.8 B

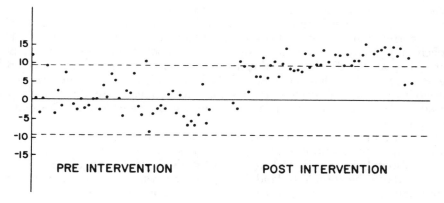

Figure 2.8 C

Figure 2.8. A: Where there is little or no serial correlation, a Shewart chart can gradually gauge the effect of an intervention, as illustrated in this time series showing the effect of praise plus a favourite activity acting to reduce the number of 'talkouts'. In this stepwise pattern there is a 'structural' shift in the mean (i.e. persistent determinism) and the normally distributed 'flux' around it. B: Time series of student performance in Ireland. In 1924 there was a policy change. C: Shewart chart constructed from Irish educational data above, after correcting for serial correlation. It shows that the policy change probably had a 'significant' effect at the 0.05 level. All three charts taken from Gottman (1981).

Stepwise time series represent pulse events which have a more lasting effect on the state variable. This pattern is most familiar as the 'punctuated equilibrium' pattern: a system is 'structurally' equilibrated with external forces, undergoes rapid change, and becomes re-equilibrated at a new level. However, there is no reason why the equilibration has to be as rigid in all systems as it is sometimes claimed to be in species. Equilibration in many cases simply means that the forces that buffet the system ('up and down' on the graph) are generally equally counterposed, offsetting, and/or cancel each other out. Of particular relevance here is the claim that species selection accounts for many stepwise morphological trends: trends are said to be manifestations of differential stepwise branches (see for example, Stanley, 1979).

An excellent way to analyse stepwise patterns is with a Shewart chart, widely used by engineers in quality control. As shown in Figure 2.8A, a Shewart chart is constructed by comparing the pre-step (pre-intervention) time series with that after the intervention. This simply involves calculating the mean and standard deviation of the pre-intervention series, setting up a 0.05 confidence interval (about two standard deviations) around the mean, and seeing if the post-intervention series drifts outside the confidence interval. Where it does, we may infer that a change has occurred due to intervention (be it long-term, as in Figure 2.8A, or short-term, as with an interrupted series, where this method can also be used to see if the 'pulse' is strongly outside the range of past flux). However, this simple procedure assumes that there is no significant serial correlation (as is true in Figure 2.8A). This points out one of the pitfalls of time series because if serial correlation is present, the confidence interval can be strongly biased. Specifically, points will tend to cluster above and below the mean, artificially spreading out the distribution of estimates of the mean around its expected value. The result is that the standard deviation will underestimate the true variability of the mean because it does not take the mean-spreading effects of serial correlation into account as it assumes independence (see Gottman, 1981, p. 59).

It is not difficult to correct for serial correlation in using the Shewart chart to determine intervention effects. This will be illustrated with a real time series, also in Figures 2.8B and 2.8C. In 1924, there was a change in policy by the Board of Education in Ireland. The question is whether this change had a significant effect on the time series of students passing the middle- and senior-level exams. It is unclear from a visual inspection of the time series itself whether post-intervention percentages are truly above the pre-intervention flux. However, before constructing the Shewart chart we must consider that the serial correlation of the series turns out to be quite significant: 0.536. Correcting for this is based on the following model:

$$x_t - \text{mean} = 0.536\,(x_{t-1} - \text{mean}) + e_t \qquad (3)$$

This AR model standardises all points relative to the pre-intervention mean and, more importantly, predicts each one on the basis of the known serial correlation. Thus, if we plot the residuals of this model (real value –

predicted value) for each point, we will have a Shewart chart that has the serial correlation effects subtracted from it (discussion and example from Gottman, 1981, p. 339). The confidence interval is thus that of the residuals, now known to be uncorrelated. Upon doing this, we find that the post-intervention effect was significant (Figure 2.8C).

Note that this Shewart chart approach is not limited to stepwise patterns. We have already noted its use in interrupted series, but trends that occur after an intervention (any event or point) may also be compared with the pre-intervention pattern. For instance, in the example just cited, there is a possible 'trend' after the intervention, as well as a possible 'structural' levelling off. Later data would be needed to be sure.

The random walk concept
Trends are, as noted, persistent directional changes in a time series. In contrast, stasis is defined as non-directional change leading to 'nowhere' (Figure 2.7). Both are intuitively obvious, but there are a number of subtleties and statistical aspects that are poorly understood by many evolutionists. Most of these revolve around the 'random walk' concept which I therefore discuss in some detail.

Raup (1977) first introduced to palaeontology Yule's (1926) observation that time series with a 'memory' could easily generate trends even if the movements were determined completely at random. That is, even if a state variable point estimate has a 50–50 chance of going up on a graph, a great deal of structure (e.g., pronounced directionality) can result if the position of each point is strongly serially correlated with the preceding point (Figure 2.9). This is simply the principle that a series of fair coin flips will often produce a number of heads or tails in a row: if each flip is retained in the memory (e.g., the variable moves up each time from the previous point), persistent trends can result. Raup (1987) and Bookstein (1987; 1988) have expanded on this, pursuing in detail the role of randomness in producing 'non-random' patterns.

In statistical terms, recall the AR equation (2) above:

$$x_t = r x_{t-1} + v_t \tag{2}$$

A 'pristine' random walk, as discussed and simulated by Raup and Bookstein, occurs when $r = 1$ (perfect memory) and the 'noise', v, is the movement up or down (in most models, it is a single increment or decrement of unit value), determined at 'random'. In time-series jargon, such random walks have non-stationarity of the mean and variance. Obviously, as the serial correlation (r) decreases, the non-stationarity (trending tendencies) will diminish since previous 'particle' movements will not be retained in the memory: each point is essentially determined anew, creating stationarity. Thus, when $r = 0$, we have a 'white noise' process, $x_t = v_t$, where v is normally distributed, with constant mean and variance.

Because random walks can often generate trends, much has been written (Raup, 1977; 1987; Raup and Crick 1981; Bookstein, 1987; 1988) about how to determine whether evolutionary trends represent simply random walks or something more consistently deterministic. The first step in determining this

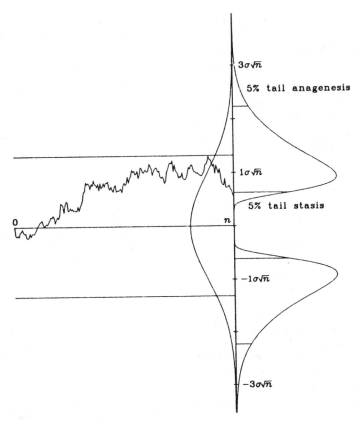

Figure 2.9. A random-walk time series produced by 20 000 'coin flips'. The horizontal band is maximum walk excursion. Expected standard deviations of a truly random are shown by tick marks, up to ±3σ (where σ is the standard deviation) representing the 99th percentile. By Bookstein's (1988) criteria, very small excursions from the original state variable value are a rare occurrence, occurring in only 5 per cent of random walks. Similarly, a pronounced 'trend' (tail anagenesis) also occurs only in 5 per cent of such walks. That is, if a time series shows an excursion as high (or low) as to fall within the 5% tail anagenesis area, it is likely not the result of a simple random walk since only 5 per cent of such walks will have such an excursion (after the 20 000 steps shown). Instead, there is a good chance that some persistent, non-random force 'of consequence' is operating to statistically bias the 'coin flips' over the long run. Modified from Bookstein (1988).

is to use the random walk as a null hypothesis. If the end-point of the time series is outside the confidence interval for a walk, then 50–50 probabilities (fully random) are unlikely to be at work: there is a statistical bias (an 'unfair' coin, as it were). Thus, as discussed by Berg (1983), if we have a series with $n = 100$ steps (e.g., 100 000 years = 100 1000-year intervals) and the chance of moving up or down (p) is exactly 0.5, the mean movement up or down is $np = 50$ vertical units each way. The standard deviation is $\sqrt{npq} = 5$, so that if, after 100 steps (100 000 years) the state variable ended up more than $+10$ vertical units (about 2 σ) either way from its original value, we should become suspicious of non-randomness (Figure 2.9). Bookstein (1987, 1988) has built more sophisticated methods from this approach. In all cases, however, note how the standard deviation (the confidence zone of Figure 2.9) increases only as the square root of n, the number of steps ('time'). Thus, the 'particle diffusion' of the random walk, while creating trends, slows down rapidly with increasing time. While equal-spaced intervals have been emphasised in theory and practice, this is one case where unequal intervals may be acceptable: the main focus is on the up/down movement, not step length. Varying step length will introduce some 'noise' but if there are enough of them, they will cancel out. Note also the role of scale again: the serial correlation of random walks can also cause 'short-term trends', that is, 'runs' on either side of the mean. Raup and Crick (1981) present a Markovian runs test for this, but it has fairly low resolution. Charlesworth (1984) presents a more powerful method.

While the above methods are used to determine whether trends can be explained by random walks, we have not yet discussed how to identify a trend. In most cases, this is obvious, via visual inspection. Most time-series texts define trends by fitting the time series with a regression:

$$x_t = a + bt + e \tag{4}$$

where the slope, b, represents the trend. As $b = 0$ is stasis, a main goal is to determine if b is significantly different from 0. Here is yet another case where serial correlation must be watched out for because, as discussed above, this will bias (underestimate) confidence intervals for the slope estimate. Another test for trend is to see if serial correlation diminishes linearly with progressive lags of the autoregression; where it does, trend occurs (Gottman, 1981).

Evolutionary meaning of trend and stasis
The serial correlation of evolutionary 'tinkering' strongly biases even random walks to create trends. Stasis is more the exception than the rule. Thus, Bookstein (1988) shows that stasis occurs in only about 5 per cent of random walks (Figure 2.9). However, I feel that the distinction between 'randomness' and determinism has been overly polarised. As stated above, randomness is simply a function of the ignorance of the observer. Even where we have a purely random walk, there are still reasons why each event occurs (e.g., Figure 2.6). Similarly, in evolutionary random walks, the randomness also has deterministic causes but they are too many and difficult to measure. Perhaps the clearest insight on this comes from Raup's (1988) comment that if a time series fits a random walk, it means that *nothing of consequence is*

driving the system. That is, no single force, pushing in one direction ('up or down') is consistently dominant. In contrast, if the series is non-random (outside the random walk confidence interval), one (or more) force is consistently dominant enough to alter the probabilities of direction ('up or down') away from 50–50. Clearly, there is a gradation in all this; the fact that we must use confidence intervals to test for random walks indicates the probabilistic nature. Note, too, that directional time series are called 'trends' whether they are within the realm of random walks (barring the few producing stasis) or in the realm of unicausal, clear-cut determinism.

What are the deterministic forces driving evolutionary time series? Raup (1977) has noted that random walks of trends are caused when biological causes of change may be so many and varied that change acts as a random walk. In anagenetic trends this may thus include a species gene pool acted upon by numerous, equally opposed (on average) environmental forces. However, recall that the state variable may also be a point estimate of cladogenetic behaviour. For instance, a random walk may represent a cladogenetic mean affected by many, varied external pressures (e.g., a symmetrical clade with equally counterposed forces of species selection). Anagenetically, genetic drift may be another possibility (Charlesworth, 1984). As the forces driving the trend become less equally opposed (on average) a directional bias will emerge; in extreme cases this will lead to single prominent 'cause' that can sometimes be identified. This brings us to the final, crucial topic.

Comparing two or more time series
Where a prominent 'force' drives a time series, we have the best situation for understanding the dynamics of state variable behaviour. We can compare the time series of the 'force' (called an *exogenous variable*) with that of the morphological state (endogenous) variable. As Bookstein (1987, p. 461) has said: 'time is unsuitable as an independent variable . . . one should instead be explaining change in terms of exogenous variables that are independently measurable'. For example, Davis (1981) provides time-series data on climate and body size in mammals over the 50 000 years which show considerable similarity in the behaviour of body size and climate, implicating the latter as a major controlling exogenous variable. Thus, even though we cannot directly prove causation *ex post facto*, parallel behaviours among time series can provide strong circumstantial evidence. Recall that the evolutionary trend need not be anagenetic. For instance, the mean or maximum of a cladogenetic pattern for size may be increasing. When plotted against the time series of, say, mean global temperature, we may find that global warming has created more habitat for the clade, promoting diversification and perhaps also increasing the number of size-increasing habitats.

Of course, given the problem of serial correlation in single time series, one should also expect to find difficulties in comparing two of them. In particular, recall that serial correlation causes an underestimate of the variability of the mean of the state variable. Thus, when we compare two state variables, we can expect similar problems with confidence intervals involving their regression. Specifically, the basic regression for comparing two time series is:

$$Y_t = a + b X_t + e_t \tag{5}$$

where Y and X are the two serial variables (X is the exogenous one). Thus, interval data between the two series are compared at the same times. Where one or both variables show strong (positive) serial correlation with time, this regression leads to an underestimate of the confidence intervals for the intercept, a, and slope, b, for much the same reason that the variable mean is underestimated: because the points are constrained, estimates of their variance will be smaller than the true variance since, as noted, the regression (ordinary least squares) calculations assume that points on the plot can vary independently.

The correction for serial correlation in such regressions is not only easy to carry out but the logic of it is easy to show, given equation (5), then

$$Y_{t-1} = a + b X_{t-1} + e_{t-1} \tag{6}$$

Subtracting (6) from (5):

$$Y_t - Y_{t-1} = b (X_t - X_{t-1}) + (e_t - e_{t-1}) \tag{7}$$

where $e_t - e_{t-1} = v_t$. If we define dY as $Y_t - Y_{t-1}$ and dX as $X_t - X_{t-1}$, then all the above reduces to:

$$dY_t = b\, dX_t + v_t \tag{8}$$

This method is called *first differencing*. Thus, one regresses differences between the variables. The key is that subtraction of the error terms (e_t, e_{t-1}) eliminates the serial correlation, producing the uncorrelated white noise needed for the assumptions of ordinary least squares regression. Unfortunately, first differencing, although common (see Raup and Crick, 1982, p. 98) is not entirely correct. In relying on the subtraction of error terms, it is assuming that serial correlation is 1. That is, recalling equation (1):

$$e_t = r\, e_{t-1} + v_t \tag{1}$$

first differencing says that $e_t - r\, e_{t-1} = v_t$, because $e_t = e_{t-1} + v_t$, that is, $r = 1$.

To remedy this, a better solution to serial correlation is *generalised differencing*, which incorporates r, permitting more accurate, regression parameters. In this case, $dY_t = Y_t - rY_{t-1}$, and $dX_t = X_t - rX_{t-1}$. The regression equation is then computed on this differenced data, as with (8), the only change being that:

$$dY_t = a' + bdX_t + v_t \tag{9}$$

where $a' = a (1 - r)$, where a is as in equation (5). The first data points of X and Y, having no previous values, cannot be differenced. Instead, they are transformed: $dY_1 = \sqrt{1 - r^2}\, (Y_1)$, and similarly for X. Note that r (calculated from equation (1)) is only an estimate of the 'true' serial correlation. It is

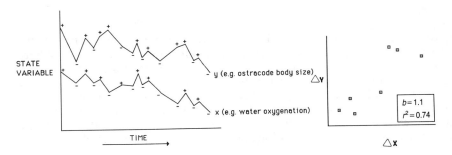

Figure 2.10.   Right: Plot and regression estimate of slope (*b*) calculated from comparing generalised differences of two time-series variables, *x* and *y*. By removing contingency effects (non-independence) of serial correlation through differencing, a truer regression estimate and comparison are obtained. In this case, *x* and *y* show a fairly strong correlation even after serial correlation effects are removed(r^2, the proportion of variation explained by the regression line, is 0.74). From the slope, we may infer that *y* increases slightly faster than *x*. Left: Plot of raw (undifferenced) time-series, *x*- and *y*-variables (e.g., ostracode body size and water oxygenation). Reyment's (1971, p. 83) runs-test method of comparing the two series involves matching the peaks and valleys (pluses and minuses). The application of a simple statistical test (see Davis, 1986) will judge if the number of matches is greater than expected by chance.

adequate for most work, especially where serial correlation is very high; however, if the worker requires maximum precision, methods to make better estimates of *r* are found in McKinney and Oyen (1989) and Ostrom (1978).

What is gained from all this labour? As summarised in Figure 2.10, the resulting regression of differences permits an unbiased direct comparison of the two time series, rather than simply looking for matches between the two in 'valleys and peaks'. The slope of the regression directly tells us how the variable *Y* changes as a function of *X* (e.g., is it isometric with slope 1? is it non-linear?). Further, the coefficient of determination (the square of the correlation coefficient) tells us what proportion of the variation is explained by the regression (relationship), that is how 'closely' the two variables are related. Even more promising is that more than two variables can be compared in this way: it is possible to perform multiple regression on a number of time series. This can lead to 'causal modelling' (i.e. path analysis—see Li, 1975) which models the causal arrows among a number of variables. This is rarely used with time-series variables but it would be intriguing to see it so applied in such areas as sea level, climate, global diversity, and so on. (Gottman, 1981, has an example using multiple-series medical data.) In some such cases, it might be better to use reduced major axis regression which, unlike least squares, assumes no single 'independent' variable, but operates symmetrically on the variables (Davis, 1986).

Finally, where the trend researcher does not wish to carry out direct regression between time series but desires more than simple visual comparison between matching 'peaks and valleys', Reyment (1971, p. 83) describes a simple runs test which tests the significance level of a given number of matches (Figure 2.10).

SUMMARY

Trends are persistent statistical tendencies in some state variable(s) in an evolutionary time series. Such variables may be point estimates (e.g., mean, maximum) of a group (e.g., cladogenetic, concerning a number of species) or a single lineage (e.g., anagenetic, concerning a number of individuals in a species). The variables may be primary (morphological) or derivatives from primary observations (e.g., taxonomic originations, extinctions). There are four broad categories of cladogenetic evolutionary trends: (accretive and non-accretive) symmetrical, and (accretive and non-accretive) asymmetrical. The asymmetrical pair are most prominent in evolutionary thought. The causes of such trends vary, but reflecting barriers and low initial states are most important in accretive asymmetrical trends, while biotic interactions (e.g., autocatakinesis) are crucial in non-accretive symmetrical trends.

As time series, evolutionary trends need to be analysed for serial correlation, that is, the non-random association of state variable values with values immediately preceding and succeeding them in time. This involves the estimation of the serial correlation coefficient. The gap-ridden nature of most fossil time series may reduce or eliminate the serial correlation. However, it also complicates the analysis of the series because most methods are based on the assumption of equally spaced data points. Interpolation, such as by autoregressive models, can help cope with this. Time-series analysis can be used not only on trends, but also interrupted, stepwise and stasis patterns observed in the fossil record.

The use of random walks as a null model has been extremely useful analytically, but has perhaps been overemphasised. I say this because 'randomness' is a confession of ignorance, so it essentially explains nothing, except that multiple forces act to buffet natural systems in unpredictable ways (which is hardly news). While it is admittedly necessary to eliminate the null possibility before searching for a 'cause of consequence', the whole notion of treating natural systems as 'random' particles must take into account the extreme scale-dependence of randomness. At sufficiently coarse (temporal and spatial) scales of observation, we may expect systems to behave randomly because so many processes are involved over the large spans of time and space observed (e.g., long-term evolution of a clade). Thus, the most informative analyses of evolutionary trends are finer-scaled (to allow isolation of determinants, acting on specific 'particles' or small groups of them). This is especially true when we can make use of exogenous time series ('independent' state variables) such as climatic indicators, which can be correlated with the biotic variable to be explained. Direct comparison among such time series is possible when corrected by generalised differencing. This is the only way to rigorously 'explain' events *ex post facto*, where no repeatable experiments are possible and direct observation is obviously lacking.

REFERENCES

Arthur, W., 1984, *Mechanisms of morphological evolution*, Wiley, New York.
Berg, H.C., 1983, *Random walks in biology*, Princeton University Press, Princeton, NJ.
Bonner, J.T., 1988, *The evolution of complexity*, Princeton University Press, Princeton, NJ.
Bookstein, F.L., 1987, Random walk and the existence of evolutionary rates, *Paleobiology*, **13**: 446–64.
Bookstein, F.L., 1988, Random walk and the biometrics of morphological characters, *Evolutionary Biology*, **9**: 369–98.
Bookstein, F., Chernoff, B., Elder, R., Humphries, J., Smith, G. and Strauss, R., 1985, *Morphometrics in evolutionary biology*, Academy of Natural Sciences of Philadelphia, Special Publication no. 15.
Bottjer, D. and Jablonski, D., 1988, Paleoenvironmental patterns in the evolution of post-Paleozoic benthic marine invertebrates, *Palaios*, **3**: 540–60.
Brace, C., Rosenberg, K. and Hunt, K., 1987, Gradual change in human tooth size in the Late Pleistocene and post-Pleistocene, *Evolution*, **41**: 705–20.
Casti, J. and A. Karlqvist (eds), 1986, *Complexity, language, and life: mathematical approaches*, Springer-Verlag, Berlin.
Charlesworth, B., 1984, Some quantitative methods for studying evolutionary patterns in single characters, *Paleobiology*, **10**: 308–18.
Cheetham, A., 1987, Tempo of evolution in a Neogene bryozoan: are trends in single morphologic characters misleading?, *Paleobiology*, **13**: 286–96.
Davis, S.J., 1981, The effects of temperature change and domestication on the body size of Late Pleistocene to Holocene mammals of Israel, *Paleobiology*, **7**: 101–14.
Davis, J.C., 1986, *Statistics and data analysis in geology*, Wiley, New York.
Eldredge, N., 1979, Alternative approaches to evolutionary theory, *Bull. Carnegie Mus. Nat. Hist.*, **13**: 7–19.
Futuyma, D.J., 1986a, *Evolutionary biology*, Sinauer, Sunderland, MA.
Futuyma, D.J., 1986b, Evolution and coevolution in communities. In D. Raup and D. Jablonski (eds), *Patterns and processes in the history of life*, Springer-Verlag, Berlin: 369–82.
Gingerich, P.D., 1985, Species in the fossil record: concepts, trends, and transitions, *Paleobiology*, **11**: 27–41.
Glass, G., Willson, V., and Gottman, J., 1975, *Design and analysis of time series experiments*, Colorado University Associated Press, Boulder.
Gottman, J., 1981, *Time-series analysis*, Cambridge University Press, Cambridge.
Gould, S.J., 1988, Trends as change in variance: a new slant on progress and directionality in evolution, *J. Paleont.*, **62**: 319–29.
Gould, S.J., Gilinsky, N., and German, R., 1987, Asymmetry of lineages and the direction of evolutionary time, *Science*, **236**: 1437–41.
Hoffman, A., 1989, *Arguments on evolution*, Oxford University Press, Oxford.
Jacob, F., 1977, Evolution and tinkering, *Science*, **196**: 1161–6.
Jersion, H., 1973, *The evolution of the brain and intelligence*, Academic, New York.

Katz, M., 1987, Is evolution random? In R. Raff, and E. Raff, (eds), *Development as an evolutionary process*, Liss, New York: 285–315.

Lande, R., 1986, The dynamics of peak shifts and the pattern of morphological evolution, *Paleobiology*, **12**: 343–54.

Levinton, J.S., 1988, *Genetics, paleontology, and macroevolution*, Cambridge University Press, Cambridge.

Li, C.C., 1975, *Path analysis—a primer*, Boxwood Press, Pacific Grove, California.

Maynard Smith, J., 1972, *On evolution*, Edinburgh University Press, Edinburgh.

MacFadden, B.J., 1986, Fossil horses from '*Eohippus*' (*Hyracotherium*) to *Equus*: scaling Cope's Law, and the evolution of body size, *Paleobiology*, **12**: 355–69.

MacDowall, D., McCleary, R., Meidinger, E. and Hay, R., 1980, *Interrupted time series analysis*, Sage, Beverly Hills, CA.

McKinney, M.L. and McNamara, K.J., 1990, *Heterochrony: the evolution of ontogeny*, Plenum, New York.

McKinney, M.L. and Oyen, C.W., 1989, Causation and nonrandomness in biological and geological time series: temperature as a proximal control of extinction and diversity, *Palaios*, **4**: 3–15.

McKinney, M.L. and Schoch, R.M., 1985, Titanothere allometry, heterochrony, and biomechanics: revising an evolutionary classic, *Evolution*, **39**: 1352–63.

Nitecki, M. (ed.), 1988, *Evolutionary progress*, University of Chicago Press, Chicago.

O'Neill, R., DeAngelis, D., Waide, J. and Allen T., 1986, *A hierarchical concept of ecosystems*, Princeton University Press, Princeton, NJ.

Ostrom, C.W., 1978, *Time series analysis: regression techniques*, Sage, Beverly Hills, CA.

Prothero, D. and Sereno, P., 1982, Allometry and paleoecology of medial Miocene dwarf rhinoceroses from the Texas Gulf Coastal Plain, *Paleobiology*, **8**: 16–30.

Raup, D.M., 1977, Stochastic models in evolutionary paleontology. In A. Hallam (ed.), *Patterns of evolution*, Elsevier, Amsterdam: 59–78.

Raup, D.M., 1987, Neutral models in paleobiology. In M. Nitecki and A. Hoffman (eds), *Neutral models in biology*, Oxford University Press, Oxford: 121–32.

Raup, D.M., 1988, Testing the fossil record for evolutionary progress. In M. Nitecki (ed.), *Evolutionary progress*, University of Chicago Press, Chicago: 293–317.

Raup, D.M. and Crick, R.E., 1981, Evolution of single characters in the Jurassic ammonite *Kosmoceras*, *Paleobiology*, **7**: 200–15.

Raup, D.M. and Crick, R.E., 1982, *Kosmoceras*: evolutionary jumps and sedimentary breaks, *Paleobiology*, **8**: 90–100.

Raup, D.M. and Sepkoski, J.J., 1982, Mass extinctions in the marine fossil record, *Science* **215**: 1501–2.

Reyment, R.A., 1971, *Introduction to quantitative paleobiology*, Elsevier, Amsterdam.

Riska, B., 1989, Composite traits, selection response, and evolution, *Evolution*, **43**: 1172–91.

Schopf, T.J.M., 1979, Evolving paleontological views on deterministic and stochastic approaches, *Paleobiology*, **5**: 337–352.

Schopf, T.J.M., Raup, D., Gould, S.J. and Simberloff, D., 1975, Genomic versus morphologic rates of evolution: influence of morphologic complexity, *Paleobiology*, **1**: 63–70.

Simpson, G.G., 1953, *The major features of evolution*, Columbia University Press, New York.

Stanley, S.M., 1973, An explanation for Cope's Rule, *Evolution*, **27**: 1–26.

Stanley, S.M., 1979, *Macroevolution: pattern and process*, Freeman, San Francisco.

Stanley, S.M. and Yang, X., 1987, Approximate evolutionary stasis for bivalve morphology over millions of years: a multivariate, multilineage study, *Paleobiology*, **13**: 113–39.

Valentine, J.W. (ed.), 1985, *Phanerozoic diversity patterns*, Princeton University Press, Princeton, NJ.

Vermeij, G., 1987, *Evolution and escalation*, Princeton University Press, Princeton, NJ.

Wilkes, M.V., 1966, *A short introduction to numerical analysis*, Cambridge University Press, Cambridge.

Wonnacott, T.H. and Wonnacott, R.J., 1984, *Introductory statistics for business and economics*, Wiley, New York.

Yule, G.U., 1926, Why do we sometimes get nonsense-correlations between time series?, *J. Roy. Stat. Soc.* **89**: 1–89.

Chapter 3

THE ROLE OF HETEROCHRONY IN EVOLUTIONARY TRENDS

K.J. McNamara

INTRODUCTION

Of the many models that have been proposed to explain evolutionary trends, few have incorporated the role of intrinsic factors, such as heterochrony, most having focused on extrinsic factors. Evolutionary trends have been interpreted as: the product of 'directional speciation' (Grant, 1963; Stanley, 1979), induced by 'mutation pressure'; 'phylogenetic drift' (Raup and Gould, 1974), induced by 'genetic drift'; 'species selection' (Stanley, 1975; 1979) induced by 'natural selection' pressures; the result of evolution towards increased specialisation in species-specific characters—(the 'effect hypothesis' of Vrba, 1980); 'environmental orthoselection', such as a consistent decrease in temperature (Futuyma, 1986); 'co-evolutionary interactions' (Futuyma, 1986), including the 'arms race' concept of Vermeij (1987); and changes in 'variance' (Gould, 1988; see Chapter 1 of this volume). None of these explanations discusses the complex interplay between intrinsic factors, such as heterochrony, and extrinsic factors, such as competition or predation pressure directing evolution along particular environmental gradients.

In recent years there has been a strengthening of the view that heterochrony (changes in timing or rate of developmental events, relative to the same events in the ancestor) plays an important role in directing morphological change along particular evolutionary pathways (Ede, 1978; Gould, 1980; Alberch, 1980; Levinton and Simon, 1980; McNamara, 1982; Maderson *et al.*, 1982; McKinney, 1988; McKinney and McNamara, 1990). As such it can be argued that heterochrony is a crucial factor in the generation of evolutionary trends. A complex nexus exists between changing timing or rates of development and changing environmental factors, which plays a critical role in channelling morphological change along particular, constrained pathways.

Evolutionary trends may range from small-scale *anagenetic* ('transformational' of Eldredge, 1979) to large-scale *cladogenetic* ('taxic' of Eldredge, 1979). Many of these anagenetic lineages have been shown to involve heterochrony (McNamara, 1982; 1988; McKinney and McNamara, 1990). In a number of subsequent chapters in this book examples of such heterochronically induced trends are described (e.g. Chapters 5, 7–9, 11 and 12). It is the aim of this chapter to show how such trends can be generated.

McKinney and McNamara (1990) have argued recently that heterochrony is perhaps the most important factor in the generation of intrinsic phenotypic change, and it is this phenotypic change which is the target for selection. This may be focused on shape or size, or on life history or behavioural strategies. Agents of selection do not focus on a random array of traits, but are constrained by the nature of organisms' developmental programmes. These may be considered as a series of constrained structural organisations. For instance, the limb morphology of living vertebrates is virtually unchanged from that in Devonian vertebrates (Darwin's 'unity of type'—see Shubin and Alberch, 1986). Such invariance in limb development has been interpreted as having arisen from historically conserved mechanisms of morphogenesis (Oster *et al.*, 1988). Thus, all limbs form under a single set of basic 'construction rules', so restricting the extent and degree of variability that can be produced. Oster *et al.*, (1988) have suggested that these rules impose a developmental order that is quite independent of adaptive pressures and other extrinsic factors. In other words developmental constraint that is a product of these construction rules imposes a pre-existing directional element, when perturbations occur. Any perturbations can only operate within the constrains of the fundamental construction rules and so result in heterochronic change with a pre-existing directional component. When combined with extrinsic directionality, such as an environmental gradient, the directions in which evolution can proceed are heavily constrained. In the long term evolutionary trends ensue.

THE CLASSIFICATION OF HETEROCHRONY

In this section a brief overview of the principal heterochronic processes is presented to aid in the understanding of some of the nomenclature that is used elsewhere in this book. The work of Gould (1977) and Alberch *et al.* (1979) in formalising the heterochronic processes has been largely followed by subsequent authors. McNamara (1986) has presented a general guide to these terms. The whole concept of heterochrony, and its role in evolution, is discussed fully in McKinney and McNamara (1990).

Heterochrony is expressed in two ways: as *paedomorphosis* and as *peramorphosis*. In the former, ancestral juvenile characters are retained by the descendant adult, whereas in the latter ancestral adult characters appear in the descendant juveniles. Heterochrony can occur within populations, and so generate intraspecific variation, or it can be expressed interspecifically.

Six heterochronic processes that involve shape changes are recognised (see also Figure 4.6). Paedomorphosis is produced by three of these processes: *neoteny*, *progenesis* and *postdisplacement*. Neoteny is a reduction in rate of morphological development. This, like other heterochronic processes, can operate at any level, from cell through organ to individual. Thus it may affect the whole organism, or only act on specific structural elements.

Progenesis occurs by early cessation of developmental events in the descendant. While precocious sexual maturation in the descendant, resulting in premature retardation, can occur and produce a maximum body size less than that of ancestor, local growth fields can also be subject to progenesis.

Postdisplacement occurs by delayed onset of growth of particular morphological structures. Should subsequent development and cessation of growth be the same in the descendant as in the ancestor, the displaced structure will attain a shape at maturity resembling that found in a juvenile of the ancestral form.

The three corresponding peramorphic processes are: *acceleration*, *hypermorphosis* and *predisplacement*. Acceleration of rate of morphological development during ontogeny will produce a peramorphic descendant. Like neoteny, acceleration can affect the whole organism, or very often it is dissociated, operating on specific structures.

Hypermorphosis occurs by extending the juvenile growth period. If this occurs by delayed onset of sexual maturation, growth allometries are extended to a larger size, and the hypermorphosis is global in its effects. Like progenesis, it can also operate just in local growth fields.

Predisplacement occurs by earlier onset of growth of structures. If subsequent development and cessation of growth are the same as in the ancestor the structure will be morphologically more advanced and larger than the equivalent structure in the ancestral adult.

HETEROCHRONOCLINES

Many fossil lineages reveal a pattern of structural change involving shape, or number, or size, such that descendant species become either 'more' paedomorphic, or 'more' peramorphic. It is less common for a structure in a lineage to show reversals, the character being, (say) paedomorphic in the first descendant, then peramorphic in the second, followed by paedomorphic in the next.

Let us first consider a series of ontogenies that change through time by paedomorphosis. The ancestral, non-paedomorphic form (termed the *apaedomorph*) undergoes a sequence of morphological changes through, for example, stages A to M, during the course of its ontogeny (Figure 3.1A). If a descendant form evolved from the apaedomorph with all or some paedomorphic characters, then the descendant will pass through fewer stages, perhaps A to K. After the evolution of this first paedomorph, another form may evolve by paedomorphosis. This second paedomorph may pass only through stages A to I. This species may ultimately give rise to a further paedomorph, that passes only through stages A to G. This pattern continues

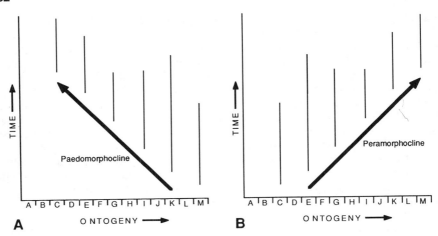

Figure 3.1. A: The paedomorphocline: a morphological gradient of progressively more paedomorphic species through time. B: The peramorphocline: a morphological gradient of progressively more peramorphic species through time. A to M represent arbitrary ontogenetic stages.

to the last species to evolve in the lineage, which is the most paedomorphic. In our theoretical lineage this last species only passes through stages A to C. The six adult morphologies in the lineage form a morphological gradient through time of M-K-I-G-E-C. These adult morphological stages follow the opposite morphological pathway to the ontogenetic development of the earliest species, the ancestral apaedomorph. This sequence of adult morphologies displaying a temporal morphological gradient of increasingly more juvenile characters has been termed a *paedomorphocline* (McNamara, 1982).

The operation of peramorphic processes will produce a similar pattern, but one that is a mirror image of the paedomorphocline. The lineage will consist of a sequence of increasingly more peramorphic species. This has been termed a *peramorphocline* (McNamara, 1982). In a peramorphocline, the ancestral (aperamorphic) species may pass through, for example, ontogenetic stages A to C during ontogeny. By the operation of any one of the peramorphic processes, a descendant form will evolve that passes through stages A to E during its ontogeny. A subsequent descendant species may arise that passes through yet more stages, A to G, and so on. The six forms in Figure 3.1B have adults that constitute a morphological gradient through time of C-E-G-I-K-M. Collectively paedomorphoclines and peramorphoclines may be termed *heterochronoclines* (McKinney and McNamara, 1990).

As has been discussed in Chapter 2, evolutionary trends are of two basic forms: anagenetic or cladogenetic. Heterochronoclines may likewise be anagenetic or cladogenetic. This is not surprising, because I believe that the vast majority of anagenetic trends are heterochronoclines. Anagenetic heterochronoclines show, on the geological time-scale, no apparent overlap

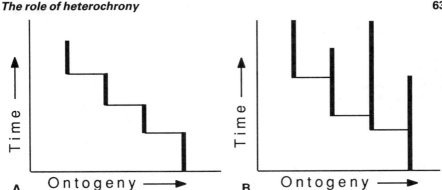

Figure 3.2. A: A stepped anagenetic paedomorphocline. B: A stepped clado-genetic paedomorphocline.

in temporal ranges between the species and there is no overall increase in numbers of species: as a new species arises so its ancestor becomes extinct (Figure 3.2A). In other words there is no increase in variance. Where there is perceptible overlap in stratigraphic (time) ranges between species in a heterochronocline, then the speciation event is cladogenetic, the ancestral species surviving as one of the branches (Figure 3.2B). In this scenario species numbers, and thus variance, will increase. Different selection pressures will determine whether the heterochronoclines are anagenetic or cladogenetic, as is discussed below.

When I first proposed the concept of paedomorphoclines and peramorpho-clines (McNamara, 1982) I interpreted these morphological gradients as being discontinuous. However, it could be argued that if the heterochronic changes are induced by (say) very small gradual changes in allometries between successive populations, then these changes could be continuous. While there is some evidence for this in some intraspecific lineages it has yet to be determined for interspecific cases. Detailed stratigraphic collecting is needed to resolve this. The apparent preponderance of discontinuous heterochrono-clines suggests that it may be that a species' genetic homeostasis or gene pool integration is so interlocked that species changes can occur only in discrete packages even when the environment is continuously graded.

The concept of heterochronoclines was originally proposed as a way to explain directional *interspecific* evolutionary trends. Examples have been described mainly in marine invertebrates, such as brachiopods (see below), echinoids (Chapter 9), trilobites (Chapter 5), ammonites (Chapter 7) and other molluscs (McNamara, 1988). In this book examples are also docu-mented in some vertebrates, such as fishes (Chapter 11) and reptiles (Chapter 12). In all of these examples the transitions are between what are interpreted, on the basis of morphological criteria, as species. Levinton (1983; 1988) has criticised my approach to explaining evolutionary trends at the species level on the basis that the changes occur without cladogenesis. He argues that

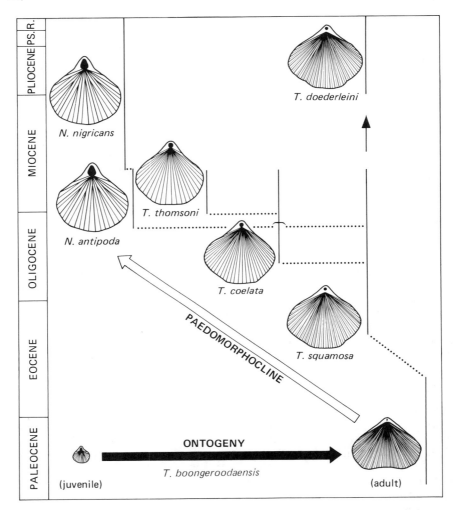

Figure 3.3. Paedomorphocline of species of *Tegulorhynchia* and *Notosaria*, illustrating temporal narrowing of shell, reduction in number of ribs, decrease in beak angle and increase in relative foramen size by progressive reduction in extent of growth of deltidial plates. These morphological changes are thought to have enabled colonisation of shallow water environments by descendant species. After McNamara (1982, Figure 3).

unless it can be shown that cladogenesis has occurred one cannot be certain that the morphotypes are actually different species. Unfortunately Levinton mistakenly interpreted the lineages that I described (McNamara, 1982) as being anagenetic patterns. While such patterns do occur, in those examples of heterochronoclines described in my 1982 paper all show appreciable temporal overlap of 'morphospecies'. Moreover, in groups, such as echinoids, it is possible to use the same morphological criteria to distinguish the fossil taxa as are used in living taxa.

Having discussed how heterochronoclines can theoretically occur, I will briefly describe an actual example. Perhaps one of the best is the *Tegulorhynchia–Notosaria* paedomorphocline (McNamara, 1982; 1983), a lineage of Caenozoic rhynchonellid brachiopods.

Fossil and living species of these two genera occur in the Indo-West Pacific region. The oldest species is the Paleocene *Tegulorhynchia boongeroodaensis*. This species underwent appreciable morphological change during ontogeny (Figure 3.3). Its shell became relatively broader; valves increased appreciably in depth, reflecting a great increase in internal volume; the commissure (the line that joins the two valves) developed a strong median fold; the rib number increased from about 25, at a shell length of 2 mm, to about 80 at a maximum shell length of 18 mm; the beak reduced in height, so increasing the umbonal angle; and the foramen became relatively smaller, as the deltidial plates joined, almost closing the foramen. In life the pedicle would have passed through this. The decrease in foramen size reflects a relative reduction in pedicle thickness through ontogeny.

The second species in the paedomorphocline is the New Zealand Late Eocene to Early Miocene *T. squamosa*. It has been suggested (Lee, 1980) that this long-ranging species may be synonymous with the living species *T. doederleini*. If, as seems likely, this is the case, then all subsequent species that evolved along the paedomorphocline did so by cladogenesis. *T. squamosa–doederleini* has fewer ribs than *T. boongeroodaensis* (only 60), a slightly narrower shell, less strongly developed median fold, larger foramen and slightly larger shell size. The descendants, *T. coelata*, which first appeared in the Middle Oligocene, and *T. thomsoni*, from the Early Miocene, continue the trends apparent between the older species: fewer ribs, narrower shell, weaker plication and larger foramen. The end members of the paedomorphocline are fossil and living species of *Notosaria*, *N. antipoda* and *N. nigricans*, respectively. The adults of these species are morphologically closest to the earliest ontogenetic stages of the apaedomorph, *T. boongeroodaensis*, in having few ribs (less than 25), relatively very narrow shells, most disjunct deltidial plates, and so largest foramen. Heterochronoclines that are described on the basis of coexisting species, such as the *Olenellus* paedomorphocline (Chapter 5 herein), are spatial heterochronoclines. These must have been formed by cladogenesis.

EFFECT OF HETEROCHRONIC PROCESSES AND GROWTH STRATEGIES ON EVOLUTIONARY TRENDS

While there have been many recent attempts to decide which particular heterochronic processes operated on fossil lineages (McNamara, 1988), many problems are encountered, not the least of which is the use of size as a proxy for time. This question has been addressed in detail by McKinney and

McNamara (1990). McKinney's (1988) concept of allometric heterochrony is one that can be used in the absence of data on age.

In the case of the *Tegulorhynchia–Notosaria* paedomorphocline, the paedomorphosis appears to have been global in its effects. It has been proposed (McNamara, 1983) that neoteny was the principal heterochronic process. A very small paedomorph. *T. sublaevis* (Thomson) that coexisted with *T. squamosa*, has been interpreted as progenetic, on the basis of its small adult size and paedomorphic characters (McNamara, 1983), but this is really allometric progenesis. While it is possible that interspecific evolution in the *Tegulorhynchia–Notosaria* paedomorphocline may have been gradual, it is more likely that the pattern of evolution within this lineage was like that depicted in Figure 3.1B, a cladogenetic paedomorphocline: long periods of morphological stasis within species, and rapid unidirectional transitions between them. This is supported by information on the stratigraphic range of *T. squamosa–doederleini*. This 'species' pair shows an extremely long period of morphological stasis, for up to 45 million years (Lee, 1980).

The main impact of differing types of heterochronic process on the generation of evolutionary trends is on the patterns of heterochronoclines. The operation of different heterochronic processes may have some influence on the rates of evolution within the heterochronoclines. Where either global progenesis or hypermorphosis operated, the chances of interspecific changes being episodic are much greater. This will occur when allometric changes during ontogeny are great, or where rates of differentiation of individual structures are high. In such cases abrupt changes in timing of maturation can produce 'punctuated' events. For instance, in trilobite paedomorphoclines, where progenesis caused size reduction combined with thoracic segment reduction, evolution of one species from another will be saltatory. In the *Olenellus* paedomorphocline (Chapter 5 herein), the apaedomorph and first paedomorph have 14 thoracic segments, but the fourth paedomorph, the last member of the paedomorphocline, only nine. In this character at least, change was episodic. The operation of global progenesis or hypermorphosis in organisms, such as arthropods, where phenotypic expression itself is episodically induced by moulting, combine to produce abrupt changes between species, and thus a pronounced *stepped heterochronocline* (in the form of *stepped paedomorphoclines* or *stepped peramorphoclines*). Where there is no overlap of species' ranges, the pattern would be a *stepped anagenetic paedomorphocline* (or *peramorphocline*) (as in Figure 3.2A). Where there is overlap between species ranges, the pattern can be described as a *stepped cladogenetic paedomorphocline* (or *peramorphocline*) (Figure 3.2B). If the lineage shows gradual directional intraspecific change (see the *Echinocyamus* lineage in Chapter 9 herein), then the evolutionary trend can be described as a *gradual anagenetic paedomorphocline* (or *peramorphocline*). Such gradual heterochronoclines are more likely to develop when heterochrony is occurring by gradual unidirectional intergeneration shifts in allometries. They are thus more likely to be seen in organisms, such as vertebrates, where growth is continuous.

The nature of the organism's growth pattern will therefore be crucial in determining patterns of evolutionary change, in particular whether evolutionary changes will be episodic or gradual. Thus different effects can be created by the same heterochronic process, depending on whether it is affecting organisms that grow continuously (e.g., vertebrates) or those whose phenotypic expression of growth (or at least the external expression of that growth) is episodic (e.g., arthropods) (see McKinney and McNamara, 1990 for a more detailed discussion).

DISSOCIATED AND MOSAIC HETEROCHRONOCLINES

Lineages, such as the *Tegulorhynchia–Notosaria* and *Olenellus* lineages, are global paedomorphoclines. That is to say, all heterochronic changes are paedomorphic. However, in other lineages only specific morphological traits are affected, or some might be paedomorphic, others peramorphic. If the traits are all peramorphic (or all paedomorphic), but produced by different processes, this is a *dissociated heterochronocline*. In other words some traits might be affected by acceleration, others by pre-displacement. Some might even be affected by more than one process. However, if some traits are affected by paedomorphic processes, while others are affected by peramorphic ones, a *mosaic heterochronocline* can be produced.

This implies that local growth fields, or suites of growth fields, may be under specific selection pressure. While the number of dissociated and mosaic heterochronoclines described in the literature is quite small, this probably reflects more a preoccupation with heterochrony that involves large-scale shape and body-size changes late in ontogeny. Thus, examples of body-size increases or reductions, and concomitant global morphological changes, are more easily recognisable than those that involve just subtle shifts of allometric coefficients or rates of differentiation of maybe only a few structures. It is probable that dissociated heterochronoclines occur more often in evolutionary trends than has hitherto been recognised.

The frequency of dissociated and mosaic heterochronoclines is dependent to some degree on the nature of the ontogenetic development of the organisms. For instance, some echinoids, such as spatangoids and holasteroids, not only undergo pronounced allometric changes during growth, but also have different coronal plates growing with different allometries (often some being positive, while others are negative). Furthermore, even different axes within a single plate may have different allometries. Through time these opposing allometries can become increasingly polarised as different heterochronic processes act on the different axes (McNamara, 1988b). The resultant structure that evolves can be quite extreme. For example, in the evolution of pourtalesiid echinoids (David, 1989), peramorphoclines occur within some lineages, such as the great increase in longitudinal growth of the coronal plates, at the expense of transverse growth, which occurs with negative allometry and forms a paedomorphocline. The extreme expression of these dissociated heterochronoclines is the holasteroid *Echinosigra*, which has an elongate, flask shape (David, 1989). In such cases, not every structure need be under individual selection pressure. Suites of characters may be under the influence of the same selection pressure, in other words there is trait covariation.

Where heterochrony acts on a two-tiered level in some colonial organisms (see Chapter 10 herein), the style of heterochrony at one level might directly influence the nature of the heterochronic processes operating at the other level. In the case of variable echinoid axial plate allometries, the reduction in plate allometry in one direction will also affect that in another direction. If not, then there would be no phylogenetic change. Developmentally it is probably also energetically more efficient if there is an overall reduction in allometric growth in one direction while another increases. Consequently, the areal extent of the echinoid plate might not change, but its shape can, quite appreciably. Thus its contribution to the functions of the organism can change.

Dommergues and Meister (1989) have described another form of extreme mosaic heterochrony where paedomorphoclines and peramorphoclines operate on the same structure within individual ammonites. Rather than different structures showing different types of heterochrony, the same structure shows different types at different stages of its ontogeny. In some harpoceratine ammonites from the Early Jurassic of north-west Europe the growth paths of the shell may follow a paedomorphocline in the early growth stages, but then in later growth stages follow a peramorphocline. Such a change occurs because of changing allometries during growth. After all, in most organisms, structures that grow allometrically do not do so at a constant rate during the entire course of ontogeny (see Chapter 2 herein).

HETEROCHRONOCLINES AND ENVIRONMENTAL GRADIENTS

For heterochronoclines to be initiated and persist, the developmentally induced morphological polarity must follow an environmental gradient. In the earlier example of a paedomorphocline in the *Tegulorhynchia–Notosaria* lineage, this is thought (McNamara, 1983) to have developed along an environmental gradient of deep to shallow water. This was attained by translation of ancestral juvenile morphological adaptations to the adult stage of descendants. Juvenile ancestral species had morphological characteristics that were adapted to function only in a certain size range in a deep water environment. They include a relatively large pedicle for attachment. Being more unstable than the adult, a relatively larger pedicle by which it attached to a hard substrate was functionally advantageous. There was virtually no real increase in pedicle thickness (as determined by the size of the foramen in the umbonal region of the shell) during ontogeny of the ancestral species, *Tegulorhynchia boongeroodaensis*, following the early juvenile growth. This juvenile ancestral pedicle size was an exaptation (see Chapter 1), for when translated into adult shells it allowed the occupation of higher hydrodynamic regimes, in shallower water. Species of *Notosaria* live in New Zealand today in the intertidal zone, whereas *Tegulorhynchia doederleini* is a deep water inhabitant. A further exaptation was the low convexity of ancestral juvenile shells, which reflects the possession of a relatively small lophophore. This is in contrast to the highly convex shell developed in ancestral adults due to positive allometric growth of the lophophore (a combined feeding and respiratory structure), necessary for the occupation of a low hydrodynamic regime in deep water. Global neoteny along the paedomorphocline resulted in the evolution of a smaller lophophore. Where current flow is much greater, a smaller lophophore is energetically more effective than a larger structure. Reduction in rib number may be of no adaptive significance at all, but merely be a covarying trait.

Many echinoid lineages (see Chapter 9 herein) show heterochronic evolution along environmental gradients of coarse- to fine-grained sediments. This may also correlate to evolution along a gradient from shallow to deep water (McKinney, 1984, 1986). Water depth and sediment grain size are common environmental gradients in the marine environment affecting invertebrate evolutionary trends (McNamara, 1988a). In the terrestrial environment, temperature and elevation are common environmental gradients along which lineages evolve. For instance, Grant (1963) recognised the importance of environmental gradients in controlling speciation in five species of the herbaceous plant *Polemonium*, species evolving along a gradient towards lower temperatures at higher elevations.

To understand heterochronocline development it is pertinent to consider each species as occupying an adaptive peak (Wright, 1932) along the environmental gradient. Wright's model comprised a landscape of adaptive peaks and valleys. The summits of peaks are occupied by the genetic 'elite', whose genotypes produce morphotypes most fitted to a particular environment. Adaptive peaks have tended to be viewed as a combination of morphological/functional characteristics of a taxon combined with the environmental regime that it occupies, the adaptive zone. Adaptive peaks are better considered solely in terms of the phenotypic character, being products of intrinsic change, while ecological niches are best regarded as the product of extrinsic factors (biotic and abiotic environment). McKinney and McNamara (1990) consider that while adaptive peaks are only as effective as the ecological niches into which they fit, exaptations (formerly known as *pre-adaptations*) imply that adaptive peaks are restricted by the phenotype. Niches become viable when an exaptation, or series of exaptations, opens up or creates the niche. Niches 'pre-exist' as potential roles in a sense, but one may also say that organism's 'create' their own niches from that potential pool. Unoccupied niches then become realised niches, while the exaptations create a new adaptive peak.

Heterochronoclines develop along a single axis in the overall adaptive landscape. The environmental gradient along which the heterochronocline develops can be considered to consist of a series of ecological niches into which the adaptive peaks slot. Each species will consist of a normal range of heterochronic phenotypes; but only some attain the adaptive peak. For example, some phenotypes within each species may show extreme development of paedomorphic traits. If these do not lie along the axis of the environmental gradient, and do not confer any adaptive advantage, then they will barely scrape the base of the adaptive peak, fail to be preferentially selected, and not contribute toward the evolution of the paedomorphocline.

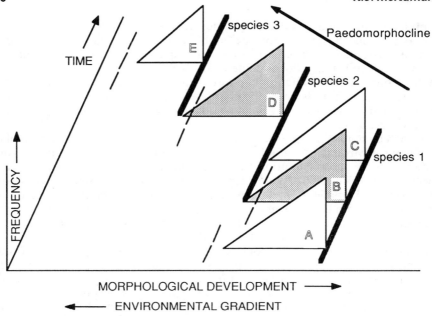

Figure 3.4.   Suggested mechanism for the development of a paedomorphocline.
Only after a population of species 1, such as population B, has paedomorphic
phenotypes able to cross an adaptive threshold and occupy a new adaptive
peak is a new paedomorphic species able to arise. Shifting adaptive thresholds
and variation in range of paedomorphic phenotypes affect the timing of
evolution of species 2. For instance, even though the range of paedomorphic
phenotypes is greater in population A than in B, the position of the adaptive
threshold is such that it is not crossed by phenotypes of population A. Extreme
phenotypes of population C of species 1 are limited by competitive exclusion.
There is no adaptive threshold. The persistence of such populations, however,
effectively blocks reverse speciation from species 2 to species 1. Selection of
extreme paedomorphic phenotypes of species 2 which can cross an adaptive
threshold result in the establishment of a further paedomorphic species,
species 3. The paedomorphocline is therefore established, and a unidirectional
evolutionary trend is established.

Other paedomorphs may reach varying heights up the adaptive peak, the
'fittest' occupying the top of the peak. Where only the ancestral species
occupies the initial adaptive peak that corresponds to the first ecological niche
along a putative environmental gradient, then extreme paedomorphs have
the potential to transcend the adaptive threshold that lies between the
existing adaptive peak and the next potential peak along the niche series.
To do this these paedomorphs must be capable of fitting into the next vacant
niche along the niche axis. If this occurs, then occupation of the next niche
and development of a new adaptive peak will produce ecological, and thus

genetic isolation. This may not necessarily be geographical, and so the speciation event need not be considered allopatric in the generally accepted sense of the term (Mayr, 1970). It could be geographically parapatric, or perhaps sympatric.

The adaptive threshold between two peaks is unlikely to remain static, mainly because the ecological niches may vary in their breadth along the axis by niche contraction and expansion. Consequently, it does not have to be mandatory for an extension of the range of paedomorphic phenotypes to occur for the adaptive threshold to be crossed (Figure 3.4). Should the adaptive threshold move closer to the earlier adaptive peak, then existing extreme paedomorphs may cross to the new adaptive peak. Normally both aspects will change over any period of time, such that a slight shift in the adaptive threshold and extension of the phenotypic range will ensure successful occupation of the next adaptive peak along the environmental gradient. A heterochronocline may therefore be considered as a series of adaptive peaks that have been established sequentially through time and which fit into a parallel series of ecological niches that lie along an environmental gradient. Each peak's position will be determined by many factors, but largely by the position of the preceding peak as well as the distance between niches. These factors themselves will be determined by the requisite morphological or ecological separation between ancestors and descendants which inhibits significant competition for resources.

LIMITS ON HETEROCHRONOCLINE DEVELOPMENT

The extent of a heterochronocline will be determined by either intrinsic or extrinsic factors, or more likely a combination of both. The limiting intrinsic factor will be the degree to which species are constrained by developmental and structural considerations. For instance, if the principal target of selection is appendage number, and so appendage number is decreasing along a paedomorphocline, then a point will be reached below which the organism will not be able to function. While it might seem that paedomorphoclines were more likely to be subjected to such constraint than peramorphoclines, the constraining factor with peramorphoclines might be organism size. If the same appendages are increasing in number, but there is no body-size increase, then there will be size constraints imposing an upper limit on appendage number, and consequently the extent of the peramorphocline. Energy and behavioural considerations will also affect the viability of such forms. However, should hypermorphosis occur, then there will be no such constraint. The dominant extrinsic factor constraining heterochronocline development is the availability of niche space along the environmental gradient.

While environmental gradients, such as temperature, sediment grain size, water depth or elevation, are generally gradational, adaptations are not so gradational, but will be applicable over part of the environmental gradient. The degree to which the adaptations range along the environmental gradients will also limit the extent of the heterochronoclines. Many adaptations also covary in suites. These are often size-related (see Chapter 4 herein). While size changes along heterochronoclines may be gradational, adaptations that are related to these size changes may not.

THE DRIVING FORCE BEHIND HETEROCHRONOCLINES

Cladogenetic heterochronoclines, where ancestral species persist following the evolution of the descendant, may be considered to be *autocatakinetic* (Hutchinson, 1959) — in other words, self-generating in a particular direction. Gould (1988) has shown that many large-scale (speciational) evolutionary trends may, in fact, be little more than an increase in variance, within a constrained framework, and thus by inference autocatakinetic. Cladogenetic heterochronoclines are a subset of this, increasing variance through a species' own dynamics. Once established, a heterochronocline can be considered to be self-generating in two ways. First, the development of a heterochronocline along a sequence of niches will, by definition, be a directional process. Second, what drives the heterochronocline in one direction is the persistence and competition from the ancestral species. For instance, in a modal distribution of heterochronic phenotypes, those phenotypes of the second species in a paedomorphocline that are on the peramorphic side of the modal phenotypic expression would be outcompeted by the paedomorphic phenotypes in the ancestral species (McNamara, 1982). These phenotypes are 'fitter' in their own ecological niche than the descendant peramorphic phenotypes would be (see Figure 3.4). This obstruction in one direction means that evolution can only occur in the other direction and is autocatakinetic.

The driving force behind anagenetic heterochronoclines, however, lies outside the heterochronocline system in the form of extrinsic selection pressure, such as predation pressure (see Chapter 9 herein). Whether this explanation applies to intraspecific anagenetic heterochronoclines also is questionable (as with the *Echinocyamus* lineage described in Chapter 9). The most likely explanation for 'movement' along such anagenetic heterochronoclines is that the ancestral morphotype occupied only part of the niche axis. Once established, the same system will apply as in cladogenetic heterochronoclines: autocatakinesis by ancestral blockage will drive the heterochronocline unidirectionally while there is temporal overlap of both morphs. Extinction of the ancestral morph may occur because of either intraspecific competition or differential survival controlled by extrinsic factors, such as predation pressure. However, if the morphological changes have no apparent adaptive significance, then explanations involving a form of 'morphogenetic drift', somewhat akin to 'genetic drift', must be invoked. Such phyletic gradualism that has an underlying morphological basis in heterochrony could also develop if the niche itself is gradually changing unidirectionally.

REFERENCES

Alberch, P., 1980, Ontogenesis and morphological diversification, *Amer. Zool.*, **20**: 653–67.

Alberch, P., Gould, S.J., Oster, G.F. and Wake, D.B., 1979, Size and shape in ontogeny and phylogeny, *Paleobiology*, **5**: 296–317.

David, B., 1989, Jeu en mosaique des heterochronies: variation et diversité chez les Pourtalesiidae (échinides abyssaux), *Geobios*, mémoire spécial, **12**: 115–31.

Dommergues, J.-L., and Meister, C., 1989, Trajectoires ontogénétiques et hétérochronies complexes chez des ammonites (Harpoceratinae) du Jurassique inférieur (Domerian), *Geobios*, mémoire spécial, **12**: 157–66.

Ede, D.A., 1978, *An introduction to developmental biology*, Wiley, New York.

Eldredge, N., 1979, Alternative approaches to evolutionary theory, *Bull. Carnegie Mus. Nat. Hist.*, **13**: 7–19.

Futuyma, D., 1986, *Evolutionary Biology*, Sinauer, Sunderland, MA.

Gould, S.J., 1977, *Ontogeny and phylogeny*, Harvard University Press, Cambridge, MA.

Gould, S.J., 1980, The promise of paleobiology as a nomothetic, evolutionary discipline, *Paleobiology*, **6**: 96–118.

Gould, S.J., 1988, Trends as change in variance: a new slant on progress and directionality in evolution, *J. Paleont*, **62**: 319–29.

Grant, V., 1963, *The origin of adaptations*, Columbia University Press, New York.

Hutchinson, G.E., 1959, Homage to Santa Rosalia or why are there so many kinds of animals? *Am. Nat.*, **93**:145–59.

Lee, D., 1980, Cenozoic and Recent rhynchonellide brachiopods of New Zealand: systematics and variation in the genus *Tegulorhynchia*, *J. R. Soc. N.Z.*, **10**: 223–45.

Levinton, J.S., 1983, Stasis in progress: the empirical basis of macroevolution, *Ann. Rev. Ecol. Syst.*, **14**: 103–37.

Levinton, J.S., 1988, *Genetics, paleontology and macroevolution*, Cambridge University Press, Cambridge.

Levinton, J.S. and Simon, C.M., 1980, A critique of the punctuated equilibria model and implications for the detection of speciation in the fossil record, *Syst. Zool.*, **29**: 130–42.

Maderson, P.F.A., Alberch, P., Goodwin, B.C., Gould, S.J., Hoffman, A., Murray, J.D., Raup, D.M., de Ricqles, A., Seilacher, A., Wagner, G.P. and Wake, D., 1982, The role of development in macroevolutionary change. In J.T. Bonner (ed.), *Evolution and development*, Springer-Verlag, Berlin: 279–312.

Mayr, E., 1970, *Populations, species, and evolution*, Harvard University Press, Cambridge, MA.

McKinney, M.L., 1984, Allometry and heterochrony in an Eocene echinoid lineage: morphological change as a by-product of size selection. *Paleobiology*, **10**: 207–19.

McKinney, M.L., 1986, Ecological causation of heterochrony: a test and implications for evolutionary theory. *Paleobiology*, **12**: 282–9.

McKinney, M.L., 1988, Heterochrony in evolution: an overview. In M.L. McKinney (ed.), *Heterochrony in evolution: a multidisciplinary approach*, Plenum, New York: 327–340.

McKinney, M.L. and McNamara, K.J., 1990, *Heterochrony: the evolution of ontogeny*, Plenum, New York.

McNamara, K.J., 1982, Heterochrony and phylogenetic trends, *Paleobiology*,

82: 130–42.

McNamara, K.J., 1983, The earliest *Tegulorhynchia* (Brachiopoda: Rhynchonellida) and its evolutionary significance, *J. Paleont.*, **57**: 461–73.

McNamara, K.J., 1986, A guide to the nomenclature of heterochrony, *J. Paleont.*, **60**: 4–13.

McNamara, K.J., 1988a, The abundance of heterochrony in the fossil record. In M.L. McKinney (ed.), *Heterochrony in evolution: a multidisciplinary approach*, Plenum, New York: 287–325.

McNamara, K.J., 1988b, Heterochrony and the evolution of echinoids. In C.R.C. Paul and A.B. Smith (eds.), *Echinoderm phylogeny and evolutionary biology*, Clarendon, Oxford: 149–63.

Oster, G.F., Shubin, N., Murray, J.D. and Alberch, P., 1988, Evolution and morphogenetic rules: the shape of the vertebrate limb in ontogeny and phylogeny, *Evolution*, **42**: 862–84.

Raup, D.M. and Gould, S.J., 1974, Stochastic simulation and evolution of morphology—towards a nomothetic paleontology, *Syst. Zool.*, **23**: 305–22.

Shubin, N. and Alberch, P., 1986, A morphogenetic approach to the origin and basic organization of the tetrapod limb, *Evol. Biol.*, **20**: 319–87.

Stanley, S.M., 1975, A theory of evolution above the species level, *Proc. Natl. Acad. Sci. U.S.A.*, **72**: 646–50.

Stanley, S.M., 1979, *Macroevolution: pattern and process*, Freeman, San Francisco.

Vermeij, G.J., 1987, *Evolution and escalation—an ecological history of life*, Princeton University Press, Princeton, NJ.

Vrba, E.S., 1980, Evolution, species and fossils: how does life evolve? *S. A. J. Sci.*, **76**: 61–84.

Wright, S., 1932, The roles of mutation, inbreeding, crossbreeding and selection in evolution, *Proc. Sixth Internat. Congr. Genetics*: 356–66.

Chapter 4

TRENDS IN BODY-SIZE EVOLUTION

Michael L. McKinney

INTRODUCTION

Body size is the central feature of any organism—physiologically, ecologically and evolutionarily. Further, it is the nexus among the levels because body size is correlated with many physiological, ecological, and life-history traits, and can be used to characterise many evolutionary patterns (e.g., large mammals tend to become extinct more easily). This should be no surprise as an individual's body size sets severe constraints on the rate of physiological (internal) processes and therefore strongly controls its relationship with the external (biotic and abiotic) environment: prey, predators, competitive abilities, thermoregulation, among many others (Peters, 1983; Calder, 1984). It also serves as a visible proxy for life-history events because changes in growth rate, maturation and death are usually manifested in body size. This too might be expected, since 'body size' is really a composite feature. It subsumes (records) the ontogenetic changes in most, if not all, morphological traits. In sum, because of its composite nature and ontogenetically changing, but central, physiological and ecological role, body size will inevitably be a major focus of changing natural selection, and therefore be evolutionarily very labile.

Palaeontologists have long appreciated the importance of body size, for at least two major reasons. First, it is an easy trait to measure. Unlike, say, the immune system, to use Raup's (1988) example, body size is readily preserved in fossils (or more correctly, some reasonably accurate proxy for body size, such as molar area, is preserved). Further, as measurable traits go, size not only is the most prominent but also unlike many, is comparable across taxa (e.g., the masses of protists and elephants may be compared). Second, the

75

most common evolutionary change observed in fossils is change in body size (Stanley and Yang, 1987; Gingerich, 1985; Boucot, 1976). This has been traditionally codified as Cope's Rule, defined as the 'widespread tendency of animal groups to evolve toward larger size' (Stanley, 1973). According to Kurtén (1953), this palaeontological rule is second in repute only to Dollo's Law of irreversibility. (Ironically, Cope never explicitly formulated the rule named after him, though it was implicit in his writings.)

In spite of its extreme popularity, Cope's Rule is broadly misunderstood, in terms of both pattern and process. In this chapter, my goal is to clarify both points. The first major section, on pattern, will build upon the pioneering work of Stanley (1973) who first clearly documented the cladogenetic nature of much long-term body-size increase, as an increase in maximum size (what Gould, 1988, has called an 'increase in variance'). In building on this, I will use later work, including MacFadden's (1986) excellent study of horses, my own compilation of size change seen in fossil studies, and incorporate the 'particle diffusion' framework discussed in Chapter 2 of this book. I believe this section will show that palaeontology has come a long way in replacing past anecdotal description of size change with solid documentation. In the second major section, I discuss the processes responsible for the patterns observed. My goal here is to replace the past *ad hoc* simplistic explanations of the pattern with explanations rooted in an emerging body of ecological studies on size. This has only recently been possible, with the rise of interest in body size by ecologists (summarised in Peters, 1983, and Calder, 1984). However, there are at least three other ecological fronts that have even more recently focused on the study of body size: quantitative genetics (Kirkpatrick, 1988), the demographic import of size-structured (as opposed to age-structured) population dynamics (Ebenman and Persson, 1988), and the role of body size in structuring food webs (Lawton and Warren, 1988). Together, the accumulating data provide a much more refined view of the role of body size in ecological (and hence evolutionary) processes. A key conclusion is that body size is subject to a much wider, more complex variety of selection pressures than appreciated in the past. In the final section of this chapter, I discuss the broad implications of these topics. Of particular importance is the emergence of 'macroecology', the study of ecosystem-level processes and their relationship to body-size evolution. For example, large-bodied animals seem to use a disporportionate share of the energy flow through an ecosystem.

However, before beginning the first major section of body size patterns of evolution, it is first necessary briefly to review just what is meant by 'body size'.

WHAT IS BODY SIZE?

Intuitively, body size is easily understood. Few people would have difficulty telling that a whale is 'bigger' than an insect. However, as with many operationally easy definitions, closer inspection of the concept reveals many complexities. In truth, 'size' (like 'shape') is a qualitative concept that is multivariate in nature. This multivariate (or composite) nature means that it is very difficult to characterise (i.e. 'measure') with single values. For example, consider the comparison of two individuals that weigh the same but differ in linear dimensions. Which is bigger? Is one linear dimension more 'important' than another?

The most common way of dealing with this problem is to create summary variables that capture the bulk of what we wish to measure. This entails the application of multivariate statistical methods, especially principal component or factor analysis, to morphometric data. Numerous authors have discussed the use of such methods relative to body size (Bookstein *et al.*, 1985; Shea, 1985; Tissot, 1988; McKinney and McNamara, 1990, Chapter 2) and there is room for only a brief, semi-technical discussion here (also see Gould, 1981, for an excellent non-technical overview).

Consider a bivariate plot of body length versus width. To compare individuals of various 'sizes' we may plot their length–width dimensions. Since growth among body parts is highly covariant, individuals larger in one dimension are usually larger in the other. Thus, a plot of many individuals will usually follow a fairly 'tight' linear trajectory, and we would say that the 'largest' individuals are at the uppermost extreme (have the highest joint values). It is a simple mental exercise to extrapolate this procedure to three or more morphological dimensions. For example, the three-dimensional homologue is a 'football'-shaped point cloud. In such a case, the longitudinal axis is the 'size' axis: it is a vector along which values of all three dimensions increase in covariant fashion. This 'first axis' thus summarises the composite change in all variables: individuals which fall at higher values (have higher 'scores' and occur near the upper end of the longitudinal axis of the point cloud) are thus 'larger' in most if not all measured variables. Thus, we may say that body size is 'a general factor which best accounts for all observed covariances among a set of distance measures taken on individuals of varying size' (Bookstein *et al.*, 1985).

There is a common misconception that this means that the first (Bookstein's 'general') axis is a 'size-only' axis. However, as Shea (1985) and Tissot (1988) have discussed at length, this is not true. The first axis incorporates shape changes that occur with general size increase. It thus does not need to represent isometry but can include size-associated positive and negative allometry, where the general vector of change does not have a slope of 1 (when all traits are 'plotted' at once). The first axis is thus more accurately called a 'size-determined' axis.

Allometry

Allometry has come to take on many meanings but it generally refers to changes in some variable, such as shape, that occur with increasing size (see, for example, the general reviews in Gould, 1966; Shea, 1985; LaBarbera, 1986; McKinney and McNamara, 1990). In the context above, such changes can be seen as those which are covariant with size increase (fall on the first axis) and those that deviate from this general covariance (fall on second, third or higher axes, which measure deviations from the first). The most visible way to view such change is to use bivariate plots, analysed with Huxley's familiar allometric power function, $y = bx^k$, where y is a trait and x is often some variable (e.g., length, area) that proxies for 'size'. Given the complexity of characterising size, as just noted, how can such a simple equation be so successfully and commonly used? Clearly it is because change is so covariant among traits: thus, separate body parts are good estimators of change in other parts and even the collective sum of parts.

With the rise of interest in body size in ecology, the term 'allometry' has also become commonly used to refer to changes in ecological relationships with body size. Thus, there is the 'allometry' of population density (which decreases with body size) and so on (Peters, 1983; Calder, 1984). I agree with LaBarbera (1989) that this use of 'allometry' is confusing. In this chapter, 'allometry' is used to denote changes *internal* to the organism that occur with size change (e.g., anatomical shape, metabolism, physiology). For changes *external* to the individual organism, such as in population density, predator or prey size, that occur with body-size increase, I use Schmidt-Nielsen's (1984) term 'scaling', as suggested by LaBarbera (1989). This should help distinguish between two qualitatively different kinds of change that occur as an organism increases in size, both ontogenetically and evolutionarily, dichotomising the changes that occur within the individual and those that reflect changes in interactions with its abiotic and biotic external environment.

Summary: selection on size, shape and timing

Aside from general examination of critical terms, the above discussion is meant to tease out three major targets of natural selection. The environment may act directly on size itself, in some cases causing shape to change as a byproduct. However, selection may also act on shape alone, with no change in body size. Finally, selection may act on the timing (and rate) of growth, either of local growth fields or of the somatic complex itself (the 'body'). In such cases (usually, rather glibly, called 'life-history' selection where body growth events are involved), indirect size selection occurs. This crucial, tripartite, suite of targets is discussed more fully in the section on process below.

PATTERNS IN BODY-SIZE EVOLUTION

As noted, body size has been the major trait analysed in most evolutionary studies (Boucot, 1976; Gingerich, 1985; Stanley and Yang, 1987). This is not only because it is prominent, easily measured, and comparable across taxa but because palaeontologists naturally focus on the traits that change the most, and size is very labile (see Flessa and Bray, 1977, for one of the few efforts to study change after correcting for size).

What patterns are discerned from this careful palaeontological documentation of size change? Much of the concern has been on examining evolutionary rates, but there is also much information about directionality. In Chapter 2 herein, I discussed two basic kinds of trend, anagenetic and cladogenetic. Both involve changes in a 'state variable' (in this case, body size); anagenetic trends occur in a single, unbranching lineage, whereas cladogenetic trends show changes in state space via creation of new species. Perhaps in part because of the past interest in evolutionary rates in specific lineages, anagenetic and small-scale cladogenetic trends have been the best documented. As discussed later, the distinction between cladogenetic and anagenetic trends is often a matter of scale. Longer time-spans tend to involve more branching while shorter scales of study focus on anagenetic 'fluctuations' in a single lineage. With this in mind, let us begin with a look at the patterns of coarsest scale and narrow down to those at finer, mostly anagenetic, scales.

Cladogenetic size trends

Any group, from a small clade to the biosphere, will result from and display cladogenesis. Thus, as shown in Figure 4.1, the biosphere as a whole has shown a cladogenetic increase in body size. In this case, diversification has produced an increase in the maximum size. Not surprisingly the general aspect is asymptotic: as maximum structural limits are reached, size increase rapidly drops.

Also documented have been body-size trends in clades of higher taxa. The results of MacFadden's (1986) excellent study of horses are shown in

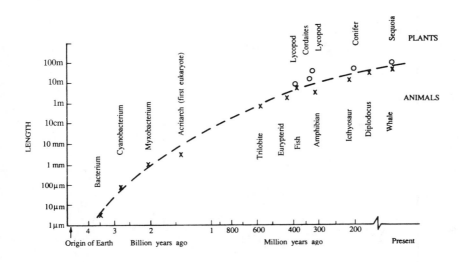

Figure 4.1. Approximate increase in maximum size of organisms. Both scales are logarithmic. Modified from Bonner (1965).

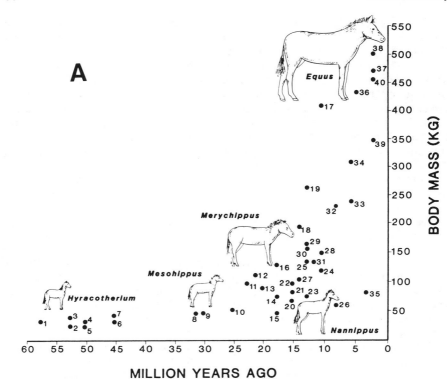

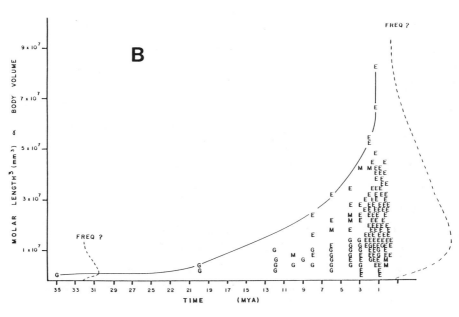

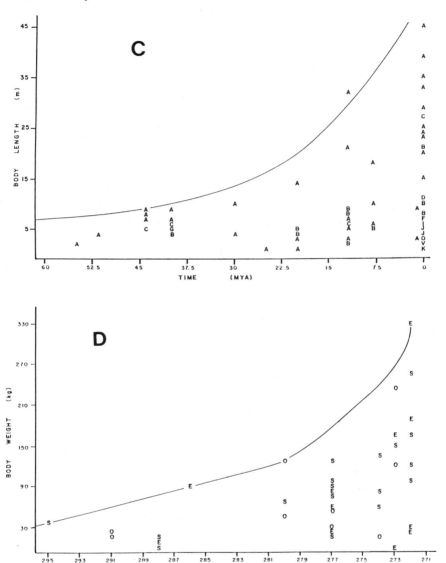

Figure 4.2. A: Body–mass averages for 40 horse species. (Modified from MacFadden, 1986.) B: Body size as approximated by molar length cubed for proboscideans. This plot, compiled by D. Walker, shows molar size of 324 individuals of three families: G = Gomphotheridae, M = Mammutidae, E = Elephantidae. Dashed frequency curves estimate size distribution at those times (many individuals occur at same point, so overprinting occurs). Data from Osborn (1936; 1942). C: Body size of cetaceans, compiled by D. Tardona.

Figure 4.2A. Examination of 40 species revealed little size change from about 57 to 25 million years ago. Thereafter, especially between about 25 to 10 million years ago there was a rapid, major diversification, with some species representing an eightfold size increase.

In addition to the horse study, Figure 4.2 shows original data collected by myself and some students. Body-size increase in proboscideans (Figure 4.2B) cetaceans (Figure 4.2C) and pelycosaurs (Figure 4.2D) is shown. The first two groups were chosen because they are among the largest mammals known on land and sea, while the last represents a well-studied non-mammalian clade from the Palaeozoic Era. In proboscideans, third inferior molar length cubed (to more closely approximate a volumetric measure of body weight) reveals that body size has increased at an exponential rate. Similar patterns are seen in cetaceans and pelycosaurs. Note that the same pattern is also seen in the various subgroups: Gomphotheridae, Mammutidae, Elephantidae in the proboscideans, baleen and carnivorous whales, and the three basic kinds of pelycosaur. In the last two sets of cases, significantly different dietary and ecological niches were involved among the subgroups.

Anagenetic size trends

The palaeontological literature has many examples of size change in a single fossil lineage. Some of the best are those of mammal size during and since the Pleistocene (e.g., Davis, 1981). Kurtén's (1960) classic study is illustrated in Figure 4.3. Invertebrate examples documented in this book include trilobites (Chapter 5), ammonoids (Chapter 7) and echinoids (Chapter 9). Gingerich (1985) compiled examples which include anagenetic size trends.

Summary: a compilation

To try to summarise the patterns seen in the literature, I have compiled a list of body-size studies (Table 4.1). It is not exhaustive but it does include many, if not most, of the major and most often cited studies on size. The sample size is small and undoubtedly biased, but a number of salient points from the table are tentatively suggested. First, most studies seem to focus on anagenetic size change. Second, the cladogenetic studies seem to find size increase or

---

Figure 4.2: continued
177 individuals are shown where letter shows amount of overprinting: A = 1, B = 2, etc. Data from numerous sources, especially many articles by Kellogg. Complete listing available from McKinney. D: Body size of pelycosaurs for three groups: E = edaphosaurs, O = ophiacodonts, S = sphenacodonts. Sample size is 51 individuals, as compiled by M. Gibson, mainly from Romer and Price (1940) and Oeson (1962). These authors estimated weight from skull size and other features. Complete references available from McKinney.

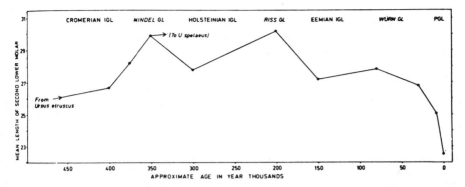

Figure 4.3.   Body-size change in the brown bear of Europe, based on length of second lower molar (in millimetres). Modified from Kurtén, 1960.

decrease in about equal numbers (six of each). In contrast, anagenetic studies seem to find a massive preponderance of size increases (92, with 28 showing no increase). However, if we ignore the data of Hallam, and then Boltovshoy (to avoid possible bias from one or a couple of large sources), the size-increase dominance is eliminated, indeed reversed (with 21 showing an increase and 25 a decrease). Clearly, the matter is unresolved as yet.

   Third, the anagenetic pattern, upon closer inspection, shows an overriding scaling influence. In studies that focus on a time-span of less than a million years, size decrease is more common (19, with three showing an increase). In contrast, those spanning over a million years show a strong dominance of size increase (89, as against nine showing no increase; with Hallam's and Boltovskoy's data removed, there are 18 showing an increase and six a decrease). Of course, a major bias is that the short-term data largely represent post-Pleistocene size decrease from climatic warming (Davis, 1981; Koch, 1986). However, there is plenty of evidence that short-term anagenetic random (not consistently preferring increase or decrease) 'flux' does occur, as revealed in subsections of many long-term anagenetic studies. For instance, in Boltovskoy's studies (Table 4.1), while many foraminifera increase from the Oligocene to Pliocene, the majority also show a reversal of this trend during the Pleistocene.

   The tentative conclusion is that long-term anagenetic studies seem preferentially to find body-size increase significantly more often than size decreases (or no change). Patterns of short-term studies (spans of less than a million years) are unclear because of sampling bias but there is certainly no evidence that size increase is more common than size decrease. Interestingly, this agrees with Bonner's (1968) assessment, which he based on evidence largely different from that used here. He specified three evolutionary size patterns, also distinguished by scaling: fast changes (1000–10 000 years) were as likely to show size increase as decrease; medium-rate changes (5–20 million years) showed predominently size increase; and slow changes over the whole span of life's evolution, representing increase in maximum size of

Table 4.1. *Body size changes (+, −, no change) reported in the literature*

Group	Changes	Span (years)	Source
Cladogenetic			
Plio–Plst radiolaria	1+,1−	3×10^6	Kellogg (1976; 1983)
Plio–Plst radiolaria	1+, 1−	4.5×10^6	Lazarus (1986)
Miocene rhinoceros	2−	few million	Prothero and Sereno (1982)
Early Cz primates	4+,2−	few million	Gingerich (1976)
Anagenetic			
Plio–Plst radiolaria	2+	6×10^6	Kellogg and Hays (1975)
Permian foraminifers	1+	many million	Ozawa (1975)
Mio–Rec foraminifers	2+,1−	8.1×10^6	Malmgren and Kennett (1981)
Mio–Plio foraminifers	1+	10^7	Malmgren *et al.* (1983)
Olig–Plio foraminifers	9+,1no	over 2×10^7	Boltovskoy (1984)
Olig–Plio foraminifers	7+,1no	over 2×10^7	Boltovskoy (1988)
Eocene ostracodes	1−	3×10^7	Reyment (1985)
Late Pz ostracodes	1−	few million	Schweitzer *et al.* (1986)
Jurassic bivalves	6+	many million	Hallam (1978; 1982)
Jurassic bivalves	30+,1−	few–many million	Hallam (1975)
Jurassic ammonites	19+	few–many million	Hallam (1975)
Jurassic ammonites	2+	few million	Raup and Crick (1981)
Mid-Cz oreodonts	2+,2−	1.2×10^7	Bader (1955)
Early Cz condylarths	3+,1−	3×10^6	Gingerich (1985)
Plio–Pleist rodents	5+	1.5×10^6	Chaline and Laurin (1986)
Pleist mammals	6−	few thousand	Davis (1981)
Pleist marsupials	8−	few thousand	Marshall and Corruccini (1978)
Pleist mammals	5−,3+	few thousand	Kurtén (1960)

phyla. This last would correspond to the biosphere trend discussed above and would be the kind of cladogenetic asymmetrical pattern (see Chapter 2 herein) seen in the clades. We might also include clade-level cladogenetic patterns in the group of medium-rate changes (as well as anagenetic patterns) since, during that time-span, asymmetrical branching will probably lead to size increase. This is discussed next, as we focus on the processes behind the patterns just described.

PROCESSES IN BODY-SIZE EVOLUTION

The general process: diffusion without drift

Gould (1988; and Chapter 1 herein) has expanded upon Stanley's (1973) original point that evolutionary size increase is largely a matter of clado-genetic 'diffusion' away from an originally small-sized ancestor. This process is shown graphically in Figure 4.4A. In terms of the above discussion, the reason for a predominance of size increase over medium to long time-spans (i.e. millions of years) is easy to see, if one assumes that the process is cladogenetic. As the descendant species evolve, they often come under selective pressures which promote size increase, either because they have migrated into such environments, or because the old ones are changing. In the shorter spans (e.g., less than a million years, but this is relative and will vary with the taxon), we would expect to observe anagenetic fluctuations that would, on average, tend to favour neither size increase nor decrease. That is, a random walk would often occur because the multiple independent variables affecting size would generate generally unpredictable directionality from one point to the next (see Chapter 2 herein).

However, while explaining the short-term anagenetic patterns, and medium- to long-term cladogenetic ones, this scenario does not explain why medium- to long-term anagenetic trends tend to show size increase. Possibly this trend is due to sampling bias of some kind—for example, workers tend to study anagenetic changes that lead to size increase—or may be explained by 'anagenetic' changes which are really cladogenetic. A more literal reading of the pattern would be that there is some long-term advantage to larger size that 'drives' the species along. McNamara (McKinney and McNamara, 1990) has documented a predator-driven system wherein echinoid species are driven by gastropod predation in such an anagenetic fashion (see Chapter 9). However, it is not likely that the widespread general trend implied by a literal reading of Table 4.1 is entirely predator-driven. Rather, some more general, intrinsic advantage of body size would be suggested. Such 'intrinsic' advantages have often been invoked in the past, with the general theme that 'bigger is better': fewer predators, more intelligence, more buffering against the environment, and so on (Stanley, 1973). If true, such reasoning would lead to the scenario shown in Figure 4.4B. This is not diffusion, but 'diffusion with drift', where drift means that each particle is subjected to a strong deterministic force (Berg, 1983; see also discussion in Chapter 2 of this volume). For example, gravity may bias random diffusion of molecules, or in this case, intrinsic size advantage would bias random diffusion of species.

However, I agree with Stanley (1973) that the 'intrinsic' advantage concept has been widely abused and oversimplified. For their particular niches, small organisms are just as well adapted as large ones. Organisms become larger under some selection regimes but there is nothing 'intrinsically' advantageous in large size, as shown by the fact that it is often selected against in many situations. Therefore, I would be inclined to reject the 'diffusion with drift' model in favour of simple diffusion (i.e. Figure 4.4A is more valid). This distinction is not at all moot but represents a fundamental difference of process. In pure diffusion, individual particles (species) are under no over-riding force to move one way or another. Evolution of larger size occurs only because random processes have acted independently. Thus, there are indeed forces acting to increase size, in some situations, but they are (on average, over the very long term) equally matched by counteractive forces towards smaller size. Nevertheless, even random diffusion is an expansive process, so that size trends occur, especially if the expansion is asymmetrical, being blocked on one side by a 'reflecting barrier' (see Chapter 2 herein).

To understand more fully both the anagenetic and cladogenetic trends in body size, a closer look at the diffusion process is required. I have isolated six major aspects, shown in Figure 4.4C. These are numbered roughly in the

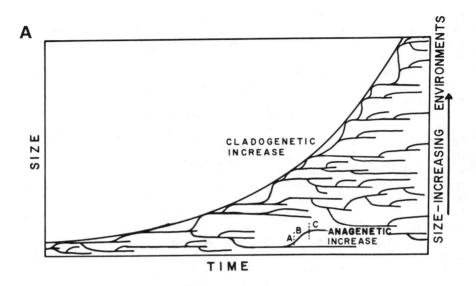

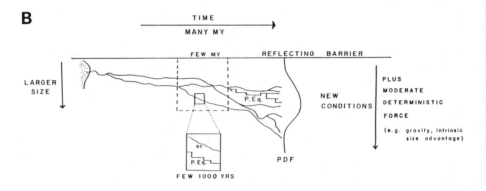

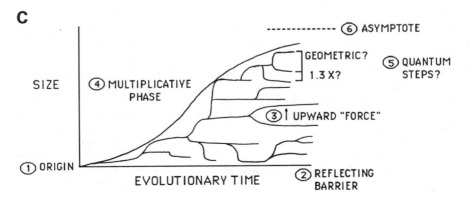

order in which they occur in the process, and in which they are discussed below. Most of these aspects represent key processes themselves, each of which is required to explain the overall process: why taxa originate at small sizes, what are the large-size selecting environments, and so on (note that no. 3 is the only one really needed to explain anagenetic change).

Origination at small size

In Stanley's (1973) original presentation of the diffusion model, he argued that taxa originate at small sizes mainly because they are less 'specialised'. In theory, large size is basically a specialisation of its own, with allometric requirements that limit the future evolution of the species. I disagree with this for the following three reasons. First, the notion that small organisms are less 'specialised' and thus have more 'evolutionary potential' is unproven and logically unsound. The basic reasoning of small size-potential is largely a subset of deBeer's (1958) famous argument that paedomorphs have more evolutionary potential. I have discussed the problems of this at length elsewhere (McKinney and McNamara, 1990, Chapter 8) and have space for only a brief review here. Among the more salient problems are that 'unspecialised' is a very subjective term and therefore unmeasurable. As a result, 'evidence' for this potential is almost always *post hoc* and involves much circular logic. In addition, the very logical basis for unspecialised potential for large size is flawed by problems of scale. Thus, while it is true that, say, a microbe has more 'potential' than an elephant in having far fewer ontogenetic contingencies to constrain its future changes (McKinney and McNamara, 1990), this is a far cry from saying that a dwarf elephant, or even a mouse, has more evolutionary potential than a large elephant.

My second and third reasons against the 'constraint' argument are positive in nature, focusing not on why it is wrong but suggesting better explanations instead. Thus, a major reason why ancestors are mostly small is simply that small organisms are much more common. As LaBarbera (1986) has pointed out in this context, body-size distribution in most clades is roughly log-normally distributed (for reasons discussed below in the final section). For instance, only about 9 per cent of mammals are 'large', with head length over 30 cm (van Valen, 1975). However, an important recent modification has been added to this basic ecological generalisation by Dial and Marzluff (1988).

Figure 4.4. A: Cladogenetic body-size diffusion via migration and speciation into environments that select for larger sizes. Note occurrence of anagenetic size change as well. B: Body-size diffusion 'with drift', wherein 'drift' refers to strong deterministic advantages favouring large size in *each* species ('particle'). This occurs in addition to migration plus speciation. The PDF is the probability density function, discussed later in the chapter. In theory, anagenetic change may be either punctuated or gradual as shown. C: Cladogenetic size diffusion model. Same idea as Figure 4.4A except that six key aspects are labelled for discussion in text.

Data from 46 different clades show that the most common body size is not the smallest (that in the first percentile), but that in the sixteenth to fortieth percentiles. Nevertheless, the same logic applies: smaller body sizes are much more common, with larger ones being progressively much less so, as the curve skews to the positive. Therefore, even if all body sizes have an equiprobable chance of beginning a clade, small ancestors will much more often be the rule.

The final reason for small-sized ancestors is that many clades originate after mass extinctions (e.g., mammals in the early Caenozoic) and larger species are more susceptible to extinction. There is both theoretical and empirical evidence for this. Pimm *et al.* (1988) have reviewed and tested the theoretical evidence, which is based on the idea that large animals have lower population densities and lower fecundity (longer generation times), so that (other things being equal) populations of large animals are more likely, respectively, to be more severely decimated and less likely to rebound from severe perturbations. Further, larger individuals require more food and other resources so that other things are usually not equal: larger-bodied individuals are often more likely to starve to death and suffer from privation. Palaeontological evidence seems to bear this out. The end-Mesozoic extinction shows selection against large-bodied forms (Bakker, 1977; Clemens, 1986) as does the Pleistocene (Martin, 1984). Barnovsky (1989) has done an exceptionally proficient job in analysing the Pleistocene data, showing clearly the preferential impact of rising extinction magnitude on large-bodied forms (Figure 4.5). Note that this last source of small ancestors is not exclusive of the above; it may well be that they operate in conjunction.

The discussion so far has focused on reasons for small originations of clades of various levels. However, if we want to account for small size origination of the biosphere (Figure 4.1), we can invoke none of the above, for there were no pre-existing organisms to provide a frequency distribution of fodder for extinction. Instead—and the answer is perhaps so obvious as to be trivial—small size of biosphere origination resulted from the basic laws of thermodynamics. Life is an open system, far from equilibrium, and in its early stages it was obviously quite small and simple since the 'ratchet' effect of replication had not yet allowed it to become as far from equilibrium as it now is. It is only through the gradual accumulation of ontogenetic contingencies ('information') that complexity has evolved (McKinney and McNamara, 1990, Chapter 8). Since size and complexity are correlated (see Bonner, 1988, for a good discussion), this growth of complexity has entailed a growth of size as well.

The reflecting barrier

In Chapter 2 herein I have discussed the nature of 'reflecting barriers' in general. In the context of the evolution of body size in clades (Figure 4.4C), this would denote limitations on the body plan of the clade's members. As with the related concept of allometric constraints or constraints of specialisation, this is a very slippery concept. For example, how small can a horse become? While such questions are difficult to answer, there are at least two

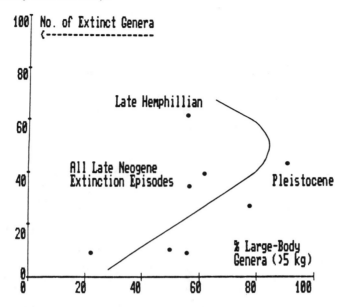

Figure 4.5. Plot showing how percentage extinction of large-bodied mammalian genera increases steadily as total number of extinct genera increases. By Pleistocene magnitudes, nearly all extinctions were of large forms. Only when overall extinction is extreme, that is, a very high number of genera are extinct (late Hemphillian) do small-bodied genera become extinct in sufficient numbers to reduce large-bodied percentage. Modified from Barnovsky (1989).

obvious avenues that suggest that there is some limit, even if we cannot precisely specify it. First, there are basic design (functional and structural) limits to any anatomical plan. A horse, for example, can only become so small before its digestive system, metabolism, bone structure or some other aspect prevents it from running, eating, or performing some other activity necessary to its existence. Second, even if radical 'down-sizing' could occur, the anatomical modifications needed to meet the scaling demands would eventually lead to deviations so novel that the organism would cease to be a 'horse' (for example).

For biosphere evolution, the nature of the reflecting barrier is clearer. There is a limit to how small an organism can be and still function as a living creature, metabolising, reproducing, and so on. The limit is apparently between the size of a virus and small prokaryotic cell.

The upward 'force'

Perhaps the most important aspect of the diffusion process (Figure 4.4C) are the causes of size increase. Simply starting at small size and limiting evolution

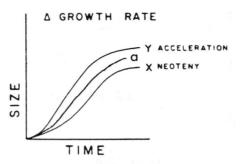

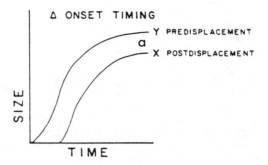

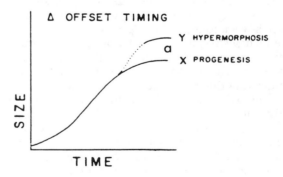

Figure 4.6.   Six major types of heterochrony in terms of size versus age plots; *a* represents ontogenetic trajectory of ancestor. Acceleration = increased rate of growth; neoteny = retarded rate. Predisplacement = early onset of growth; postdisplacement = late onset. Hypermorphosis = late offset of growth; progenesis = early offset. Modified from McKinney (1988).

in one direction is not enough: there must also be some process 'pushing' upward. Of course this 'force' is the two-fold Darwinian process of 'random' mutation plus environmental selection. Couched in dialectical terms, this is 'internally' (intra-organismic) produced (size) variation sorted by external conditions.

Turning first to the ('internal') mechanisms of production of size change, body size generally has a heritability of over 50 per cent (Atchley, 1983). While there is room for much ecophenotypic plasticity, body size shows much compensatory growth, making up for uneven conditions and reaching some 'target' size (Atchley, 1984). However, there is an important dichotomy in body-size genetics. On the one hand, it is typically a polygenic trait—for example, mouse size has over 100 loci (LaBarbera, 1986). On the other hand, there are single gene mutations that can have dramatic size effects, usually by interfering with the complex interplay in the hormone production-response system. For instance, the African human pygmy results from neoteny (slow growth) from such a mutation (Shea, 1988, and personal communication). This dichotomy would seem to have major implications for evolutionary rates since single gene mutations could cause rapid size change.

Whatever the genetic basis, the developmental process of size change is virtually always the result of heterochronic (rate or timing) changes in somatic growth. A classification of the six basic heterochronic ways of changing size is shown in Figure 4.6 (see the Glossary on pages 351–6 of this volume). This is discussed fully in McKinney and McNamara (1990), but to summarise, there are only three ways to become larger: growing faster in a constant unit of time (acceleration), growing for a longer period (hypermorphosis), or beginning growth sooner (predisplacement), with the illustrated opposite processes occurring for size decrease. Such 'global' (whole-body) heterochronies need not (indeed probably do not) occur exactly as shown in the 'pristine' curves of Figure 4.6: growth changes often involve a mixture of, say, growing longer and faster as well (illustrated in McKinney, 1988). A major (relatively undiscussed) reason for this is that ontogenetic growth curves are usually multiphasic so that delays or accelerations in one phase have complex, cascading effects in later phases (for a full discussion, see McKinney and McNamara, 1990, Chapters 2 and 7).

Ecological forces: direct size selection

The second process in the 'upward' force of phylogenetic size increase is selection acting on the heterochronically produced variants. (In large part, this is an internal–external dichotomy: internal—genetic-developmental—mechanisms versus external selection of their products.) In addition to a better understanding of the genetic and developmental mechanisms, our knowledge of ecological selection is also much improved. Indeed, this section is probably the most critical to understanding the primary factors operating in evolution of body size. We may consider these external selective forces as the specific, sometimes identifiable, fine-scale deterministic processes that 'buffet' and propel the random diffusion of species (particles) through morphospace (see Chapter 2 herein). Only in large scale does such diffusion appear 'random'. Here, we are focusing on forces propelling species through that (large) subsection of morphospace subsumed under 'body size'.

To start, let us take a brief overview and dispel some of the simplistic misconceptions about size selection advanced in the past. I will cover four basic flaws of past discussions on size increase: supposed intrinsicality of size benefits; ignoring multiple environmental pressures on size; ignoring size change caused by indirect size selection (on growth rate and shape); and ignoring the fact that selection acts not only on adult size (and timing and shape) but on all ontogenetic phases, which will often affect adult size.

First, as noted, there are no intrinsic advantages to larger body size. Many benefits of large size have been suggested: greater competitiveness, fewer predators, larger relative brain size among them (listed more fully in Stanley, 1973, and Schwaner, 1985). This 'bigger is better' logic completely ignores the blatant fact, discussed above, that small organisms are much more abundant and do quite well at what they do (and in fact often persist longer geologically —see LaBarbera, 1986). By more finely subdividing the environment they essentially take advantage of conditions that favour their size: specialising on high-quality foods, using microhabitats for cryptic evasion, and so on (references in Dial and Marzluff, 1988). The key to understanding size increase, then, is to isolate the environmental conditions that favour it, as I will do shortly. In addition to the intrinsic fallacy itself, some of the past logic for it has been flawed. For instance, Rensch (e.g., 1959) was a strong proponent of a link between larger organisms and larger relative brain size from allometric extrapolation of brain/body growth. Yet our present, much better, understanding of brain ontogeny shows that it can be 'decoupled' from somatic growth and that the allometry can change (see, for example Riska and Atchley, 1985; McKinney and McNamara, 1990, for review). In short, it is possible to create large brains without scaling up body size. Another example is the often suggested benefit that larger organisms suffer less predation. Yet, as discussed in the zooplankton example in the next paragraph, predators sometimes prefer larger prey.

Another major past error is the preoccupation with 'one' particular advantage, as if, even granted the importance of local environmental selection (non-intrinsicality), there was one overriding 'cause' of large size. Thus, as reviewed by Roff (1981), most workers have focused on three areas of size advantage: metabolic benefits (physiological efficiencies, such as storing heat, lower locomotor costs); fewer predators; and competitive benefits. Yet, as argued by Mayr (1983), there are always multiple pressures acting on the size of an individual, often in complex fashion, so that trade-offs must be teased apart by the investigator to see which ones are actually determining size and the relative importance of each. A classic example was documented in zooplankton by Brooks and Dodson (1965). They showed that in the absence of predators, body size in plankton increased significantly. The reason is that large-bodied plankton are more efficient feeders but predators prefer larger-sized prey. It is obviously hopeless to try to infer such fine-scale causes from the fossil record but, surprisingly, even neontologists have been very superficial in studies of size selection (but see Monaghan and Metcalfe, 1986, who

Table 4.2. *Environmental correlates fostering larger body size, in (very) roughly descending order of decreasing importance. (A) = abiotic selection, (B) = biotic. This is non-exhaustive, intended only as a sampler of the more prominent selection pressures*

Correlate	Group	Source
(B) Regularly abundant food	All	See text*
(B) Low-nutrient food	All	See text*
(A) Cool ambient temperature	Warm-blooded	See text*
(A) Warm ambient temperature	Cold-blooded	Stevenson (1985)
(B) Prey size	Predators	Peters (1983)
(B) Predation	Marine organisms	Gerritsen (1982)
(A) Seasonality	All	Lindstedt and Boyce (1985)
(B) Sex selection	Many	Woolbright (1983)
(B) Female fecundity	Many	Shine (1988)
(A) Post-arboreality	Vertebrates	Taylor *et al.* (1972)
(A) Water density	Plankton	Malmgren and Kennett (1981)
(A) Nutrient starvation	Benthic foraminifers	Hallock (1985)

* References are too numerous to list here.

present one of the few more complete analyses). This is changing, especially with the rise of controlled size selection experiments and the application of quantitative genetics to body size, developmental timing, and other traits (discussed briefly below).

Also amenable to these methods are the third and fourth deficiencies of past size discussions: that body size is often affected by environmental selection not only on size itself (but also on developmental rates and shape which can affect final adult size), and that all phases of ontogeny (size, timing and shape) are under selection. Obviously, then—and this is the main point of this section so far—the crucial selection process leading to body-size change is much more complex than has been implied by past discussions. I can hardly describe all of these in a single chapter; rather, my goal is to outline the major explanatory advances recently made in ecology and new directions of future work. I turn first to a discussion of some of the multiple factors operating to affect size in particular environments, beginning with direct selection. This is followed by a brief discussion of indirect size selection, on timing and shape, including preadult ontogeny. Finally (in this part on 'upward forces'), I cover briefly some of the most promising ongoing work of controlled study of environmental selection affecting size.

Table 4.2 gives a compilation of some of the main (abiotic and biotic) selection variables that favour larger-sized individuals. The first mentioned, regularly abundant food, is considered the most important by many ecologists. (By 'important' I mean most widespread as a key determinant in many groups.) To cut a long story short, larger animals have a two-way advantage

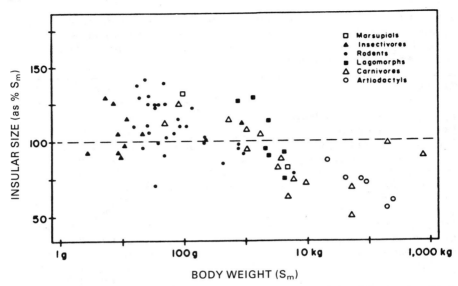

Figure 4.7. Size of island faunas as a function of mainland body size (S_m). Smaller mainland forms tend to increase in size on islands, from competitive release, while larger forms decrease in size due to reduced resources. Modified from Lomolino (1985).

over smaller ones when there is competition for predictable resources (referred to as 'competitive asymmetry'). Larger individuals show *exploitative* asymmetry in being able to eat foods that smaller ones cannot, while the reverse is not true; being able to consume more per feeding, leaving less for others; and, in general, having foraging advantages such as lower locomotor costs per unit mass. Schoener (e.g., 1983) has been a major proponent in documenting this (also see Wilson, 1975). In addition, larger individuals show *interference* asymmetry in that they can actively impede the feeding of smaller forms (Persson, 1985). These asymmetries are not limited to mobile feeding but are also important in sessile feeders (Sebens, 1982). This is not to say that both advantages always operate together, as Persson (1985) has discussed, but, under many conditions they can, singly or together, explain the competitive benefits of larger size.

Evidence for this reasoning was provided by an elegant study of island faunas by Lomolino (1985). As shown in Figure 4.7, the size of island faunas increases then decreases as a function of mainland size. Lomolino cogently argues that this is because the reduced abundance and variety of resources on islands is insufficient to support the greater absolute needs of normally larger species. Hence, they must become smaller (which explains much of the dwarfism of large animals seen in the record—see Sondaar, 1977—restricted not necessarily to islands but to any resource-poor environment). Lister (1989) has recently provided an exceptionally well-documented example, showing that red deer isolated on islands in the last interglacial became reduced to one-sixth their body weight in less than six thousand years. In contrast, normally small species increase in size because the removal (or size decrease) of their larger-sized competitors provides 'competitive release' to sizes as large as the island can support. This example shows why the resources need be not only predictable but also abundant and varied to produce large animals: while large size lowers relative metabolic rate (energy burned per unit mass) the absolute needs of each individual are greater.

The second entry in Table 4.2, low-nutrient food, is also widespread, being a main correlate of body size in many kinds of organism. The driving force here is the fact that the length and capacity of the digestive tract scales quite closely with body size (best documented quantitatively by Demment and van Soest, 1985, for mammalian herbivores). Therefore larger organisms can eat foods that require more processing (but, conversely, are also much more abundant). Thus, larger ungulates are generally grazers of grasses and other low-nutrient plants while smaller ones are forest-dwelling browsers of higher-quality buds (Janis, 1982; Demment and van Soest, 1985). Similarly, large primates (e.g., gorillas) eat lower-quality foods such as leaves, while progressively smaller ones eat progressively more higher-nutrient foods of fruits, nuts and insects (Jungers, 1985).

Going down Table 4.2, ambient temperatures have opposite effects on warm- and cold-blooded organisms. The well-known Bergmann's Rule describes the tendency for cooler climates to increase mammalian body size, for reasons of surface/volume efficiency (documented so well by Kurtén, 1960, and then by Davis, 1981, and Koch, 1986). In contrast, ectotherms show a tendency to increase size in warmer, especially stable, climates. This is because in climates with pronounced seasonality, ectotherms can survive only by finding shelters in microhabitats during seasonal flux. In warmer, equable climates, large individuals can survive and are selected for because they can retain (externally assimilated) body heat for longer periods, staying active much longer, putting them at great advantage (e.g., competitive and predatory) over smaller individuals (Stevenson, 1985). This 'inertial' homeothermy may well explain the large size of reptiles during the warm equable Mesozoic Era. It may also explain why some might have thrived in temperate areas since unseasonality is more important for larger size than the ambient temperature itself.

Other correlates on Table 4.2 (in descending order) include the selection on predators to increase in size as their prey does and the selection on prey to increase in size when predators prefer smaller individuals. Radical seasonality of food availability promotes size increase because larger individuals can survive longer on body stores. Mate-size preference, especially by females, may well affect average species size. In environments where reproductive output is a dominant selective force, body size of females may increase because such output increases with size. Groups that spend more time on the ground after arboreality (e.g., apes) tend to increase in size because arboreality promotes smaller size for locomotor and other reasons (e.g., the relative energy per unit mass expended by a mouse to run up a tree is about one-eighth that of chimpanzee—see Taylor *et al.*, 1972). Increasing water density increases body size in plankton for reasons of buoyancy. Nutrient-poor environments promote size increase in some benthic foraminifera because they foster a long growth phase with algal symbionts.

Begon (1985) has discussed how size selection is also important in onto-genetic stages. Thus, while most would view the correlates in Table 4.2 as affecting final adult size, we should recall that they also often affect juvenile size. Of course, such selection pressures will change with ontogeny, especially if there is significant size change. Werner and Gilliam (1984) have used the term *ontogenetic niche* to denote each of such successive changes. In cases where the ontogenetic niches are not removed enough from the adult, juveniles may be forced to compete with adults of the same species. This may be an important force driving the diffusion process as shown in Figure 4.4: descendant species may be driven to larger sizes to avoid or minimise competition with the ancestor. Bonner (1988) discusses this 'competitive escape' via size increase. McKinney and McNamara (1990) refer to this as 'autocatakinetic' (self-driving) cladogenesis since the earlier species provide a selective force pushing 'outward' of themselves (see Chapter 3 herein). Ecologists refer to this as a competitive 'juvenile bottleneck', when juveniles of a large species must compete with smaller species for resources (Ebenman and Persson, 1988).

Ecological forces: indirect size selection and analysis

It has rarely been noted by explainers of evolutionary size increase that selection on developmental timing (or its derivative, rate) *and/or* shape can lead to size change by indirect selection. Examples of the former occur in environments that either are subject to unpredictable catastrophes or provide constantly available superabundant resources. As Begon (1985) discusses, such environments remove the competitive advantages of large body size: catastrophic selection eliminates individuals regardless of size and in super-abundant conditions there is no competition. Such environments are thus 'size-neutral' and selection therefore favours not individuals which invest resources and energies in somatic development but those which invest them in gonadal (reproductive) output: generally more offspring as soon as possible (see Begon *et al.*, 1986, for good general discussion). Thus the selection is largely on reproductive timing; the sooner sexual maturation occurs, the greater one's fecundity and therefore reproductive fitness. Size is basically irrelevant.

Many readers will recognise this last scenario as containing elements of the well-known r–K theory. K-strategists live in stable, abundant environ-ments (Table 4.2) where competition (and hence size) is important, while r-strategists live in unpredictable and/or superabundant ones. This r–K notion has, of course, been the subject of much debate and few would deny that any simple dichotomy oversimplifies nature: competition/no competition, size selection/non-size selection, somatic/gonadal investment, stable/unstable environments, and so on. However, as I have tried to show here, the r–K continuum is not so much incorrect as it is incomplete. Thus, as long as we restrict the discussion to competition and the lack of it as determinants of size and timing, the predictions of the $r-K$ theory are generally accurate. However, it is incomplete in that the $r-K$ theory simply does not (at least directly) address many other factors that can affect body size or timing. For example, most of the other entries in Table 4.2 simply are not variables on the $r-K$ continuum: ambient temperature, for example. It also is incomplete in that, even in dealing with factors on the continuum, it focuses only on certain stages in the organism's ontogeny, that is, only those 'life-history' events that are deemed 'important': birth, maturation, death. However, every point in an individual's life-span (at least before reproduction) is equally important to fitness.

Another one of the most basic oversimplifications of the $r-K$ theory (in the context of body size), is its widely cited association of larger size with slow growth and delayed maturation. These three are, in turn, associated with stable, K-selecting regimes. As a result, many workers (especially palaeontologists, who usually lack ontogenetic age data) assume that larger individuals grow more slowly and mature later. On a very coarse level, these generalisations are true. The problem is that on a finer level, that of interspecific variation, which is after all the level at which ecological and micro-evolutionary processes work, the generalisation is practically worthless. Bigger animals often grow much faster than smaller ones and can mature at the same time or even sooner (e.g., gorillas grow faster than chimpanzees and mature at the same time—Shea, 1983). Numerous examples are becoming known (see Jones, 1988, for a review). This is illustrated in Figure 4.8A which shows that on a 'bacterium to elephant' basis it does indeed take longer to 'build' and replicate an elephant, for instance. However, as shown by Wootton's (1987) excellent work (Figure 4.8B), a fine-scale view shows that, for any given size, maturation may occur at many different ages. Conversely, larger body sizes often mature earlier (and hence must grow faster to be bigger, since growth stops earlier) than smaller ones. Such 'vertical allometries' (Calder, 1984) represent 'local' selection and developmental processes that operate within the overall 'structural' ('mouse to elephant') constraints. The main point is that, in such cases, the differences reflect selection on timing of maturation and rate of growth as well as (or perhaps instead of) size itself. Many mammals are the same size but reach that size at different rates and in different times.

As an example of such indirect size selection on timing or rate, consider humans and gorillas. Both are considerably larger than their close relative, the chimpanzee, but gorillas simply grow faster and mature at the same time while humans grow only at about the same rate as the chimpanzee, becoming larger because we mature much later (in about 13 compared with only 8 years; see McKinney and McNamara, 1990). Clearly the rate and timing differences in size are related to different selection pressures: human cultural adaptations, such as a large brain and extended learning stages, require extended (delayed) growth stages (brain growth is difficult to accelerate; only a prolonged foetal stage, where all neurons are generated, can increase neuronal number—see McKinney and McNamara, 1990). In contrast, the large gorilla size is largely tied to folivory and it is advantageous to become large as soon as possible: the gut allometry discussed above means that leaf digestion is

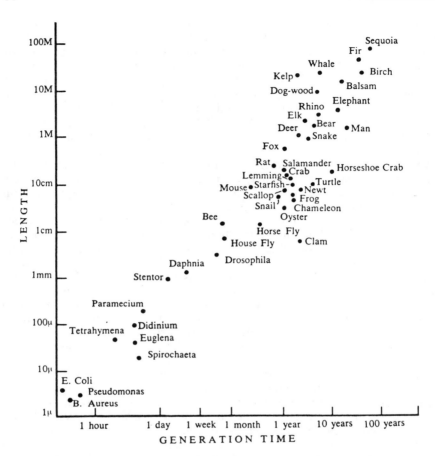

Figure 4.8. A: Organism length as a function of generation time; both scales are logarithmic (from Bonner, 1965).

more effective at larger size. (Thus, even though leaves and other low-grade vegetation are superabundant so that there is little competition, a classically 'size-neutral' situation noted above, the digestive benefits of large size, along with perhaps other pressures, favour fast growth. This illustrates once again the often intricate interplay of forces behind size change.) One interesting implication of this is that our own large body size (we are the second largest primate) is at least in part, and perhaps largely, due to selection on rate and timing of development; to cultural animals, being smarter contributes much more to fitness than being larger.

A final, palaeontologically very important, example of size selectional interplay is discussed by Vermeij (1989, p. 348): 'in marine invertebrates, large body size is usually linked to rapid growth, which in turn is associated

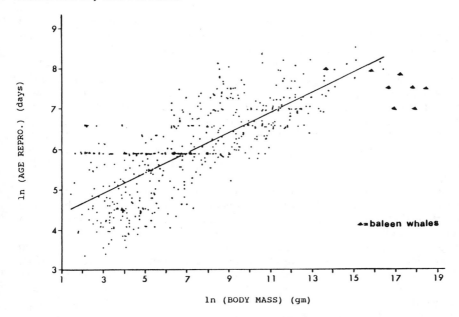

Figure 4.8.   B: Age of reproduction in mammals as a function of body mass; both scales logarithmic. Note large variation, even among related forms, within coarse correlation (from Wootton, 1987).

with abundant food'. If correct, this important generalisation, aside from giving us information (albeit very general) on the size–age problem, indicates that large marine invertebrates may usually result from selection on size (and not timing or shape): as noted above, stable, abundant (but not superabundant) regimes promote competition wherein significant somatic investment is advantageous. This can be visualised for competing suspension feeders, detritivores and predators, among others. Note the major break with the oversimplified *r–K* of the past in that the larger, more competitive size is attained rapidly, again disagreeing with the fallacy discussed above that larger means slower-growing and/or delayed maturation. Indeed, aside from organisms in which learning and parental care is important for competitiveness, it seems likely that large size is, in general, more likely to result from faster growth than prolonged growth. That is, where large size is favoured, it may often be the case that the sooner one achieves this, the better. This badly needs empirical data, but I suggest that our predilection for associating largeness with slower, prolonged growth (this is also wrong since slowness, or rate, can be decoupled from offset time, or duration, of that rate— see McKinney and McNamara, 1990) results not only from the 'bacterium to elephant' scaling fallacy, but from our past emphasis on vertebrates (especially mammals and primates) in theorising about this.

Aside from indirect size selection, yet another aspect is usually overlooked in the past simplified explanations of selection operating to change size—that rate and size trade-offs operate throughout ontogeny and each stage is subject to a variety of different selective pressures and genetic constraints. Thus, predators will select prey by size at any ontogenetic stage, regardless of the growth rates used to attain the size. Yet since size is the sum of growth increments during previous age intervals (see Lynch and Arnold, for a thorough quantitative review), there are many 'routes' to attaining a size, with the optimum rate and timing route being selected for. Thus, for a juvenile stage where there is competition with other species (juveniles or adults), selection will operate to promote rapid rate of growth during that stage.

The second kind of indirect size selection occurs when some 'shape' (e.g., organ) that correlates with body size is selected for and 'drags along' body size to attain the optimum shape size. For instance, there is a strong correlation between body size and antler size in deer, so that some workers have suggested that the large body size of the extinct 'Irish Elk' may have resulted from selection (e.g., male combat) acting on the antlers alone (Gould, 1974). However, such inferences are very difficult to prove and, given the many facets of physiology and ecology (internal and external processes, respectively) affected by body size, it seems unlikely that such instances of shape 'pulling' size along are common (literally, a case of the tail wagging the dog). In addition, many traits ('growth fields'—see McKinney and McNamara, 1990) can be 'decoupled' from other fields fairly easily so that such 'correlated response to selection' (as termed by quantitative geneticists, discussed shortly) is reduced. In other words, the genetic (and developmental) covariance structure among traits is itself subject to selection (recently reviewed by Clark, 1987). Bonner and Horn (1982) have discussed some problems and examples of isolating size, shape or timing as the main target of selection. They conclude that examples of shape selection determining body size are rare and relatively special. Nevertheless, where it occurs, such as in a size-neutral environment (e.g., as noted above) where one organ is of overriding importance to fitness, indirect size selection would occur.

In sum, selection on timing, rate or shape can indirectly alter body size. When we consider that such selection is operating throughout ontogeny, and involves many factors acting at once, it is evident that complete explanations of evolutionary change in body size are unlikely to be as simple as suggested in the past. Indeed, in some groups, selection has been for (ecophenotypic) body-size plasticity itself (Ebenman and Persson, 1988), complicating the rate/time problem even more. Figure 4.9 attempts graphically to summarise this complex interplay and how it culminates in the deceptively simple manifestation most familiar to palaeontologists: the allometric plot. Notice that environmental gradients are often translated into morphological (size, shape, age) gradients. These have been well documented by McNamara (1982) and are called 'heterochronoclines' (see also McKinney and McNamara, 1990; and Chapter 3 of this volume). Yet given the just-noted complexity of interaction between selection and the ontogenetic factors, how could such consistency be realised? Apparently, it is because the ontogenetic

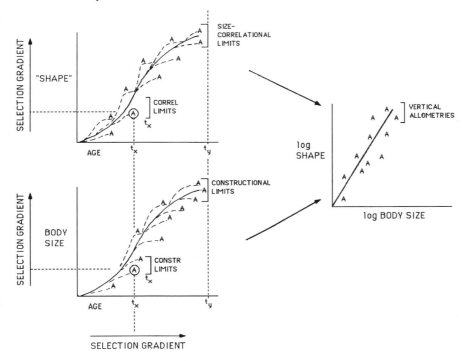

Figure 4.9. Schematic relationships among body size, 'shape' (e.g., organ length, but which, like body size, may be a multivariate summary variable) and age. Both size and shape are originally functions of age. Graphs on the left show how selection may act on shape, body size, and/or age to alter one, two, or all three parameters. For instance, in the species Ⓐ, selection favours the size and shape shown, to be attained at time t_x, as opposed to some other age. Other As represent other, related species that follow similar trajectories, with minor heterochronic variations in offset, rate, and so on (Figure 4.6). 'Constructional limits' denote maximum and minimum body size that can be attained at a given age with this group's basic developmental pattern (*sensu* Figure 4.8). Similar to this are 'size-correlational limits', which are limits imposed on organ size/shape by the organ's inability to fully decouple its growth from that of body size (correlated traits). Note how size, shape, and age selection all follow a general gradient, reflected general vectors of grain size, energy levels, and so on, in virtually all environmental parameters, and coadaptation of morphology to them. The loss of age data obscures virtually all of this complex and crucial selectional/developmental interplay, rendering a simple allometric plot (As represent average adult values, i.e. static allometries). In spite of much lost information, vertical allometries remain, as testimony to the size and shape variation selected for, permitted within the constructional and correlational limits imposed by rates and timing intrinsic to developmental processes of the group. Even so, without age data, we cannot infer how any given A developed because a large number of age/size/shape permutations could have produced any given A.

factors are coadapted as a suite to environmental factors that are also correlated among themselves. For example, many of McNamara's (1982) heterochronoclines run along onshire–offshore (i.e. selection) gradients that have covarying vectors of directionality in many parameters (see Chapter 9 herein). Thus, proceeding offshore, the following vectors occur: grain size decreases, water depth increases, water temperature decreases, water energy level decreases, water pressure increases, light decreases, and so on.

Clearly, the only way fully to tease apart the various forces in the complex interplay is to have information on ontogenetic change in these four basic parameters: size, shape, age and environment. This permits the study of the growth of individuals, as a population in an environment of known, preferably controllable, conditions. A reading of the growing literature on this work, and the difficulties of measuring selection and genetic/developmental change even under such conditions should convince anyone that palaeontologists have an extremely difficult job in explaining size change in long-dead animals, where we know so little about environmental conditions and usually lack a knowledge of an individual's age, such that determination of growth rates is lacking. Nevertheless, a brief review of how to explain size change where better information is available is extremely useful in telling us what we do not know and how we might go about making more educated inferences than those of the past.

Simply stated, the 'evolution of size and growth can only be understood through knowledge of the selection processes acting on the population and the patterns of genetic variation that are present' (Kirkpatrick, 1988). In terms of the four parameters noted above, size, shape, and age represent the manifestation of genetic variation and selection in the environment. Without delving into the equations of quantitative genetics (which are well described by Kirkpatrick, 1988), we can say that the phenotypic mean of a trait is the sum of two basic effects: the additive covariance among traits and the selection gradient. The main point is that to analyse how the phenotypic mean (e.g., body size) evolves we must have measurements of the traits under study (e.g., size and some correlated shape or organ traits) in a series of related individuals and their offspring. From this we can determine the covariance among traits, by standard statistical methods, based on matrix manipulations. Clearly some of this kind of information is not accessible to the palaeontologist (e.g., whether individuals are related, and longitudinal study of a series of generations). However, it does show the importance of knowing ontogenetic ages in trying to characterise even fossil populations (e.g., using methods discussed by Jones, 1988), and perhaps the potential of analysing trait covariances in cross-sectional biometrics (from dead individuals representing different ontogenetic stages). The second determinant, selection gradients, can also be measured directly in living populations. The selection gradient is defined as the regression of relative fitness on the phenotypic value of that trait (Lande and Arnold, 1983). Again we see the problem for fossil reconstruction, in determining the 'fitness' of a trait relative to some incompletely reconstructed environment. Nevertheless, again there is also probably room for improvement. For instance, size distributions of fossils could reveal stage-specific mortality patterns; this would be even more important if ontogenetic age were known through growth-line analysis, for example. Individuals with certain phenotypes might have greater mortality (see, for example McKinney *et al.*, 1990).

While the methods outlined above may seem involved already, they are not adequate for the study of selection on growth trajectories. In order to really understand the evolution of body size, what we must actually study is the evolution of growth trajectories. The application of quantitative genetics has been explored by Kirkpatrick (1988). His technique involves measuring the size of an individual at different ages and treating each measurement as if it were a trait. The mean phenotype (growth trajectory in this case) is then calculated in the manner discussed above and a selection gradient is calculated. Again, workers on fossils are at a disadvantage but it seems reasonable that where ontogenetic age could be determined, cross-sectional data from numerous individuals might be substituted for his longitudinal method, if appropriate allowances were made. In addition to methodological insight, Kirkpatrick's (1988) conclusions are of interest to body-size evolution. He finds that only a limited number of growth trajectories can be realised because patterns of genetic and ontogenetic covariation among sizes at different ages constrain evolution. That is, individuals that are large at one age tend to be relatively large at the next, and so on. As there is much plasticity among groups in many trait covariances, this raises many interesting questions about the evolution of body size in fossil lineages. Do certain groups show less size covariance among stages than others? If so, is this lack of constraint manifested in higher rates of size evolution? Or greater size change? Given age data in fossils, it seems likely that palaeontologists could at least make some further progress towards understanding these kinds of key questions. For instance, growth increments in a number of fossilised adult individuals might be used to reconstruct their size at various earlier ontogenetic ages. From these, the covariance (degree of constraint) among growth trajectories could be estimated and compared with rates of body-size evolution in the group (the role of ontogenetic constraint in phylogenetic change).

Multiplicative cladogenesis and step size

Returning to Figure 4.4 (especially 4.4C), we turn away from the driving force behind size increase and look at the broader aspects of the pattern itself. The 'multiplicative phase' represents a time of relatively rapid size increase in the clade (see also Figure 4.2). I suggest that this phase occurs because of cladogenetic dynamics: as the number of branches accumulates, the number of potential lineages available to undergo branching increases rapidly because branching is a multiplicative process. Thus, as with binary cell fission, there is an initial lag phase followed by a 'snowball effect'. More rigorously, consider that each branch has an equal chance of branching into two more branches, where p is this constant probability of branching. Then:

$$N_{t+1} = N_t + pN_t \tag{1}$$

where N is the number of branches at time t. The point of this simplified illustration (e.g., it ignores branch extinction probabilities) is to show that the amount of branching is largely a multiple of the number of pre-existing branches. Therefore, if there is much branching (high value of p), clade growth will be more geometric than arithmetic—for example, where $p = 1$ and all branches split, clade size will double during each time interval $(t, t+1)$. This is a good example of positive feedback, common in many replicating biological processes (DeAngelis *et al.*, 1986). Note that not all branches contribute to the size-increasing pattern ('go up'). By the reasoning above, only about half of the branches (but it could just as well be less, under different assumptions) result in species larger than the ancestor. However, the same multiplicative pattern applies since the upward-moving proportion is roughly constant (i.e. a constant proportion of increasing multiples). Essentially, we are talking about increasing the skewed 'tail' of the size distribution noted above, along with adding many smaller species well within the ancestral size range, too.

This interpretation is at variance with that suggested by Hayami (1978) who also described a logistic curve of size increase, using data from Jurassic bivalves and ceratopsian dinosaurs. However, he ascribed the rapid growth phase to multiplicative body growth: size increases in multiples of previous sizes. In part, this is because he focused on anagenetic size increase, but the same principle applies, except that in the cladogenetic model the ancestor lives on. As shown in Figure 4.4, the implication is that each size increase occurs as a 'step', where step size increases (roughly geometrically) with each upward step. Note the key distinction between this model and the one above: in the earlier model, multiplicative increase results from increasing multiples of the number of steps (i.e. number of branches upward); in Hayami's model the increase is in multiples, not of number of steps, but of step size.

Aside from the positive evidence advanced in favour of my model, I cite two negative lines of evidence that seem to disfavour Hayami's. First, it is based on flawed reasoning. Hayami states that the logistic pattern is shown because lineages start off at a 'suboptimal' size and move towards some optimal size. The leveling off (asymptote—see Figure 4.4) occurs because as it nears optimal size, selection pressure decreases. This logic (originally presented by Stanley, 1973) assumes that early, small ancestors are less well adapted than their descendants. It is probably unprovable either way, but there is no reason to assume that *Protoceratops* (to use Hayami's example) was 'suboptimal' relative to *Triceratops*. As discussed in detail above, the body size of any organism is a result of many selective factors finely tuning size, age and shape to the organism's particular environment. If ceratopsian dinosaurs (or any lineage) became larger it is because the optimum ('adaptive peak') moved along, from environmental change. The second problem with Hayami's model is more empirical: given the heterochronic ease with which rapid size change can be carried out (McKinney and McNamara, 1990;), there is no reason for Hayami's bivalve and dinosaur lineages to take millions of years to go through the rapid growth stage. Even aside from the developmental biology of size change, fine-scale fossil studies (Davis, 1981; Koch, 1986) show that phylogenetic size can closely track environmental changes, changing rapidly within a few thousand years (see especially Lister's, 1989, documentation of red deer size, reduced to less than 20 per cent in less than six thousand years). The point is that for the multiplicative phase to take so long, multiples of species-level processes (cladogenesis) are more likely to be involved than are multiples of ontogenetic processes (morphogenesis).

A final aspect of the multiplicative phase is the possibility that step size was constant, at multiples of about 1.3. This arises because of the highly cited, and debated, proposal of Hutchinson (1959) that a doubling in weight (i.e. a 1.3-fold increase in linear dimensions such as length) is necessary for niche-separation of related or competing species. This became ecological dogma for over twenty years, and many workers (among them Lovtrup *et al.*, 1974) did indeed seem to find peaks in body-size distributions at intervals of roughly 1.3. If it were true, then we would expect to find steps of 1.3 multiples in Figure 4.4C where selection had acted to minimise competition among the differentiating species. However, since 1979, a number of empirical and theoretical studies have demonstrated the unlikelihood of Hutchinson's size ratio. In a thorough statistical analysis, Roth (1979) showed that the 1.3 multiple peaks in size are not at all clearly demonstrated. Simberloff (1983) showed that many inanimate objects are characterised by 1.3 ratios. As noted by Tonkyn and Cole (1986), this is probably because the human mind categorises objects into regularly-sized categories: thus, ratios of adjacent categories will always be slightly larger than 1. Perhaps most damaging of all is the work of Eadie *et al.* (1987) which showed that the 1.3 size ratio is an artefact of log-normal distributions, such as characterise body-size distributions in most clades, as noted above. Nevertheless, this should not completely detract from the importance of size in influencing competitive relationships. As Werner and Gilliam (1984) have pointed out, increasing size during growth will have a major effect on the changing (ontogenetic) niche occupied by the individual, even if no simple law of incremental thresholds applies.

In conclusion, the multiplicative phase of clade size increase results from the positive feedback inherent in cladogenesis. Further, there is no evidence that branching occurs in quantum increments of size change, either at a constant 1.3 (or doubling, if weight) or at geometrically (or other multiple) increasing increments.

The clade size asymptote

The final aspect of the model (Figure 4.4C) to explain is why the multiplicative process of size increase slows down and begins to level off. As noted, Hayami (1978) ascribed this to approaching an optimum size. However, the cladogenetic view here assumes that each species in the clade is equally well 'optimised' to its environment (the correlates discussed above) so that his reason does not apply. A better possibility is that the clade is approaching not an optimum but a limit: the structural/functional maximal size limit for the clade's basic body plan. For instance, to use the horse example above, how large can a horse become and still function as a horse, or even at all? While this idea is appealing, and at first seems testable, it is very difficult to determine the upper size limit of a group. For instance, Schmidt-Nielsen (1984), in his discussion of the largest land mammal to exist, the 30-tonne *Baluchitherium* (now re-named *Indrichotherium*), estimates that the bone structure could have withstood a static load of 280 tonnes. Yet this proves very little because static loads provide no information on response to stress from an active, moving animal. Thus, Schmidt-Nielsen (1984, p. 6) concludes that such upper size limits simply cannot be known.

An alternative explanation for the asymptote is that it results from the diffusive nature of cladogenesis. In Chapter 2 herein, I noted that the maximum limit of diffusion (standard deviation from the starting point) increases only as the square of time ($\sqrt{npq}$, where n = number of steps, or time increments). Thus, if we plot the maximal limit (the outermost particle) through time, we obtain a curve very like the one of Figure 4.4C, because, as the square of time, the outer limit levels off after a fairly rapid increase. This reflects the dynamics of diffusion: each particle has an equal chance of going up or down so that, through time, the number of particles towards the top becomes progressively more rarefied, with fewer and fewer near the maximum to extend the envelope of the group as a whole. If we consider that species are particles (see Chapter 2), they may show random behaviour if there are enough multiplicative, independent forces operating on each species. If this is even roughly accurate, the asymptote represents not a structural limit, intrinsic to the clade, but the simple fact that diffusion has not had enough time to proceed to still greater sizes. Could it be that, given more time, even larger land mammals might have evolved? Even if so, the declining-square law of diffusion dictates that such very large species would evolve very rarely. This may represent yet another case where random models may be applied to evolutionary processes (Raup, 1988).

SYNTHESIS: ORIGINATION, EXTINCTION, AND ECOLOGY OF BODY SIZE

Overview of origination and extinction

Body size (a multivariate concept) has evolved in two ways: via anagenesis and cladogenesis. Fossil studies have focused on anagenetic evolution, showing a preponderance of trends to size increase rather than decrease, at least in the long term. This is difficult to explain, except as an artefact or bias. I reject the argument (Stanley, 1973; Hayami, 1978) that species originate at suboptimum size (because large ones are 'allometrically constrained' and evolve towards more optimal (i.e. 'larger') sizes), for numerous ecological reasons. Two major ones are: that there is no evidence that species originate at 'suboptimal' sizes (all species are well adapted), rather they track adaptive peaks (see Chapter 2 herein); and that there is no evidence for the vague notion of 'allometric constraint' when applied to anything except (micro-evolutionarily irrelevant) mouse versus elephant scales. Shorter-term anagenesis seems to show the expected random 'flux' (no clear preference for increase or decrease).

Large-scale (e.g., biosphere- and clade-level) size changes involve clado-genetic diffusion. This diffusion is asymmetrical to larger size because the diffusion starts near the smaller limit of body size so that there is 'nowhere to go but up'. Rejecting again the allometric constraint logic for small-sized origin, two other reasons remain: that small organisms survive major perturbations better (many clades originate after such events), and that most organisms in a clade are small to begin with. Following small-size origination, diffusion to environments that foster larger size occurs. Many abiotic and biotic factors select for larger size, with individuals often under many selective influences at once so the optimum body size is a trade-off among them. Among those influences discussed, resource competition is a major promoter of larger body size, as captured by the $r–K$ continuum. Indirect size selection, such as on reproductive timing (also part of the $r–K$ continuum), and shape may also be important factors in size. Because of multiple influences, and their effects on rate and size itself throughout ontogeny, explaining the selective forces in body-size evolution is probably more complex than historically indicated. Cladogenesis is a multiplicative process so that, as diffusion proceeds, a rapid growth phase occurs, followed by a levelling off. Size increases during this process probably do not occur in quantum steps. The cause of the levelling off is not clear, but one or both of the following appear likely: a structural/functional size limit is approached for the clade body plan; and diffusion (being a square root law process) occurs relatively more slowly as time goes on, and the process is eventually halted anyway when mass extinctions occur.

This whole scenario is reviewed in Figure 4.10. Figure 4.10A shows that the probability density function (PDF), that is, size distribution increases in two ways through time: maximum body size in the clade (length of tail), and the overall number of species (height of PDF), including the number of smaller species, which grows during the branching as well. Figure 4.10B shows that body-size evolution of a clade (and indeed the biosphere and many collective units) may be a sequence of diffusion processes (e.g., migration) broken up by major perturbations that 'reset the clock'. (The term 'perturbation' is purposely used instead of mass extinction because some clades may be affected during non-mass extinction times, i.e. small extinction events.) Note that such perturbations could theoretically have two basic effects, manifested in the two alternative post-extinction PDFs shown. Where selection against large size does not actively occur, the post-extinction PDF is basically a diminutive version of the pre-extinction PDF, indicating that each body-size class had an equal chance of extinction. (Because they are so few in number, very large classes, in the PDF tail, probably die out even if no selection occurs; e.g., if three exist and 90 per cent of all species die out at the event, then there is a 73 per cent ($0.9 \times 0.9 \times 0.9$) chance that all three will die out.) Thus, even without active size selection, major perturbations will in that sense 'select against' large size. However, should active selection against large size occur (for the many reasons cited above), the post-extinction PDF (as shown) will be more peaked toward the small-sized end. (As with true PDFs, the two post-extinction curves are scaled to have the same area under them.)

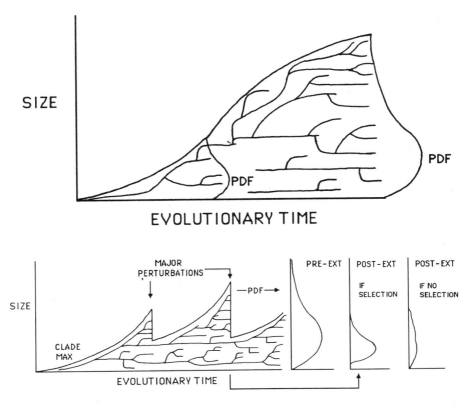

Figure 4.10. A: PDF = probability density function (i.e. size–frequency distribution) of cladogenetic diffusion process: at any given point in clade evolution, small species are much commoner than large ones. B: Cladogenetic size diffusion is punctuated by episodes of major extinction events that usually eliminate the large-sized species (i.e. the PDF 'tail'). It is uncertain whether this is because of selectivity against large species (schematically shown on the post-extinction curve on the left) or because there are simply fewer large species (post-extinction curve on the far right, produced by random decimation of pre-extinction PDF).

An interesting aspect of this whole scenario is that body-size trend reversals occur not because of reversals of diffusion, or more properly in the selective regimes that led to size increases, but because of a 'crash' in the whole system. In addition, buried within the major outlines of Figure 4.10 are still other dynamics that are illustrated in Figure 4.11. We see here that during 'normal' (background) times of stability, small body sizes tend to have higher extinction rates. While not shown here, this is because small species are the most endemic, with many often having greatly reduced geographic ranges compared with larger species (Brown and Maurer, 1989, Figure 2). This makes them much more susceptible to the vagaries of local habitat destruction (the leading cause of extinction) that characterise background times (McKinney, in preparation). Also, large species may have a competitive advantage in their dominance of resources during stable times (Brown and Maurer, 1986), contributing to higher small-size extinction rates as well. In contrast, large species have higher extinction rates during environmental disturbances (Figure 4.11) for the reasons cited above. Perhaps of most interest is the pattern of origination rates relative to size: smaller-sized species have higher ones than large species (Figure 4.11). The reasons, discussed by Dial and Marzluff (1988; 1989) are that smaller species have a number of traits (e.g., life-history ones such as early maturation) that make them better colonisers, opportunists and so on, to take advantage of background fluctuations in resources and local extinctions; and that smaller species subdivide the environment more finely so that there are more niches for new species to inhabit. These combine to make per-species origination higher for smaller species. However, one might add that there are more smaller species to begin with so that, again, the multiplicative nature of small size encourages higher rates of those size classes per unit time. Finally, note how the net result of this extinction–origination interplay results in the well-known log-normal size distribution discussed throughout this chapter (Figure 4.11).

The interpretations of Figure 4.11 are partly consistent with documented palaeontological data: groups that have high origination rates also tend to have high extinction rates (Stanley, 1979; in press). In this case, the high rates are attributed to small taxa. However, Bonner (1988) has interpreted Stanley's (1979) data to mean that large-bodied groups have higher extinction (and origination) rates. Again, this appears to be a problem in scales of comparison. Stanley's rates involve comparisons of such diverse groups as foraminifers, corals, mammals and so on. In this sense, the 'small' organisms, such as foraminifers, do indeed have lower rates than 'large' ones like the mammals. There are many possible reasons for this: greater behavioural complexity and lower dispersal of mammals, for instance (Stanley, in press). However, such gross taxonomic comparisons are not clearly applicable to the clade-level (e.g., genus, family, order) patterns discussed here. The extinction and origination of foraminifers versus mammals may have little in common with the same processes generating large versus small horses. Evidence for this is shown by van Valen's (1975) calculations that large mammals have lower background extinction rates than small mammals. The ecological arguments presented by Dial and Marzluff (1988) and Pimm *et al.* (1988) are not only rooted in solid ecological generalisation, but they are much more relevant to originations and extinctions at the (intra-class) clade level.

Ecology of body size

The cladogenetic evolution of larger body sizes has important implications beyond the phylogenetic level, affecting ecosystem structure itself. For instance, Lawton and Warren (1988) have discussed the role of body size as

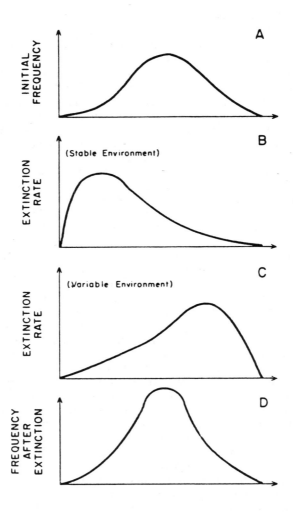

Figure 4.11.   Model showing how size-specific extinction and speciation may produce classic circum-'log-normal' distribution of body size found in most clades (F). Barring extinction or reflecting barier, diffusion and origination would lead to a normal (or any arbitrary) size distribution (A). However, selective extinction (after diffusion and origination) in stable environments (C) and variable environments (D) leads to higher extinction rates of small- and large-sized species, respectively, leading to (D). However, D is not realised because of the final differential macroevolutionary property: small species speciate faster (E). Thus, diffusion and origination lead to (F), when differences in extinction and origination rate are accounted for. From Dial and Marzluff (1988).

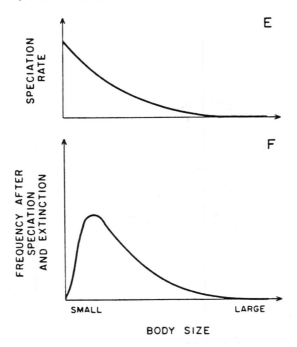

Figure 4.11. *continued*

the major structuring element of food webs, the heart of ecosystemic processes. Similarly, Dickie *et al.* (1987) have reviewed the role of body-size distributions in controlling energy-flow rates through ecosystems, through rates of metabolism, locomotion and other allometrically scaled processes. While it is too early to prove conclusively the ecosystemic effect of increasing body size through time, it appears that it increases the complexity of interactions, adds trophic levels, and probably retards rates of energy and matter flow through ecosystems. This last occurs not only because of more interactions and added trophic levels, but because larger organisms have relatively lower mass-specific rates of energy and matter use (Peters, 1983; Calder, 1984).

Another major size-related ecological trend has been well documented in a series of papers by Brown and Maurer (1986; 1989; Maurer and Brown, 1988). They show that increasing body size, in addition to altering the structural complexity and rate of resource flow in ecosystems, has led to a major change in the way those resources have been partitioned: large species have come to utilise a disporportionate share of ecosystem resources (Figure 4.12). The reasons for this are not entirely clear. It may be that the competitive advantages of large individuals (discussed above) permit larger species to monopolise resources (Brown and Maurer, 1986). However, it is also true that smaller species often do not directly compete with larger ones,

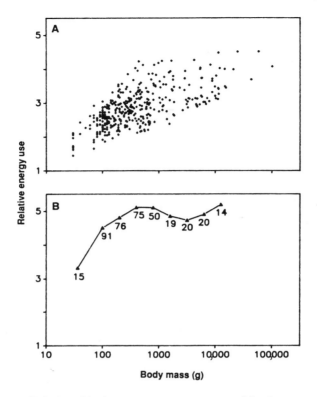

Figure 4.12.   Relationship between energy use and body mass (log scale) for North American land birds. A: Average values for species. B: Values summed for all of the species in equal-sized log (body-size) categories. Numbers are number of species in each size class. Large birds use more energy than small ones. From Brown and Maurer (1989).

becoming highly specialised (i.e. needing high-nutrient foods) to meet higher mass-specific metabolic food needs and other scaling changes. Thus, it may be that energetic dominance of larger-sized species may be based on their ability to utilise more kinds of foods, of lower quality (e.g., the digestive tract allometry discussed above). In this sense, the evolution of larger body size results in the evolution of greater utilisation of available energy.

Prospects

The evolution of body size, both anagenetic and cladogenetic, is becoming better understood, as the amount of information on both pattern and process increases. Documentation of fossil patterns has improved greatly over the last 20 years, although much more needs to be done. Especially useful would be thorough studies of cladogenetic size change in a variety of clades of different taxa and at different taxonomic levels. Documentation of fossil patterns outside of body-size studies *per se* are also likely to yield insight into body-size evolution, because body size is such a pervasive trait. For instance, the well-known nearshore–offshore trend of nearshore clade origination and later offshore diversification (Bottjer and Jablonski, 1988) may be integrated with the body-size observations offered herein: small-sized originations in near-shore (higher-energy, less predictable) environments are consistent with ecological theory, as is later diversification ('diffusion') offshore into stabler environments, which would foster larger-bodied species (McKinney, 1986; see also the discussion in McKinney and McNamara, 1990).

Further insights are likely from the improving understanding of the developmental and ecological processes underlying fossil patterns. Hetero-chrony and other ontogenetic mechanisms are becoming increasingly better understood (McKinney and McNamara, 1990; and Chapter 3 of this volume). Ecological processes, such as selection on size versus life history at different ontogenetic stages, are more difficult to tease apart. Nevertheless, much progress is being made, such as by the application of quantitative genetics to both controlled and natural conditions.

REFERENCES

Atchley, W., 1983, Some genetic aspects of morphometric variation. In J. Felsenstein (ed.), *Numerical taxonomy*, Springer-Verlag, Berlin: 346–63.

Atchley, W., 1984, Ontogeny, timing of development, and genetic variance-covariance structure, *Am. Nat.*, **123**: 519–40.

Bader, R.S., 1955, Variability and evolutionary rate in the oreodonts, *Evolution*, **9**: 119–40.

Bakker, R., 1977, Cycles of diversity and extinction. In Hallam, A. (ed.), *Patterns of evolution*, Elsevier, Amsterdam: 431–78.

Barnovsky, A.D., 1989, the Late Pleistocene event as a paradigm for widespread mammal extinction. In S. Donovan (ed.), *Mass extinctions: processes and evidence*, Belhaven, London: 235–54.

Begon, M., 1985, A general theory of life-history variation. In R. Sibly and R. Smith (eds), *Behavioural ecology*, Blackwell, Oxford: 91–7.

Begon, M., Harper, J., and Townsend, C., 1986, *Ecology*, Sinauer, Sunderland, MA.

Berg, H.C., 1983, *Random walks in biology*, Princeton University Press, Princeton, NJ.

Boltovskoy, E., 1984, On the size change of benthic Foraminifera of the bathyal zone during the Oligocene–Quaternary interval, *Revista Española Micropaleont.* **16**: 319–30.

Boltovskoy, E., 1988, Size change in the phylogeny of Foraminifera, *Lethaia*, **21**: 375–82.

Bonner, J.T., 1965, *Size and cycle*, Princeton University Press, Princeton, NJ.

Bonner, J.T., 1968, Size change in development and evolution, *J. Paleont.*, **42**: 1–15.

Bonner, J.T., 1988, *The evolution of complexity*, Princeton University Press, Princeton, NJ.

Bonner, J.T. and Horn, H.S., 1982, Selection for size, shape and developmental timing. In Bonner, J.T. (ed.), *Evolution and development*, Springer-Verlag, Berlin: 259–77.

Bookstein, F., Chernoff, B., Elder, R., Humphries, J., Smith, G., and Strauss, R., 1985, *Morphometrics in evolutionary biology*, Academy of Natural Science of Philadelphia, Special Publication 15.

Bottjer, D. and Jablonski, D., 1988, Paleoenvironmental patterns in the evolution of post-Paleozoic benthic marine invertebrates, *Palaios*, **3**: 540–60.

Boucot, A., 1976, Rates of size increase and of phyletic evolution, *Nature*, **261**: 694–6.

Brooks, J.L. and Dodson, S., 1965, Predation, body size, and composition of plankton, *Science*, **150**: 28–35.

Brown, J.H. and Maurer, B., 1986, Body size, ecological dominance, and Cope's Rule, *Nature*, **324**: 248–51.

Brown, J.H. and Maurer, B., 1989, Macroecology: the division of food and space among species on continents, *Science*, **243**: 1145–50.

Calder, W., 1984, *Size, function, and life history*, Harvard University Press, Cambridge, MA.

Chaline, J. and Lauren, B., 1986, Phyletic gradualism in a European Plio-Pleistocene *Mimomys* rodent lineage, *Paleobiology*, **12**: 203–16.

Clark, A.G., 1987, Genetic correlations: the quantitative genetics of evolutionary constraints. In V. Loeschcke (ed.), *Genetic constraints on adaptive evolution*, Springer-Verlag, Berlin: 25–45.

Clemens, W.A., 1986, Evolution of the terrestrial vertebrate fauna during the Cretaceous–Tertiary transition. In D.K. Elliot (ed.), *Dynamics of extinction*, Wiley, New York: 63–85.

Davis, S.J., 1981, The effects of temperature change and domestication on the body size of Late Pleistocene to Holocene mammals of Israel, *Paleobiology*, **7**: 101–14.

DeAngelis, D., Post, M. and Travis, C., 1986, *Positive feedback in natural systems*, Springer-Verlag, Berlin.

deBeer, G., 1958, *Embryos and ancestors*, Clarendon Press, Oxford.

Demment, M. and van Soest, P., 1985, A nutritional explanation for body size patterns of ruminant and non-ruminant herbivores, *Am. Nat.*, **125**: 641–72.

Dial, K.P. and Marzluff, J., 1988, Are the smallest organisms the most diverse?, *Ecology*, **69**: 1620–4.

Dial, K.P. and Marzluff, J., 1989, Nonrandom diversification within taxonomic assemblages, *Syst. Zool.*, **38**: 26–37.

Dickie, L.M., Kerr, S. and Boudreau, P., 1987, Size-dependent processes underlying regularities in ecosystem structure, *Ecol. Monogr.*, **57**: 233–250.

Eadie, J.M., Brockhoven, L. and Colgan, P., 1987, Size ratios and artifacts: Hutchinson's rule revisited, *Am. Nat.*, **129**: 1–17.

Ebenman, B. and Persson, L. (eds), 1988, *Size-structured populations*, Springer-Verlag, Berlin.

Flessa, K. and Bray, R., 1977, On the measurement of size-independent morphological variability, *Paleobiology*, **3**: 350–9.

Gerritsen, J., 1982, Size efficiency reconsidered: a general foraging model for free-swimming aquatic animals, *Am. Nat.*, **123**: 450–67.

Gingerich, P.D., 1976, Paleontology and phylogeny: patterns of evolution at the species level in Early Tertiary mammals, *Am. J. Sci.*, **276**: 1–28.

Gingerich, P.D., 1985, Species in the fossil record: concepts, trends, and transitions, *Paleobiology*, **11**: 27–41.

Gould, S.J., 1966, Allometry and size in ontogeny and phylogeny, *Biol. Rev.*, **41**: 587–680.

Gould, S.J., 1974, The evolutionary significance of 'bizarre' structures: antler size and skull size in the 'Irish Elk', *Megaloceras gigantans, Evolution*, **28**: 191–220.

Gould, S.J., 1981, *The mismeasure of man*, Norton, New York.

Gould, S.J., 1988, Trends as changes in variance: a new slant on progress and directionality in evolution, *J. Paleont.*, **62**: 319–29.

Hallam, A., 1975, Evolutionary size increase and longevity in Jurassic bivalves and ammonities, *Nature*, **258**: 493–7.

Hallam, A., 1978, How rare is phyletic gradualism and what is its evolutionary significance? Evidence from Jurassic bivalves, *Paleobiology*, **4**: 16–25.

Hallam, A., 1982, Patterns of speciation in Jurassic *Gryphaea, Paleobiology*, **8**: 354–66.

Hallock, P., 1985, Why are larger Foraminifera large?, *Paleobiology*, **11**: 195–208.

Hayami, I., 1978, Notes on the rates and patterns of size change in evolution, *Paleobiology*, **4**: 252–60.

Hutchinson, G.E., 1959, Homage to Santa Rosalia or why are there so many kinds of animals? *Am. Nat.*, **93**: 145–59.

Janis, C., 1982, Evolution of horns in ungulates: ecology and paleoecology, *Biol. Rev.*, **57**: 261–318.

Jones, D.S., 1988, Sclerochronology and the size versus age problem. In M.L. McKinney (ed.), *Heterochrony in evolution: a multidisciplinary approach*, Plenum, New York: 93–110.

Jungers, W. (ed.), 1985, *Size and scaling in primate biology*, Plenum, New York.

Kellogg, D.E., 1976, Character displacement in the radiolarian genus *Eucyrtidium, Evolution*, **29**: 736–49.

Kellogg, D.E., 1983, Phenology of morphologic change in radiolarian lineages from deep-sea cores: implications for macroevolution, *Paleobiology*, **9**: 355–62.

Kellogg, D.E. and Hays, J.D., 1975, Microevolutionary patterns in late Cenozoic Radiolaria, *Paleobiology*, **1**: 150–60.

Kirkpatrick, M., 1988, The evolution of size in size-structured populations. In B. Ebenmann and L. Persson (eds), *Size-structured populations*, Springer-Verlag, Berlin: 13–28.

Koch, P.L., 1986, Clinal variation in mammals: implications for the study of chronoclines, *Paleobiology*, **12**: 269–81.

Kurtén, B., 1953, On the variation and population dynamics of fossil and Recent mammal populations, *Acta Zool. Fennica*, **76**: 1–122.

Kurtén, B., 1960, Rates of evolution in fossil mammals, *Cold Spring Harbor Symposium: Quantitative Biology*, **24**: 205–15.

LaBarbera, M., 1986, The evolution and ecology of body size. In D. Raup and D. Jablonski (eds), *Patterns and processes in the history of life*, Springer-Verlag, Berlin: 69–98.

LaBarbera, M., 1989, Analyzing body size as a factor in ecology and evolution, *Ann. Rev. Ecol. Syst.*, **20**: 97–117.

Lande, R. and Arnold, S.J., 1983, The measurement of selection on correlated characters, *Evolution*, **37**: 1210–26.

Lazarus, D., 1986, Tempo and mode of morphologic evolution near the origin of the radiolarian lineage *Pterocanium prismatium, Paleobiology*, **12**: 175–89.

Lawton, J.H. and Warren, P., 1988, Static and dynamic explanations for patterns in food webs, *Trends Ecol. Evol.*, **3**: 242–5.

Lindstedt, S. and Boyce, M., 1985, Seasonality, fasting endurance, and body size in mammals, *Am. Nat.*, **125**: 873–8.

Lister, A.M., 1989, Rapid dwarfing of red deer on Jersey in the Last Interglacial, *Nature*, **342**: 539–43.

Lomolino, M., 1985, Body size of mammals on islands: the island rule re-examinated, *Am. Nat.*, **125**: 310–16.

Lovtrup, S., Rahemtulla, F., and Hoglund, N., 1974, Fisher's axion and the body size of animals, *Zoo. Scr.*, **3**: 53–8.

Lynch, M. and Arnold, S., 1988, The measurement of selection on size and growth. In Ebenman, B. and Persson, L. (eds), *Size-structured populations*, Springer-Verlag, Berlin: 47–59.

MacFadden, B.J., 1986, Fossil horses from '*Eohippus*' (*Hyracotherium*) to *Equus*: scaling, Cope's Law, and the evolution of body size, *Paleobiology*, **12**: 355–69.

Malmgren, B.A. and Kennett, J.P., 1981, Phyletic gradualism in a late Cenozoic planktonic foraminiferal lineage, *Paleobiology*, **7**: 230–40.

Malmgren, B.A., Berggren, W., and Lohman, P., 1983, Equatorward migration of *Globorotalia truncatulinoides* ecophenotypes through the late Pleistocene: gradual evolution or oceanic change? *Paleobiology*, **9**: 377–389.

Marshall, L. and Corrucini, R., 1978, Variability, evolutionary rates, and allometry in dwarfing lineages, *Paleobiology*, **4**: 101–19.

Martin, P.S., 1984, Prehistoric overkill: the global model. In P.S. Martin and R. Klein (eds), *Quaternary extinctions*, University of Arizona Press, Tucson: 354–403.

Maurer, B.A., and Brown, J.H., 1988, Distribution of energy use and biomass among species of North American terrestrial birds, *Ecology*, **69**: 1923–32.

Mayr, E., 1983, How to carry out the adaptationist Program?, *Am. Nat.*, **121**: 324–34.

McKinney, M.L., 1986, Ecological causation of heterochrony: a test and implications for evolutionary theory, *Paleobiology*, **12**: 282–9.

McKinney, M.L., 1988, Classifying heterochrony: allometry size, and time. In M.L. McKinney (ed.), *Heterochrony in evolution: a multidisciplinary approach*, Plenum, New York: 17–34.

McKinney, M.L. and McNamara, K.J., 1990, *Heterochrony: the evolution of ontogeny*, Plenum, New York.

McKinney, M.L., McNamara, K.J. and Zachos, L., 1990, Heterochronic hierarchies: application and theory in evolution, *Historical Biology*, **3**: 269–87.

McNamara, K.J., 1982, Heterochrony and phylogenetic trends, *Paleobiology*, **8**: 130–42.

Monaghan, P. and Metcalfe, N., 1986, On being the right size: natural selection and body size in the herring gull, *Evolution*, **4**: 1096–9.

Olson, E.C., 1962, Late Permian terrestrial vertebrates, U.S.A. and U.S.S.R., *Trans. Am. Phil. Soc.*, **52**: 1–44.

Osborn, H.F., 1936, *Proboscidea*, Volume 1, American Museum Press, New York.

Osborn H.F., 1942, *Proboscidea*, Volume 2, American Museum Press, New York.

Ozawa, T., 1975, Evolution of *Lepidolina multiseptata*, *Mem. Fa. Sci. Kyushu Univ., Series D*, **23**: 117–64.

Persson, L., 1985, Asymmetrical competition: are larger animals competitively superior?, *Am. Nat.*, **126**: 261–6.

Peters, R.H., 1983, *The ecological implications of body size*, Cambridge University Press, Cambridge.

Pimm, S., Jones, H., and Diamond, J., 1988, On the risk of extinction, *Am. Nat.*, **132**: 757–85.

Prothero, D. and Sereno, P., 1982, Allometry and paleoecology of medial Miocene dwarf rhinoceroses from the Texas Gulf Coastal Plain, *Paleobiology*, **8**: 16–30.

Raup, D.M., 1988, Testing the fossil record for evolutionary progress. In M. Nitecki (ed.), *Evolutionary progress*, University of Chicago Press, Chicago: 293–318.

Raup, D.M. and Crick, R., 1981, Evolution of single characters in the Jurassic ammonite *Kosmoceras*, Paleobiology, **7**: 200–15.

Rensch, B., 1959, *Evolution above the species level*, Columbia University Press, New York.

Reyment, R.A., 1985, Phenotypic evolution in a lineage of the Eocene ostracod *Echinocythereis*, *Paleobiology*, **11**: 174–94.

Riska, B. and Atchley, W., 1985, Genetics of growth predict patterns of brain size evolution, *Science*, **229**: 668–71.

Roff, D., 1981, On being the right size, *Am. Nat.*, **118**: 405–22.

Romer, A.S. and Price, L.W., 1940, *Review of the Pelycosauria*, Geological Society of America, Special Paper 28.

Roth, V.L., 1979, Can quantum leaps in body size be recognized among mammalian species? *Paleobiology*, **5**: 318–36.

Schmidt-Nielsen, K., 1984, *Scaling: why is animal size so important?* Cambridge University Press, Cambridge.

Schoener, T., 1983, Field experiments on interspecific competition, *Am. Nat.*, **122**: 240–85.

Schweitzer, P., Kaesler, R., and Lohmann, G.P., 1986, Ontogeny and heterochrony in the ostracode *Cavellina* Coryell from Lower Permian rocks in Kansas, *Paleobiology*, **12**: 290–301.

Schwaner, T., 1985, Population structure of black tiger snakes. In G. Grigg, R. Shine, and H. Ehmann (eds), *Biology of Australasian frogs and reptiles*, Royal Zoological Society of New South Wales: 35–46.

Sebens, K.P., 1982, Competition for space: growth rate, reproductive output, and escape in size, *Am, Nat.*, **120**: 189–97.

Shea, B.T., 1983, Allometry and heterochrony in the African apes, *Am. J. Phys. Anthrop,*, **62**: 275–89.

Shea, B.T., 1985, Bivariate and multivariate growth allometry: statistical and biological considerations, *J. Zool. Soc. (London)*, **206**: 367–90.

Shea, B.T., 1988, Heterochrony in primates. In M.L. McKinney (ed.), *Heterochrony in evolution: a multidisciplinary approach*, Plenum, New York: 237–68.

Shine, R., 1988, The evolution of the large body size in females: a critique of Darwin's 'Fecundity Advantage' model, *Am. Nat.*, **131**: 124–31.

Simberloff, D., 1983, Sizes of co-existing species. In D. Futuyma and M. Slatkin (eds), *Coevolution*, Sinauer, Sunderland, MA: 404–30.

Sondaar, P., 1977, Insularity and its effect on mammal evolution. In M. Hecht, P. Goody, and B. Hecht (eds), *Major patterns in vertebrate evolution*, Plenum, New York: 671–707.

Stanley, S.M., 1973, An explanation for Cope's Rule, *Evolution*, **27**: 1–26.

Stanley, S.M., 1979, *Macroevolution: pattern and process*, Freeman, San Francisco.

Stanley, S.M., in press, The general correlation between rate of speciation and rate of extinction. In R. Ross and W. Allmon (eds), *The causes of evolution*, University of Chicago Press, Chicago.

Stanley, S.M. and Yang, X., 1987, Approximate evolutionary stasis for bivalve morphology over millions of years: a multivariate, multilineage study, *Paleobiology*, **13**: 113–39.

Stevenson, R.D., 1985, Body size and limits to the daily range of body temperature in terrestrial ectotherms, *Am. Nat.*, **125**: 102–17.

Taylor, C.R., Caldwell, S., and Rowntree, V., 1972, Running up and down hills: some consequences of size, *Science*, **178**: 1096–7.

Tissot, B., 1988, Multivariate analysis. In M.L. McKinney (ed.), *Heterochrony in evolution: a multidisciplinary approach*, Plenum, New York: 35–52.

Tonkyn, D.W. and Cole, B., 1986, The statistical analysis of size ratios, *Am. Nat.*, **128**: 66–81.

van Valen, L., 1975, Group selection, sex, and fossils, *Evolution*, **29**: 87–94.

Vermeij, G.J., 1989, Geographical restriction as a guide to the causes of extinction: the case of the cold northern oceans during the Neogene, *Paleobiology*, **15**: 335–56.

Werner, E.E. and Gilliam, J., 1984, The ontogenetic niche and species interactions in size-structured populations, *Ann. Rev. Ecol. Syst.*, **15**: 393–425.

Wilson, D.S., 1975, On the adequacy of body size as a niche difference, *Am. Nat.*, **109**: 769–84.

Woolbright, L.L., 1983, Sexual selection and size dimorphism in anuran amphibia, *Am. Nat.*, **121**: 110–19.

Wootton, J.T., 1987, The effects of body mass, phylogeny, habitat, and trophic level on mammalian age at first reproduction, *Evolution*, **41**: 732–49.

PART TWO

EVOLUTIONARY TRENDS
IN INVERTEBRATES

Chapter 5

TRILOBITES

R.A. Fortey and R.M. Owens

INTRODUCTION

Trilobites are common fossils in Palaeozoic rocks, and, with their complex morphology and rapid temporal variation, are appropriate subjects for the investigation of evolutionary trends. Trends have been identified from time to time (pioneered by Kaufmann, 1933) but they have never been summarised. With the possible exception of the ostracodes, trilobites are the only arthropodan group with a fossil record adequate to investigate trends at the species-to-species level. Some of these are capable of being explained in functional terms; many happen several times during the history of the group. An exhaustive compendium of all those examples that have been claimed as trends in trilobite evolution would extend this chapter beyond a reasonable length. Instead, we have selected examples which are representative of the different kinds of change observed. The sources of literature are disparate, and sometimes obscure, and so we have encapsulated the important points in the figures. For morphological terminology, the reader is referred to the *Treatise on Invertebrate Paleontology*, Part O (Moore, 1959).

Definition of trends

The term 'trends' has been used rather loosely to describe virtually any evolutionary change. For the purposes of this account two categories are recognised: those within species-to-species phyletic sequences, and repeated trends which are represented by the periodic recurrence of specific morpho-types from different phylogenetic sources. The former type of lineage has

121

been recorded from more or less confacial sequences, spanning no more than a few million years, and the change involved is frequently of a kind which recurs in different stocks and at different times in the Trilobita. The case of loss of eyes is a typical example. For the moment we leave aside the question of whether such trends can or cannot be interpreted within the adaptationist paradigm. The second kind of trend approximates to that described as 'iterative' evolution in classical palaeontological texts. In many cases the actual tree recording the derivation of the morphotype has not been constructed species by species, and instead the end product is recognised from its homoeomorphic resemblance to other, unrelated taxa. Paradoxically, precisely those characters which are of most interest in the discussion of trends are the ones which are the greatest trouble to cladistic analysis of phylogenetics (Wiley, 1981). The stuff of trends is homoplasy, whereas the recognition and removal of homoplasous characters is the particular virtue of cladistic analysis. While it undoubtedly helps to have a phylogenetic classification available to comment on taxonomic aspects of trends, it is not a *sine qua non* as it is in the case of extinction patterns (Briggs *et al.*, 1988; Patterson and Smith, 1987). One can talk about repeated trends without having a fully resolved classification, because those that are similar in kind may occur in taxa which are, quite obviously, taxonomically separate. Eye loss, for example, occurred in Agnostida in the early Cambrian, and Proetida in the Devonian and Carboniferous, and one can describe these trends without knowing how these groups fit into trilobite classification as a whole. However, when we attempt to summarise *patterns* of trends (for example, the number of families giving rise to different morphotypes) it becomes important that the families cited are natural units. We have selected examples where there are no major taxonomic problems known to us.

TRENDS IN PHYLETIC SERIES OF SPECIES

The examples described comprise case histories which have been derived from stratigraphically controlled series of collections, mostly from continuous sequences, and for which the taxonomy is up to date. Although subtle change in more than one character is usually involved in the trend, in the chosen examples it is predominant in a single and striking exoskeletal character. The kinds of change are also those which have happened on more than one occasion during trilobite history. In the explanation for Figure 5.1, we have also indicated a time calibration for the lineages. This may be somewhat approximate, being based on average duration of biozones, but it is of the right order of magnitutude.

Eye loss

Trilobites were primitively equipped with well-developed eyes (Fortey and Whittington, 1989); loss of eyes is a derived character. Blind trilobites appeared repeatedly during the 300-million-year history of the group, from many different phylogenetic sources: eye loss constitutes a typical trend. It occurred independently in many taxa, including Agnostida, Conocoryphacea (itself a polyphyletic group), Trinucleacea, Shumardiidae, Dindymeninae, Dalmanitidae, Phacopidae, Pliomeridae, Prosopiscidae and Proetida; in very many of these instances the mechanism species-by-species leading to eye loss is not known in detail. But an example where eye loss has been claimed to have been observed in a phylogenetic series is in the Upper Devonian benthic tropidocoryphine proetide *Pterocoryphe*, the ancestor of the blind *Pteroparia* (Clarkson, 1988; Feist and Clarkson, 1989) (Figure 5.1A). This example shows reduction and ultimately loss of the eye accompanied by modest changes in sutural outline through a series of four species over about 3 million years. The changes are claimed as gradual, although each of the four species is represented as a distinct entity. So much of the rest of the trilobite remains similar that it is obvious that loss of the eye does not entail major genetic change. This is in accord with the common and polyphyletic occurrence of eye loss. It is worth mentioning that the way in which the eyes are lost does seem to vary among different lineages: in Trinucleacea dorsal sutures are lost, and in a few forms a small eye remains 'marooned' on the fixed cheek (Hughes *et al.*, 1975, Plate 5, Figure 64). In Phacopacea the eyes either migrated forwards and are reduced and finally lost close to the anterior margin (e.g., Figure $5.2A_{3, 4}$; $5.2C_{2, 3}$; $5.2G_{2, 3}$), or the number of lenses on the eye is reduced (e.g., Figure 5.2E, $5.2F_2$); Chlupáč (1977) has given a detailed summary of examples in the Phacopidae.

Anterior migration of hypertrophied eyes

The opposite extreme of eye development is in those trilobites with hypertrophied, globular eyes—which occupy most, or even all, of the available area of the fixed cheeks. Such trilobites include those with pelagic habits (Fortey, 1985). A trend which occurs in trilobites with this morphology is a progressive anterior migration of the visual surfaces of the eyes, which become effectively more forwardly directed (Figure 5.1B). In its extreme development, this process permits the enlarged eyes to become fused together to form a single, gigantic eye. At least seven lineages which express this trend are known, and it has its analogue among the living hyperiid crustaceans. Curiously, we have only been able to detect it in Ordovician trilobites. The example illustrated is a time-morphological series in *Pricyclopyge*, a common genus in deeper water facies in the Ordovician of Europe. The stages in the sequence have conventionally been dubbed with subspecific names, and it has been claimed that the series is gradualistic. At any one horizon there is variation in a population in the distance between the eyes. There is some 3–5 million years between the first appearance of *Pricyclopyge* with wide-apart eyes and the first occurrence of forms in which they touch. *Pricyclopyge* is thought to have been a mesopelagic trilobite (Fortey and Owens, 1987), capable of utilising low light intensities. Having thousands of lenses in each eye, it must have been sensitive to small movements of prey or predator species in the Ordovician ocean.

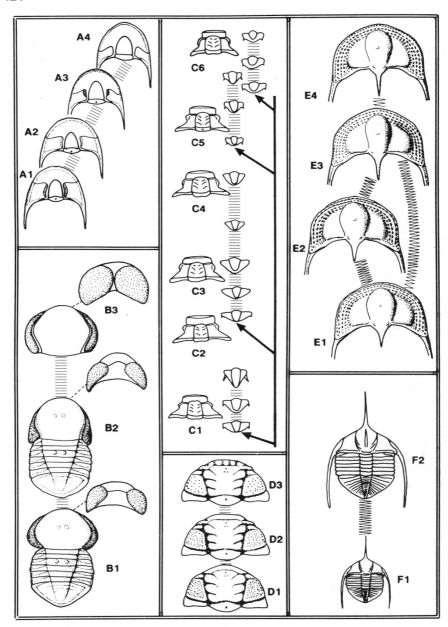

Figure 5.1.   Species-to-species trends in trilobites.

Olenid trilobites

The family Olenidae comprises a diverse array of trilobites ranging in age from late Cambrian to late Ordovician. Olenids were adapted to nekto-benthic life under conditions of low oxygen tension. They tend to be found to the exclusion of other groups in black, unbioturbated mudstones and 'stink-stones'. Oxygen-poor environments often preserve a relatively continuous stratigraphic sequence. The deposits of the 'Olenid sea' basin in the Upper Cambrian of Scandinavia have yielded a sequence of olenid species which have been studied in great detail (Henningsmoen, 1957). Kaufmann (1933) described the succession of *Olenus* species in the earlier part of the Scandi-navian olenid sequence (Figure 5.1C). Quantitative analysis of large samples showed Kaufmann that a series of *Olenus* species (recognised especially on cranidial characters) appeared suddenly in the succession, and were characteristic of particular stratigraphic intervals. However, during the history of each of these species there were changes in the pygidial shape (development of spines, width) which proceeded relatively rapidly.

Figure 5.1. – cont.
A_1–A_4: Eye reduction, finally leading to blindness in a lineage of Upper Devonian (Frasnian) tropidocoryphines, over about 3 million years. A_1 *Ptero-coryphe languedociana* Feist; A_2 *P. progrediens* Feist; A_3 *P. oculata* Feist; A_4 *Pteroparia coumiacensis* Feist (after Clarkson, 1988).

B_1–B_3: Enlargement and anterior merging of hypertrophied eyes in *Pri-cyclopyge* in the Ordovician (late Arenig and Llanvirn), over about 5 million years. B_1 *P. binodosa eurycephala* Fortey and Owens; B_2 *P. binodosa binodosa* Salter; B_3 *P. synophthalma* (Klouček).

C_1–C_6: Pygidial length changes in *Olenus* species in the Upper Cambrian (*Olenus* Zone) over about 5 million years (after Kaufmann, 1933).
C_1 *O. gibbosus* (Wahlenberg); C_2 *O. transversus* Westergård;
C_3 *O. truncatus* (Brünnich); C_4 *O. wahlenbergi* Westergård;
C_5 *O. attenuatus* (Boeck); C_6 *O. dentatus* Westergård.

D_1–D_3: Development of cranidial coaptative structures in *Placoparia* species in the Ordovician (Llanvirn and Lower Llandeilo), over 12–15 million years. D_1 *P. (P.) cambriensis* Hicks; D_2 *P. (Coplacoparia) tournemini* (Rouault); D_3 *P. (C.) borni* Hammann (after Henry and Clarkson, 1975).

E_1–E_4: Changes in pitting in the fringe of *Onnia superba* subspecies in the Ordovician (Caradoc, Onnian Stage) over about 1.5–2 million years. E_1 *O. superba cobboldi* (Bancroft); E_2 *O. superba creta* Owen and Ingham; E_3 *O. superba superba* (Bancroft), early form); E_4 *O. superba superba* (Bancroft), late form (after Owen and Ingham, 1988).

F_1–F_2: Increase in size and in number of pygidial pleural ribs in *Cnemidopyge* species in the Ordovician (Llandeilo Series), over about 4 million years. F_1 *C. nuda* (Murchison); F_2 *C. bisecta* (Elles) (diagrams by courtesy of Dr P.R. Sheldon).

Horizontal bars represent apparently unidirectional lineages; zigzag lines, those in which character 'reversals' have been identified or suggested.

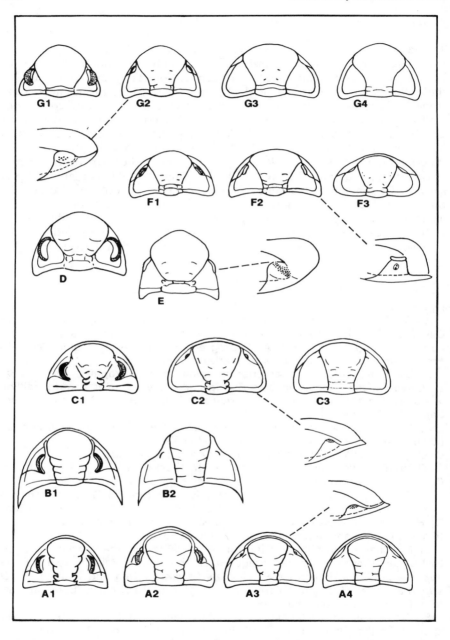

Figure 5.2 Repeated trends towards eye reduction and blindness in Phacopacea from the early Ordovician to late Devonian.

Kaufmann postulated the appearance of each species as derived from a 'conservative stock' not present in the sections he studied, and this is how it has been represented in Figure 5.1C. However, one might regard the appearance of new taxa instead as examples of punctuational change—rapid replacement by species which had evolved as peripheral isolates as in Eldredge's (1971) allopatric model. This example, although most thoroughly studied, is not amenable to functional interpretation.

Structures concerned with enrolment

Nearly all post-Cambrian trilobites could and did enrol tightly; some Cambrian forms were also capable of enrolment, but specimens in the enrolled state are distinctly rare. Because enrolment involves the co-operation between different parts of the exoskeleton, changes in enrolment style often entailed co-ordinated changes in both cephalon and pygidium. A prong developing on the pygidial margin might be matched by an appropriately shaped groove on the cephalic doublure, for example. Such specialised *coaptative* structures have been the subject of detailed work in several trilobites, and there is a trend towards a more intimate fit between the cephalic margin and the pygidium, plus, in some cases, the tips of the thoracic segments. We know of six examples, of which that illustrated is the Ordovician trilobite *Placoparia* (Figure 5.1D). Successive species of *Placoparia* allow for a progressively closer fit of the pygidial spines over the border in front of the glabella. This results in the appearance, and then emphasis, of a series of grooves in the border corresponding with the tips

---

Figure 5.2 – cont.
The selected examples do not necessarily represent evolutionary lineages, but illustrate the kinds of successive change that occur. A, *Ormathops*, Ordovician, Arenig and Llanvirn series: A_1 *O. borni* Dean; A_2 *O. atavus* (Barrande); A_3 *O. llanvirnensis* (Hicks); A_4 *O. nicholsoni* (Salter). B, *Mucronaspis* and *Songxites*, Ordovician, Ashgill Series: B_1 *M. mucronata* (Brongniart); B_2 *Songxites* sp. nov. C, *Phacopidella* and *Denckmannites*, Silurian, Wenlock and Ludlow series: C_1 *P. glockeri* (Barrande); C_2 *D. volborthi* (Barrande); C_3 *D. caecus* Schrank. D, *Phacops* (*Phacops*) *degener* Barrande, Devonian, Emsian Stage, a typical phacopid with well-developed eyes, the kind that probably lies at the beginning of lineages that gave rise to E, F and G. E, *Phacops* (*Prokops*) *prokopi* Chlupáč, Devonian, Pragian Stage. F, *Nephranops*, Devonian, Frasnian and early Famennian stages: F_1 *N. miserrimus* (Drevermann); F_2 *N. incisus incisus* (Römer); F_3 *N. incisus dillanus* (Richter and Richter). G, Devonian, Frasnian and Famennian species: G_1 *Phacops granulatus* (Münster); G_2 *Cryphops acuticeps* (Kayser); G_3 *Trimerocephalus mastophthalmus* (Reinh. Richter); G_4 *Dianops griffithides* (Richter and Richter). C_3 after Schrank (1973), F_1–F_3 and G_1–G_4 after Richter and Richter (1926). Others prepared from photographic illustrations in various sources.

of the pygidial spines (Henry and Clarkson, 1975). The change takes about 10–12 million years for completion. According to Henry (1980) the change between successive taxa along this lineage is punctuational, so that a particular stage (i.e. species) in the series persists unchanging through a stratigraphic interval before being rapidly succeeded by the next species. Of the available examples of this trend *Placoparia* is particularly apposite because the other features of the exoskeleton remain virtually unchanged through the species series. In this case it would be difficult to claim that the trend towards a more effective cephalic–pygidial interlocking mechanism was a mere incidental to some other more important change. Even more complex coaptations than those between cephalon and pygidium are possible: for example in Remopleurididae, and again in Cybelinae, a special boss (the rhynchos) on the hypostome became involved in enrolment. It is interesting to speculate whether changes so complex could be under the control of a single gene. Among living arthropods enrolment (conglobation) is familiar as a protective response in the isopods.

Trinucleid fringes

Trinucleidae is a large family of trilobites confined to Ordovician rocks. The cephalon is bordered by a pitted *fringe*, a complex structure in which opposing pits are present on the upper and lower lamella of which it is comprised (Hughes *et al.*, 1975). The trinucleid fringe evolved very rapidly, so much so that trinucleids form the basis of zonal schemes of local or even intercontinental use in the Ordovician. General trends have been described — for example, early and primitive trinucleids have numerous, irregularly arranged pits, while later and advanced ones have fewer and more ordered pits. At the species-to-species level there are a number of examples which show systematic trends in size, arrangement or number of pits. A recently published example (Owen and Ingham, 1988) from the genus *Onnia* is shown in Figure 5.1E; the components of the trend were recognised as subspecies of a single species by the authors, and the changes apparently occur faster than the other trends given here. They are rather subtle also, being primarily concerned with details of pitting, the rest of the cephalic features remaining very similar. Despite much speculation in the literature there is little agreement about how the trinucleid fringe functioned. But because phyletic changes occur commonly, and are fixed within a species, it seems very likely that the fringe had a particular function, and that natural selection operated to direct the changes. So far as we know the pitted fringe is unique to trilobites, where it occurs in trinucleids, dionidids and harpids; there is no plausible recent analogue.

Pygidial ribbing

A trend towards an increase in the number and expression of the pygidial ribs has been recorded in a number of trilobite lineages. A recent detailed study based on exhaustive collecting in the Ordovician of Wales (Sheldon, 1988) has provided quantitative evidence for this trend. One example from here, the genus *Cnemidopyge*, is shown in Figure 5.1F. Gradual increase in the number of pygidial ribs is accompanied by increase in mean size of the adults in the population. Sheldon's study showed that a number of lineages of unrelated trilobite genera underwent similar, presumed phyletic trends at the same time, through the same rock sections. Conventional taxonomy had given different specific names to respective end members of the trend in some cases. Small-scale shifts (including short-lived reversals) from bed to bed but overprinted with an overall directional change supported the notion of gradualistic change. Because more than one character is usually involved in this trend it is perhaps less easily interpreted than some of the others described above. We also note that the reverse, a progressive *decrease* in number of pygidial ribs, has been described by Engel and Morris (1983) for *Weberiphillipsia* species from the Carboniferous of Australia.

ADAPTIVE EXPLANATIONS FOR TRENDS: FACT OR WISHFUL THINKING?

With this sample of trilobite trends in mind it is worth looking briefly at the problems of explaining them in adaptive terms. It is wise to be cautious about such explanations (Gould and Lewontin, 1979), lest by an excess of adaptationist zeal the better examples are damned with the bad. However, Levinton (1988) is surely right in pointing to cases where 'adaptive improvement' in time series of taxa is more than wishful thinking. The examples we have chosen from the trilobites illustrate the problems well. They can be ranked from trends in which an improvement in adaptation seems a reasonable claim, to those in which the reason for the trend is obscure.

In the case of trends in structures involved with enrolment, especially those in *Placoparia* in which other characters are stable, it would take an unusually sceptical mind to see the changes that occur during the trend as other than an improvement in the tightness of fit between cephalon and pygidium. It would be rather like claiming a simultaneous change in both lock and key as fortuitous. This trend is unequivocally function-related. In the case of forward migration of hypertrophied eyes we have to make at least one assumption, and that is that the former pelagic habits of these trilobites are well established (Fortey, 1985). Enough is known about the visual mechanism of trilobites to be sure that the increased anterior spread of lenses which happened in these lineages would increase the acuity of vision forwards, which appeals as an adaptation in a free swimming animal. Hence the trend becomes sensibly explicable, without, we believe, burdening the reasoning with *ad hoc* hypotheses. The tendency to loss of eyes is more difficult, if only because loss of a structure is hard to prove as a positive adaptation. However, secondary blindness is well known in living arthropods, especially 'cave blindness', and it is known to happen easily, because cave-blind species have sister taxa with normal sight. As with the trilobites, it is a trend that happens polyphyletically. One could argue whether redundancy constitutes something

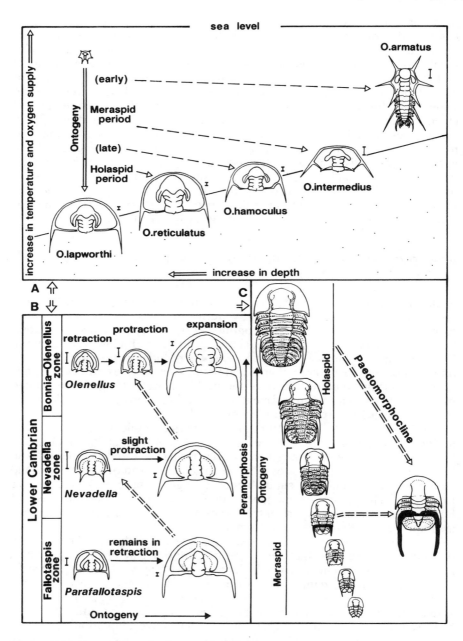

Figure 5.3   Heterochronic changes in trilobites.

which can be selected for, but the rarity of cave forms which have retained their eyes argues that blindness confers some advantage, or perhaps that the necessity for reliance on other senses 'switches off' the normal developmental programme leading to the formation of eyes. In many of the trilobite cases secondary blindness is associated with deep water benthic habitats. This applies particularly to cases where closely related trilobite taxa have normal vision, as in the phacopacean examples (Figure 5.2). In all these occurrences an adaptational explanation seems to contribute to our understanding and suggest further fruitful questions. At the opposite extreme, perhaps, come the small changes in the trinucleid fringe in *Onnia* species. We do not understand the function of the fringe, and the changes to it during the trend are not explicable. An adaptational explanation cannot necessarily be assumed.

Between the *Onnia* case and the well-supported ones there are the olenid trends and the pygidial ribbing trends. Quite a lot is known about the low-oxygen olenid palaeoenvironment, and the palaeoenvironment of the Ordovician sediments in central Wales from which pygidial rib changes have been documented may not have been dissimilar. However, there is no obvious connection between changes that are observed in the trends and any environmental factor. It might be argued that an increase in pygidial ribs was the dorsal expression of a ventral increase in limb number, and hence an increase in respiratory exites as a response to poorer oxygenation—but there is, as yet, no proof of this. In summary, it can be said that trilobite trends known from species-level phylogenies vary between those in which it would seem perverse to look beyond the morphology for an adaptive explanation, through those in which a reasonable adaptive explanation can be proposed if certain mode of life assumptions are accepted, to those in which an adaptive explanation is inaccessible. Finally, the examples discussed divide about evenly between those in which punctuational change between species is described, and those in which change is described as gradualistic.

Figure 5.3 – cont.
A: Paedomorphosis as an explanation for the origin of certain *Olenellus* species, with possible environmental controls (after McNamara, 1978; 1982). B: Peramorphic evolution of olenellid genera in the early Cambrian. Broken arrows indicate achievement of progessive glabellar grades during ontogeny (after McNamara, 1986). C: Possible origin of *Acanthopleurella* from *Conophrys* by paedomorphosis. Development is arrested after release of the fourth (macropleural) segment to achieve the number of segments of the former. Note that direct development of *A. grindrodi* Groom from *C. salopiensis* Callaway is *not* implied, and the examples are to show only the possible mechanism.

HETEROCHRONIC TRENDS

The examples of species-to-species trends just described concern the appearance of new structures, or at least the modification of existing ones. Another kind of trend relates to the timing of events in developmental history. Trilobites are good subjects for the study of such heterochronic trends, because the ontogeny of many species is known from the larval (protaspis) stage to the adult. Hence statements about shifts in developmental timing can be supported by facts rather than surmise, given plausible ancestor–descendant sequences of species. McNamara (1978; 1981; 1982; 1983; 1986) has made a particular study of heterochrony in trilobites, and our examples are mostly gleaned from his 1986 review. Heterochronic change was a potent source of speciation in the group. Because ontogenetic changes were often profound, heterochronic trends provided one way to introduce radical morphological change by what may have been a simple genetic control on development timing.

Some six different kinds of heterochronic change were recognised by McNamara (see Chapter 3 herein). Figure 5.3 shows three examples of common kinds of heterochronic trend in trilobites, which are supported by plausible series of species or genera. Paedomorphic trends are those in which ancestral juvenile characters are retained in the adult. Of the different kinds of paedomorphosis, *progenesis* (the precocious onset of sexual maturity) is frequently associated with trilobites having especially small adult size. The early Ordovician genus *Acanthopleurella* (Figure 5.3C) is probably the smallest trilobite of all; it is plausibly derived heterochronically from a genus like *Conophrys*, itself small, of which the development has long been familiar. Mature *Conophrys* had six thoracic segments, whereas *Acanthopleurella* had only four (Fortey and Rushton, 1980). Because thoracic segments in trilobites are released progressively during trilobite ontogeny, before achieving a stable mature number, the secondary attainment of a smaller number of segments is an indicator of paedomorphic processes in operation. A similar reduction in thoracic segment number happened in the transition from *Olenellus (Olenellus)* to *Olenellus (Olenelloides)* in the Lower Cambrian (Figure 5.3A), the latter also retaining a whole suite of 'immature' characteristics, such as marginal cephalic spines. Despite their very different appearance these two olenellids are considered to be closely related. Additionally, a whole series of progressively paedomorphic *Olenellus* species *(O. reticulatus, O. hamoculus, O. intermedius)* are characterised by progressively reduced morphological development as compared with *O. lapworthi*. The ontogeny of *O. lapworthi*, is, as it were, writ large in this series of olenellid species (Figure 5.3A). This kind of complex array of paedomorphosis is by no means uncommon in Cambrian trilobites, and probably later ones as well, and provides problems for the systematist.

The opposite sense of heterochronic process is *peramorphosis*, in which the adult form of the putative ancestor is retained in the juvenile of the descendant. If the development is extended this can lead to hypermorphosis (in the opposite sense of progenesis) but there are not many trilobite examples of this trend. Fortey and Owens (1987) described one, the nileid

Illaenopsis, from the early Ordovician of south Wales. Most nileids are 5–10 cm long, but *Illaenopsis* could grow to twice this size or more. That this was produced by a size displacement of the whole ontogeny is shown by the fact that an immature pygidium of *Illaenopsis* (a transitory pygidium still retaining an unreleased thoracic segment) is the largest known for any trilobite, over 1 cm long (Fortey and Owens, 1987, Figure 69). Commoner is a peramorphocline in which progressively younger species incorporate the features of their supposed ancestry in their ontogeny (Figure 5.3B). McNamara (1986) discussed several examples of this kind, of which the one selected is of interest as it concerns some of the very earliest trilobites in the Cambrian. Several changes are involved, including the shape of the glabella and its distance from the anterior cephalic margin. In a stratigraphic suite of genera the ontogenetic series of the youngest, *Olenellus*, includes in sequence the morphological states of the two, progressively older genera, *Nevadella* and *Parafallotaspis*. If this example is correctly interpreted it does seem that heterochronic trends were of some importance in generating new morphologies near the beginning of the trilobite history. Although most of the worked examples concern taxonomic novelty of a relatively low order, it is worth remembering that the Order Agnostida appears very early in the Cambrian, and that agnostids show several features which could be interpreted as progenetic. It is possible that the *initiation* of this major group was produced by heterochronic change (e.g., from an olenellid), although its subsequent history is complex and has little in common with that of the trilobites as a whole. It has also been claimed that early Cambrian trilobites exhibited a high degree of intraspecific variation; heterochronic trends operating at a low taxonomic level would have contributed to this.

The adaptive role of heterochronic trends is imperfectly understood. Because several morphological features are often involved the examples cannot be interpreted in the straightforward way that applied to changes in the coaptative devices concerned with enrolment (above). For example, what are we to make of the relatively advanced genal spines in the olenellid trend in Figure 5.3A? Such advancement of the genal spines happened several times in trilobite history (in the Ordovician family Remopleurididae, for example), and it seems reasonable to expect that advanced genal spines could be selected for. But no unequivocal explanation for this feature exists, and it is equally possible that such structures were simply a by-product of heterochronic change, fixed in morphogenetic sequence, and that the real business of selection was happening elsewhere in the morphology. McNamara (1978) proposed a plausible relationship between palaeoenvironment and the olenellid species known (Figure 5.3A). This does not, however, address the question of the adaptedness of the changes observed, which is speculative. Many cases of trilobite heterochrony incorporate changes in size, as in the example of *Acanthopleurella*. Because size differentiation is common among living, closely related species, as a response to partition of resources—for example, feeding on a different particle size—it is not an unreasonable supposition that this was the case also in the trilobites (Robison, 1975; Fortey and Rushton, 1980). A progenetic daughter species might have been able to exploit a smaller-sized food than its parent. The new species may have

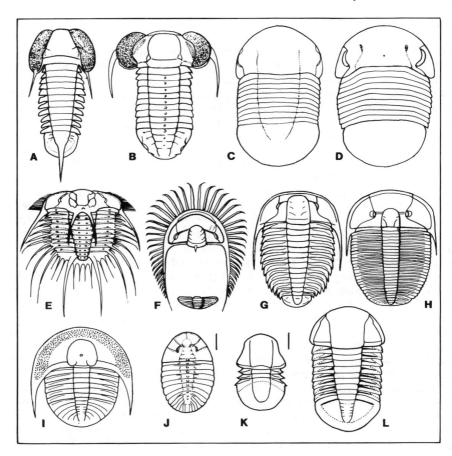

Figure 5.4.   Distinctive morphologies occurring in the Trilobita. A, B, Pelagic: *Opipeuter inconnivus* Fortey and *Carolinites genacinaca* Ross; C, D, Illaeni-morph: *Illaenus sarsi* Jaanusson and *Bumastus barriensis* Murchison; E, F, Marginal cephalic spines: *Odontopleura ovata* Emmrich and *Bowmania americana* (Walcott); G, H, Olenimorph: *Olenus micrurus* Salter and *Aulaco-pleura koninckii* (Barrande); I, Pitted fringe: *Dionide levigena* Fortey and Owens (see also Figure 5.1E, *Onnia superba* (Bancroft)); J, K, Miniaturisation: *Schmalenseeia amphionura* Moberg and *Thoracocare minuta* (Resser) (see also Figure 5.3C, *Conophrys salopiensis* Callaway and *Acanthopleurella grindrodi* Groom); L. Atheloptic: *Illaenopsis harrisoni* (Postlethwaite and Goodchild) (see also Figure 5.1A, *Pterocoryphe oculata* Feist and *Pteroparia coumiacensis* Feist; Figure 5.2A, *Ormathops llanvirnensis* (Hicks) and *O. nicholsoni* (Salter); Figure 5.2B, *Songxites* sp.; Figure 5.2C, *Denckmannites volborthi* (Barrande) and *D. caecus* Schrank; Figure 5.2F, *Nephranops incisus incisus* (Römer) and *N. i. dillanus* Richter and Richter; Figure 5.2G, *Trimerocephalus mastophthalmus* (Reinh. Richter) and *Dianops griffithides* (Richter and Richter)). Magnifications range from × 0.5 to × 2.5 approximately except for J and K, where the scale

arisen parapatrically, and might thereafter have been able to coexist with the parent by competitive exclusion. However attractive such explanations might seem—and they may go some way towards explaining how several closely related species can coexist in the same fossil bed—they are very difficult to test by independent evidence. Heterochronic trends exist in the Trilobita, but the 'driving force' behind them is not readily accessible.

REPEATED MORPHOTYPES

For a general look at evolutionary trends in the Trilobita, we can define a series of advanced morphological types, and look at the timing of their appearance and phylogenetic origins. Certain complex or specialised morphologies recurred during the history of trilobites (Figures 5.4 and 5.5). The close resemblance between unrelated forms makes it likely that they occupied similar niches. The fact of their separate derivation means that we can think of a trend leading to the morphotype in question even if we do not know the process species by species. As an example, Figure 5.2 shows the appearance of reduced-eyed phacopides on several occasions during the Ordovician to Devonian. Other exoskeletal features, including those of the glabella, co-aptative structures and the pygidium, show that these instances have no direct phylogenetic connection: this is a repeated trend from separate ancestors. It is important to know the phylogenetics to the extent that we need to be certain we are dealing with apomorphic character condition, and we need enough knowledge of relationships to resolve the convergences into their separate phyletic origins. It is not absolutely necessary to know the mode of life implied by the convergences, because the *pattern* can be surveyed regardless. This notion of trends goes back to Simpson's (1953) view of adaptive zones, which can be more or less fully occupied. The eight different morphotypes chosen (Figures 5.4, 5.5) may not provide an exhaustive description of the adaptive possibilities of trilobites, but they are distinctive enough to give an indication of times when iterative trends predominated, and when they were in abeyance.

One of the commonest supposed trends in trilobite evolution requires a brief mention here. This is effacement—the tendency for the dorsal furrows to become obliterated. A fully effaced trilobite can be left with virtually no dorsal features (*Rhaptagnostus*). More usually, the glabellar furrows are subdued and the axial furrows become shallow. Some typical examples are shown in Figure 5.7. Effacement is so common a tendency in trilobites that it can happen in virtually any family. This suggests that there is no single, simple explanation for it—for example, it appeared with equal facility in deep water olenids, and shallow shelf styginids. Most likely there were several different

Figure 5.4 – cont.
bars represent I mm. Illustrations after the following: A, B: Fortey (1985); C: Jaanusson (1957); D, E, H: Moore (1959); F: Ludvigsen (1982); G: Rushton (1985); I: Fortey and Owens (1987); J: Whittington (1981); K: McNamara (1986).

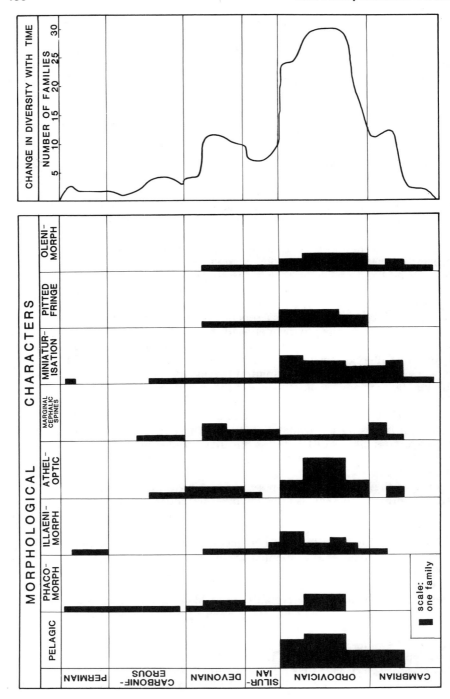

reasons why effacement happened. Some effaced trilobites were well stream-lined (Fortey, 1985); others, such as styginids or *Plethometopus*, were not. Some were large; others, like the agnostids, were small. Effacement happened in different trilobite stocks from the Lower Cambrian to the Permian. It would not, therefore, be wise to list effacement as if it were a single trend.

Trilobite morphotypes

Pelagic (Figures 5.4A and 5.4B)
Characters are hypertrophied eyes, reduced pleurae, occurrence in offshore facies, or, if epipelagic, in all biofacies. Fortey (1985) uncertainly included some Cambrian forms which are given the benefit of the doubt in Figure 5.5.

Phacomorph (Figure 5.2, 5.6)
Defined by Fortey and Shergold (1984) for nearly isopygous convex trilobites, with tuberculate sculpture in many species, prominent eyes, a forward-expanding and tumid glabella, with posterior glabellar furrows (if any) forming a collar-like structure at the base of the glabella; pygidium with strongly furrowed pleural fields and prominent axis. Phacomorphs with well-developed eyes are typical epeiric shelf inhabitants; there are several opinions on their life habits, ranging from predators to grazers. Figure 5.6 shows a striking convergence between unrelated Ordovician, Devonian and Permian phacomorphs.

Illaenimorph (Figures 5.4C and 5.4D)
Highly effaced trilobites on which the cephalic axis, and often the entire dorsal axis, is not sharply differentiated from the pleural areas by a change in slope; eyes posteriorly placed; typically the sagittal convexity of the head exceeds that of the thorax and pygidium. They have been associated with 'reefs', or with partial burial in sediment in the 'bumastoid stance', with the cephalon resting on the sea-bed, and the thorax and pygidium buried in the sediment in a vertical position. This has been interpreted as a feeding position, or associated with algal grazing. There is no general agreement on life habits of the distinctive illaenimorph trilobites.

Atheloptic (Figures 5.4L; $5.2A_3$; $5.2C_{2,3}$; $5.2F_{2,3}$; $5.2G_{2-4}$)
The *Pteroparia* species series (Figure 5.1) and the phacopaceans (Figure 5.2) quoted above are examples of this trend. Atheloptic trilobites (Fortey and Owens, 1987) are those in which the eyes are reduced, although their close relatives are known to have normal eyes. Hence major clades characterised by blindness like agnostids are excluded. Atheloptic trilobites tend to be

Figure 5.5. Numbers of occurrences and change in diversity of eight common trilobite morphologies plotted against time.

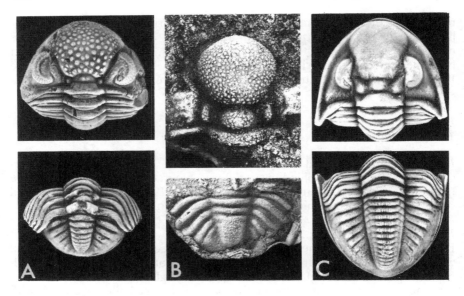

Figure 5.6. Three phacomorph trilobites belonging to different orders. A: *Phacops schlotheimi* (Bronn) (Phacopida), × 2, Devonian, Eifelian, Eifel, West Germany (National Museum of Wales 84.31G.77 and 84.31G.78). B: *Norasaphus* (*Norasaphites*) *vesiculosus* Fortey and Shergold (Asaphida), cranidium × 5, pygidium × 2, Ordovician, Arenig Series, central Australia (Commonwealth Palaeontological Collection, Bureau of Mineral Resources, Canberra, nos 22677 and 22687). C: *Ditomopyge decurtata* (Gheyselinck) (Proetida), × 2, Permian, Wolfcampian, Kansas (US National Museum, no. 145322). Note the common presence of median and lateral preoccipital lobes, and the forwardly expanded frontal lobe.

found together in former deep water habitats marginal to palaeocontinents, or more widely at times of major transgressions.

Marginal cephalic spines (Figures 5.4E and 5,4F). A curious and distinctive morphology in which the cephalon is fringed by a dense array of spines. We do not know what these were for, although Clarkson (1969) has suggested that in odontopleurids the downwardly-directed spines may have helped orientate the cephalon and hypostome during feeding. Horizontally disposed spines may have had a defensive role.

Miniaturisation (Figures 5.3C; 5.4J; 5.4K). This has been mentioned, briefly, under progenetic changes. Miniature trilobites have a mature size of a few millimetres. We make the assumption that miniaturisation is a comparable adaptation regardless of its several separate phylogenetic origins.

Pitted fringe (Figures 5.1E; 5.4I). The distinctive pitted fringe was discussed above (page 128), being developed on trinucleids and several other trilobites. It is likely that the fringe developed from a single row of pits, which are not uncommon in the cranidial border furrows of trilobites belonging to many families. Primitive trinucleaceans such as *Myinda* have single pit rows of this kind. The term 'fringe' has to be limited to the presence of several rows of pits.

Olenimorph (Figures 5.4G; 5.4H). We discussed olenids in the context of the species-to-species trend in *Olenus*. Olenimorphs had thin exoskeletons, laterally extended thoracic pleurae, which were also narrow (exsagittally) and often crowded, and usually there is a multiplication in the number of free segments. The cheeks were usually caecate. These features are regarded as

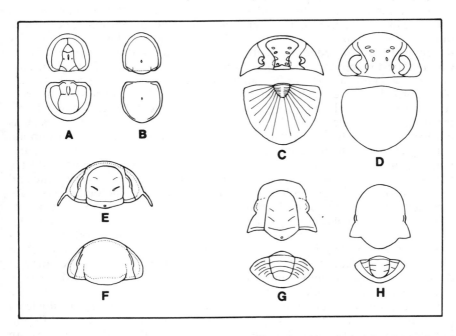

Figure 5.7. Examples of non-effaced and effaced pairs of taxa from unrelated groups of trilobites. A, B: Diplagnostidae. A, *Pseudagnostus* (*Pseudagnostus*) *ampullatus* Öpik; B, *Rhaptagnostus bifax* Shergold. Both from late Cambrian, Queensland, Australia. C, D: Styginidae. C, *Planiscutellum planum* (Hawle and Corda); D, *Rhaxeros pollinctrix* (Lane and Thomas), after Lane and Thomas, 1978. From Silurian of Czechoslovakia and Queensland, respectively (*Bumastus barriensis*, Figure 5.4D, continues the sequence). E, F: Olenidae. E, *Psilocara comma* Fortey; F, *P. patagiatum* Fortey. Both from the early Ordovician of Spitsbergen. G, H: Plethopeltidae. G, *Plethopeltis saratogensis* (Walcott); H, *Plethometopus glaber* Westrop. From late Cambrian of New York State and western Canada, respectively.

adaptations for life in a dysaerobic environment. Although Olenidae are typical, olenimorphs were recruited from several other families. The trinucleacean *Seleneceme* was an olenimorph; similarly the proetacean *Aulacopleura*.

The plot of the occurrence of these morphotypes through time is shown in Figure 5.5. This provides a measure of the frequency and taxonomic origins of these trends. It is possible to identify certain of the morphotypes back to the early Cambrian, but in general there is an increase throughout the Cambrian in both the variety of morphotypes represented, and in the number of families from which they are recruited. By any criterion the Ordovician was the acme of appearance of trends leading to the different morphotypes, with several families involved in all the categories. At the Ordovician–Silurian boundary there was a dramatic drop, this doubtless being a reflection of the disappearance of a number of trilobite clades at this horizon, which is recognised as a major extinction event in most groups of marine organisms. The pelagic morphotype disappeared at the end of the Ordovician, and we cannot identify its reappearance subsequently. This was a trend that was not renewed. There was a modest reactivation of some of the trends in the earlier Devonian. Late in the Devonian, at the end-Frasnian extinction event, we record our last olenimorph and pitted fringe. In the Carboniferous there are still examples of several of the trends, but they are being recruited from fewer and fewer families; all examples from the Permo-Carboniferous are proetaceans. It is necessary to beware, however, of possible 'Lazarus taxa'—for example, aulacopleurids with marginal spines disappear during the Famennian but reappear in the early Carboniferous, and it is considered probable that they persisted through the Famennian in a site as yet unknown. Even near the end of the history of the group in the Permian several morphological types are represented—these were confined to the inshore palaeogeographic sites.

This history reveals an early phase when a few families produced examples of the morphotypes, followed by a time when many families produced trends towards the morphotypes. Several taxa with the same adaptation but in different families could exist at the same time. The Ordovician was a time of dispersed palaeocontinents, and it is possible that the taxonomic richness in trends then was the product of parallel evolution in different sites. But while palaeogeographic differentiation may account for some of this Ordovician peak, it is also true that a *single site* may yield examples from different families with the same morphotype. For example, in the Ordovician of South Wales (Fortey and Owens, 1987) a deep water shale yielded three different families of pelagics, and at least four atheloptics. So the trilobite end products of these trends were able to subdivide a given environment without competitive exclusion. The richness of the Ordovician in the morphological trends must tell us something about the capacity of trilobites to exploit a variety of habitats at that time, and is not just about palaeogeography. After the Ordovician there is taxonomic loss, but with the exception of pelagics the remaining families threw up examples of each morphotype into the Devonian. After this, proetides alone continued, but even from this phylogenetically limited stock some of the morphotypes were reproduced.

In general, this approach to trends shows clearly the flexibility of the trilobite exoskeleton, and provides a way of evaluating the timing of that most nebulous of concepts, evolutionary success.

ACKNOWLEDGEMENTS

We thank Mrs D.G. Evans and Mrs L.C. Norton for their skilful drafting of the text figures.

REFERENCES

Briggs, D.E.G., Fortey, R.A. and Clarkson, E.N.K., 1988, Extinction and the fossil record of the arthropods. In G.P. Larwood (ed.), *Extinction and survival in the fossil record*, Clarendon Press, Oxford: 171–209.

Chlupáč, I., 1977, The phacopid trilobites of the Silurian and Devonian of Czechoslovakia, *Rozpr. ustred. Ust. geol.*, **43**, 1–172.

Clarkson, E.N.K., 1969, A functional study of the Silurian odontopleurid trilobite *Leonaspis deflexa* (Lake), *Lethaia*, **2**: 329–44.

Clarkson, E.N.K., 1988, The origin of marine invertebrate species: a critical review of microevolutionary transformations, *Proc. Geol. Assoc.* **99**: 153–70.

Eldredge, N., 1971, The allopatric model and phylogeny in Paleozoic invertebrates, *Evolution*, **25**: 156–67.

Engel, B.A. and Morris, N., 1983, *Phillipsia–Weberiphillipsia* in the early Carboniferous of eastern New South Wales, *Alcheringa*, **7**: 223–51.

Feist, R. and Clarkson, E.N.K., 1989, Environmentally controlled phyletic evolution, blindness and extinction in late Devonian tropidocoryphine trilobites, *Lethaia*, **22**: 103–21.

Fortey, R.A., 1985, Pelagic trilobites as an example of deducing the life habits of extinct arthropods, *Trans R. Soc. Edinb.*, **76**: 219–30.

Fortey, R.A. and Owens, R.M., 1987, The Arenig Series in South Wales, *Bull. Br. Mus. Nat. Hist. (Geol.)*, **41**: 69–307.

Fortey, R.A. and Rushton, A.W.A., 1980, *Acanthopleurella* Groom 1902: origin and life habits of a miniature trilobite, *Bull. Br. Mus. Nat. Hist. (Geol.)*, **33**: 79–89.

Fortey, R.A. and Shergold, J.H., 1984, Early Ordovician trilobites, Nora Formation, central Australia, *Palaeontology*, **27**: 315–66.

Fortey, R.A. and Whittington, H.B., 1989, The Trilobita as a natural group, *Historical Biology*, **2**: 125–38.

Gould, S.J. and Lewontin, R.C., 1979, The spandrels of San Marco and the Panglossian paradigm: a critique of the adaptionist programme, *Proc. R. Soc. London*, ser. B, **205**: 581–98.

Henningsmoen, G., 1957, The trilobite family Olenidae, *Skr. Norske Vid.-Akad. Oslo. 1. Mat-Nat. Kl.*, **1**: 1–303.

Henry, J.-L., 1980, Trilobites ordoviciens du Massif Armoricain, *Mém Soc. géol. miner. Bretagne*, **22**: 1–250.

Henry, J.-L. and Clarkson, E.N.K., 1975, Enrollment and coaptations in some species of the Ordovician trilobite genus *Placoparia, Fossils Strata*, **4**: 87–96.

Hughes, C.P., Ingham, J.K. and Addison, R., 1975, The morphology, classification and evolution of the Trinucleidae (Trilobita), *Phil. Trans. R. Soc. London*, ser. B, **272**: 537–607, 10 pls.

Jaanusson, V., 1957, Unterordovizische Illaeniden aus Skandinavien, *Bull. Geol. Instn. Univ. Uppsala*, **37**: 79–165.

Kaufmann, R., 1933, Variations statistische Untersuchungen über die 'Artabwand-lung' und 'Artumbildung' an der Oberkambrischen Trilobitengattung *Olenus* Dalm., *Abh. Geol.-Paläont. Inst. Univ. Griefswald*, **10**: 1–54.

Lane, P.D. and Thomas, A.T. 1978, Silurian trilobites from NE Queensland and the classification of effaced trilobites. *Geol. Mag.*, **115**: 351–8.

Levinton, J., 1988, *Genetics, macropaleontology and evolution*, Cambridge University Press, Cambridge.

Ludvigsen, R., 1982, Upper Cambrian and Lower Ordovician trilobite biostratigraphy of the Rabbitkettle formation, western District of Mackenzie, *Contr. Life Sci. Div. R. Ont. Mus.*, **134**: 1–188.

McNamara, K.J., 1978, Paedomorphosis in Scottish olenellid trilobites (Early Cambrian), *Palaeontology*, **21**: 53–92.

McNamara, K.J., 1981, Paedomorphosis in Middle Cambrian xystridurine trilobites from northern Australia, *Alcheringa*, **5**: 209–24.

McNamara, K.J., 1982, Heterochrony and phylogenetic trends, *Paleobiology*, **8**: 130–42.

McNamara, K.J., 1983, Progenesis in trilobites, *Spec. Pap. Palaeontology*, **30**: 59–68.

McNamara, K.J., 1986, Heterochrony in the evolution of Cambrian trilobites, *Biol. Rev.*, **61**: 121–56.

Moore, R.C. (ed.) 1959, *Treatise on Invertebrate Paleontology Part O, Trilobita*. Geological Society of America and University of Kansas Press, Lawrence.

Owen, A.W. and Ingham, J.K., 1988, The stratigraphic distribution and taxonomy of the trilobite *Onnia* in the type Onnian Stage of the uppermost Caradoc, *Palaeontology*, **31**: 829–55.

Patterson, C. and Smith, A.B., 1987, Is the periodicity of extinctions a taxonomic artefact? *Nature*, **330**: 248–52.

Richter, R. and Richter, E., 1926, Die Trilobiten des Oberdevons: Beiträge zur Kenntnis devonischer Trilobiten IV, *Abh. preuss. geol. Landesanst.*, N.F., **99**: 314 pp.

Robison, R.A., 1975, Species diversity among agnostoid trilobites, *Fossils Strata*, **4**: 219–26.

Rushton, A.W.A., 1985, Palaeontological notes. In P.M. Allen and A.A. Jackson, *Geological Excursions in the Harlech Dome*. Classical areas of British Geology, British Geological Survey, HMSO, London.

Schrank, E., 1973, *Denckmannites caecus* n. sp., ein blinder Phacopide aus den höchsten Thüringen Silur, *Z. geol Wiss.*, **1**: 347–51.

Sheldon, P.R., 1988, Trilobite size-frequency distribution, recognition of instars, and phyletic size changes, *Lethaia*, **21**: 293–306.

Simpson, G.G., 1953, *The major features of evolution*, Columbia University Press, New York.

Whittington, H.B., 1981, Paedomorphosis and cryptogenesis in trilobites, *Geol. Mag.*, **118**: 591–602.

Wiley, E.O., 1981, *Phylogenetics, the theory and practice of phylogenetic systematics*, Wiley Interscience, New York.

Chapter 6

BIVALVES

Arnold I. Miller

INTRODUCTION

On a global scale, the ongoing diversification of the class Bivalvia has proceeded at a remarkably continuous per-taxon rate since the group's initial radiation during the Ordovician. While the exact path of this diversity trend and the evolutionary processes that produce it continue to be debated among palaeobiologists (see Gould and Calloway, 1980; Miller and Sepkoski, 1988), its orderly structure suggests that there are emergent class-level properties of evolutionary significance shared historically by all bivalves. Indeed, a perusal of any detailed description of the group (see, for example, Cox *et al.*, 1969; Pojeta, 1987a) reveals, not surprisingly, a variety of morphological characteristics common to all bivalve species. However, just as evident from these descriptions is a considerable degree of morphological and ecological diversity. The hierarchical web of evolutionary trends that must be responsible for this variability belies an underlying complexity that seems difficult to reconcile with such a uniform class-level diversification.

The purpose of this chapter is to review a number of pervasive global, morphological, environmental and ecological trends recognised in the history of the Bivalvia. Evaluation of these trends, in turn, provides an opportunity to consider several associated issues of general importance, including:

1. *The relationship between evolutionary trends and life habit.* Among bivalves, it is apparent that life habit has played a major role, possibly transcending phylogeny, with respect to recognised paths of diversification.

2. *The hierarchical relationship between trends recognised within individual lineages and large-scale, macroevolutionary transitions.* With respect to certain morphological innovations, parallel trends are recognised in several lineages; in turn, these same trends are thought to have fuelled global transitions of substantial importance.
3. *The space dimension as an evolutionary factor.* The diversification of bivalves is not just a temporal process but also a spatial one. Palaeo-environmental transitions appear to accompany major temporal transitions and, thus, may serve as important arbiters with respect to evolutionary trends.
4. *The degree to which evolutionary trends are driven in particular directions by extrinsic 'forcing' mechanisms, either biotic or abiotic.* Net changes in the physical/biological environments that bivalves encounter may yield associated evolutionary transitions at any hierarchical level.
5. *The relationship between the variety of evolutionary trends and pathways among bivalves, and their synoptic, class-level history.* Does the class-level pattern have intrinsic meaning, or is it simply the consequence of a collage of evolutionary trends distributed randomly in space and time?

CLASS-LEVEL GLOBAL DIVERSIFICATION

A semi-logarithmic graph of bivalve global generic diversity through the Phanerozoic (Figure 6.1A) shows two particularly distinctive attributes. First, the per-genus diversification rate remained essentially unchanged through most of the Phanerozoic, suggesting that bivalves have undergone slow, but steady exponential diversification throughout their history. This pattern was recognised by Stanley (1977; 1979; 1985), and was taken as an indication of the inelastic nature of this group, in contrast to the more elastic patterns exhibited by other taxa. With respect to global diversity, Stanley noted that among elastic taxa, interspecific competition dampens what might otherwise be much higher rates of diversification, except during mass extinctions. By reducing the number of competitors in ecospace, 'mass extinction removes the major constraint on diversification. Following a sudden extinction, we would expect diversity to rebound rapidly' (Stanley, 1979, p. 284). In contrast, interspecific competition among inelastic taxa such as bivalves is relatively weak and might not dampen diversification in the first place. Thus, for an inelastic taxon, 'mass extinctions will not necessarily be followed by an increase in the rate of speciation . . . before and after mass extinctions, they tend to expand at some typical value of R [the net rate of diversification]' (Stanley, 1979, p.284).

A second attribute of the semi-logarithmic graph indicates, however, that bivalve diversification has been more elastic than suggested by Stanley. Specifically, bivalves exhibited dramatically increased diversification rates during three intervals: the initial Ordovician radiation, and following the Late Permian and Late Cretaceous mass extinctions. Utilising a coupled (i.e. interactive) logistic model, Miller and Seposki (1988; Figure 6.1B) showed that this 'combined' pattern is probably a consequence of interaction with other taxa that has served to impede the diversification rate of bivalves throughout most of their history. The diversification rate that bivalves could potentially exhibit has generally been dampened substantially; the only exceptions were periods when the summed global diversity of all taxa was relatively low (the Ordovician) or had been reduced markedly (mass extinctions). This overall pattern more closely conforms with Stanley's description of elastic, rather than inelastic, diversification.

Among many palaeontologists, it had traditionally been contended that bivalves gradually rose to prominence during the mid- to Late Palaeozoic as they outcompeted articulate brachiopods, whose diversity declined slowly during this interval. However, Gould and Calloway (1980) demonstrated that the apparent negative correlation through time between bivalve and brachiopod diversity curves was not statistically significant. They ascribed an upturn in the total generic diversification rate of bivalves to the Late Permian mass extinction, suggesting that the post-Palaeozoic bivalve radiation was a direct consequence of relative success (or luck) in weathering the event(s), rather than some manifestation of competitive superiority over brachiopods (see Gould, 1985). But it is apparent that, despite its short-term effect on bivalve global diversification, the Late Permian mass extinction did not substantially alter the long-term pattern. Following recovery from the extinction, the per-genus diversification rate returned to its pre-extinction level (Figure 6.1A). The post-Palaeozoic upturn in the total (i.e. untransformed) diversification rate is simply the expectation of an exponential system (Miller and Sepkoski, 1988), and a coincident biological event, such as a mass extinction, need not be invoked to account for this upturn.

This is not to suggest that bivalves were directly outcompeting brachiopods over the long-term. In a series of studies, Thayer (1979; 1981; 1983; 1985; 1986) showed that in some instances, bivalves exhibit a competitive advantage over brachiopods (see also Steele-Petrovic, 1979), but that in other cases the opposite is true. It is difficult, if not impossible, to demonstrate that any single higher taxon enjoys a long-term competitive advantage over another. Moreover, throughout the Phanerozoic, it is probable that bivalves have interacted with a wide variety of organisms; to ascribe their global pattern of diversification to interaction with a single group seems absurd. As noted by Eldredge (1987), the traditional comparison of brachiopods and bivalves is probably a consequence of their superficial morphological resemblance. With respect to the Palaeozoic diversification of bivalves, it would make just as much (or as little) sense to compare them directly with stalked crinoids, rugose and tabulate corals, trepostome bryozoa, or the variety of other organisms that were diverse and abundant on Palaeozoic sea-floors.

LIFE HABITS: THE HIERARCHY OF EVOLUTIONARY TRENDS

Throughout their history, bivalves have exhibited a wide variety of adaptations for life above, at and below the sediment–water interface. These life habits, and adaptations of shell form associated with them, were summarised

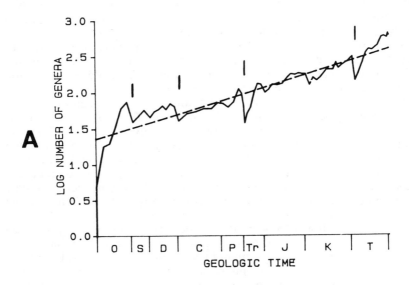

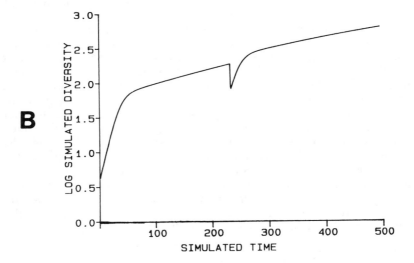

Figure 6.1. A: Semi-logarithmic graph of bivalve global generic diversity during the Phanerozoic. The dashed line is a least-squares fit of a simple exponential function to all the data. Vertical tick marks delineate four intervals of mass extinction including, most prominently, the Late Permian and Late Cretaceous events (from Miller and Sepkoski, 1988). B: Semi-logarithmic graph of simulated global diversity for bivalves, utilising a couple logistic model

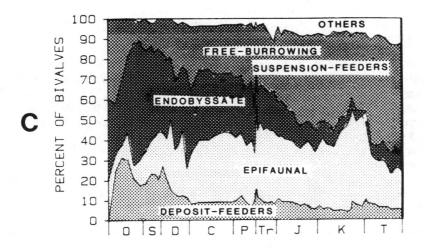

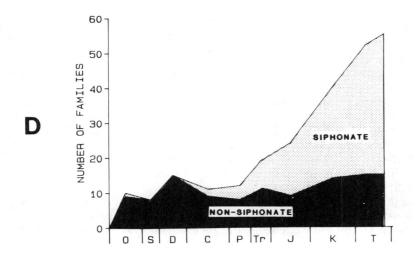

Figure 6.1 – cont.
developed by Miller and Sepkoski (1988). C: Graph depicting changes during
the Phanerozoic in the relative percentages, among bivalves, of the four
principal ecologic groups. D: Graph depicting the relative diversities, during the
Phanerozoic, of non-siphonate and siphonate free-burrowing bivalves (after
Stanley, 1977).

most effectively in a pair of classic studies by Stanley (1970; 1972). While Stanley delineated several life-habit groups, the majority of adult fossil bivalves can be divided into four major ecologic categories:

1. *Epifaunal*. Generally sedentary bivalves that usually attach to substrates with a byssus or through cementation, and live at or above the sediment–water interface.
2. *Endobyssate*. Generally sedentary bivalves that attach to solid objects with a byssus and live partially or completely buried below the sediment–water interface.
3. *Free-burrowing suspension-feeders*. Fully infaunal, unattached (as adults) suspension-feeding bivalves.
4. *Free-burrowing deposit-feeders*. Fully infaunal, unattached (as adults) bivalves that probably rely on deposit-feeding as their primary means of obtaining food.

The relative global generic diversities through the Phanerozoic of these four categories are summarised in Figure 6.1C, based on life-habit assignments made to nearly 2000 of the genera included in the global diversity curve (Figure 6.1A). These assignments were accomplished utilising information and interpretations provided directly in several references (Yonge, 1939; 1967; Kauffman, 1967; 1978; Stanley, 1968; 1970; 1972; 1979; Cox *et al.*, 1969; Bretsky, 1970; Kennedy *et al.*, 1970; Newell and Boyd, 1970; Morton, 1970; Pojeta, 1971; Runnegar, 1974; Thomas, 1975; Morris, 1978; Skelton, 1978; Hoare *et al.*, 1979; Jablonski and Bottjer, 1983; Seilacher, 1984; Kriz, 1984), as well as principles of the relationships between shell form and life habits outlined by Stanley (1970, 1972; further details of assignment procedure are provided in Miller, 1986).

The oldest documented bivalves, from the Lower Cambrian (Runnegar and Pojeta, 1974; Pojeta, 1975; 1987a; Jell, 1980; Runnegar and Bentley, 1983), are thought to have been shallow infauna. There is a stratigraphic hiatus, representing some 30 million years, before the next appearance of bivalves in Lower Ordovician strata. As they radiated during the Ordovician, all four principal life-habit groups became well represented (Figure 6.1C; Pojeta, 1971; 1978; 1987a, 1987b).

During the Ordovician, the proportion of bivalves that were free-burrowing suspension-feeders declined as byssate forms radiated substantially. However, in the mid- to Late Palaeozoic, this pattern reversed; an unprecedented diversification of free-burrowing suspension-feeders began that continues to the present day (Figure 6.1C). Stanley demonstrated that these global transitions are actually manifestations of evolutionary trends that occurred polyphyletically in several evolutionary lineages. Specifically, the Early to mid-Palaeozoic diversification of byssate bivalves is associated with the neotenous retention in adults of the juvenile byssus present in ancestral forms (Yonge, 1962). A trend from free-burrowers through endobyssate individuals to fully epibyssate forms has been recognised, in whole or part, in lineages of several different bivalve orders including the Mytiloida, Modiomorphoida, Arcoida, Pterioida, Veneroida, and Pholadomyoida

(Stanley, 1972; in some instances, Stanley has also recognised the 'reverse' trend, from byssate forms to free-burrowers). These transitions involved a suite of well-established morphological modifications to shell form.

The mid- to Late Palaeozoic radiation of free-burrowing suspension-feeders was fuelled by the advent of siphons associated with mantle fusion (Yonge, 1948), which permitted the invasion of habitats well below the sediment–water interface (Figure 6.1D; Stanley, 1968; 1977). While most siphonate suspension-feeders have been veneroids, some were pholado-myoids, and at least one was a trigonioid (Stanley, 1968; 1977; Runnegar, 1974; Newell and Boyd, 1975). Thus, as with the evolution of endo- and epibyssate bivalves, the development of siphons was apparently polyphyletic, arising in three different bivalve orders in the mid- to Late Palaeozoic.

The evolutionary mechanisms associated with the development of adult byssate forms and the advent of siphons both suggest that, in the evolution of bivalves, it is common for large-scale transitions to represent the sum total of parallel changes recognised within several individual lineages. This kind of relationship, where local-scale evolution within individual lineages ultimately produces a global-scale macroevolutionary trend, can be viewed as hierarchical.

Clearly, heterochrony has played a major role in fuelling global transitions through this hierarchical pathway. The evolution of the adult byssus is but one example. Another prominent case involves the apparent development, in several bivalve families, of morphological attributes conducive to symbiotic relationships of bivalves with sulphide-oxidising bacteria. Reid and Brand (1986) evaluated adaptations for this form of symbiosis among lucinaceans and recognised its association with several paedomorphic changes to the siphons, gills and guts. Moreover, they suggested that similar trends accompanied the development of bacterial symbiosis in other lineages. Although global diversity has yet to be measured through the Phanerozoic for bivalves exhibiting bacterial symbiosis, Reid and Brand provide convincing evidence of its potential importance as a macroevolutionary pattern.

Other examples of heterochrony among bivalves include: the paedo-morphic retention in adult cockles belonging to the species *Cardium fittoni* of juvenile spines and reduced costae number associated with an ancestral species (Nevesskaya, 1967); and the paedomorphic retardation of coiling relative to size in a lineage of *Gryphaea* (Hallam, 1968). While these cases were apparently more isolated than adult byssal development or adaptations for bacterial symbiosis, they nevertheless serve to illustrate the probable general importance of heterochrony as an evolutionary mechanism among bivalves.

The evolution of byssate adults and of siphonate forms also point to the over-riding importance of life habit in the history of bivalves. Parallel trends associated with global transitions provide unifying threads with respect to evolutionary patterns in even disparate, long-since diverged monophyletic groups. The ubiquity of parallel trends during bivalve evolution is indicative of the degree to which phylogenetic membership has been transcended. Among bivalves, the polyphyletic development of ecologically and globally significant morphological attributes appears to be the rule, rather than the exception.

ENVIRONMENTAL OVERPRINTS ON BIVALVE DIVERSIFICATION

Until recently, the prevailing view among evolutionary palaeobiologists had been that the evolution of bivalves, particularly during the Palaeozoic, was relatively static in space; much of the radiation of bivalves into habitats away from marginal, nearshore settings was ascribed, once again, to the Late Permian mass extinction (see Miller, 1988). However, on the basis of several palaeo-ecological investigations of Palaeozoic fossil assemblages (for example, Yancey and Stevens, 1981; Boardman *et al.*, 1983; 1984a; 1984b; Kammer *et al.*, 1986; Frey, 1987a; 1987b; Miller, 1989), it is now evident that the Palaeozoic history of bivalves was spatially dynamic. There were marked transitions in their palaeo-environmental distributions throughout the Palaeozoic, including incursions into deep water. These transitions were summarised and quantified in a time-environment diagram, shown here in Figure 6.2A; details of its construction are provided by Miller (1988; time-environment diagrams for several life-habit groups, and for terrigenous and carbonate subsets of the data, are also provided therein). Utilising a data base of fossil assemblages primarily from North America, this diagram depicts the average generic diversity of bivalves as a percentage of total diversity (percent diversity) in habitats ranging from nearshore (zone 1, on the right-hand side of the diagram) to deep water (zone 6, on the left-hand side) through the Palaeozoic. Bivalves were absent from the unshaded zones. Successively darker-shaded contours enclose zones where bivalves comprised, respectively, less than 10, 10–20, 20–30, and greater than 30 per cent of total generic diversity.

Patterns on the time-environment diagram illustrate the degree to which bivalve distributional patterns changed through the Palaeozoic. While bivalves were most consistently diverse nearshore, they became established offshore and in deep water, albeit at limited diversity, during their Ordovician radiation. Through the remaining Palaeozoic, their percentage diversity in deep water increased slowly and irregularly to levels comparable to nearshore values.

Direct comparison of these environmental transitions with the bivalve global generic diversity curve (Figure 6.2A) reveals a compelling relationship; the timing of environmental transitions appears to correspond closely with changes in global diversity. On a global scale, the most substantial Palaeozoic pulse of diversification took place during the Ordovician. Diversification during the remaining Palaeozoic was slower and more irregular. Correspondingly, bivalves rapidly became established in habitats from nearshore to offshore during the Ordovician, but their subsequent spatio-temporal expansion was also more sluggish. Miller (in press) utilises a coupled logistic model that incorporates a space dimension to demonstrate further the relationship between global diversity and environmental transitions, and to suggest that the interactions apparently associated with bivalve global diversification (Miller and Sepkoski, 1988) were also partly responsible for observed palaeoenvironmental trends.

Environmental overprints on bivalve diversification were not limited to the Palaeozoic. Jablonski and Bottjer (1983) found that on soft substrates, 'Palaeozoic-type' epifaunal organisms, including oysters and inoceramids (plus other non-bivalve elements), became restricted to offshore environments during the Jurassic and Cretaceous, while a more 'modern' infaunal biota that included a variety of free-burrowing bivalves diversified and expanded closer to shore. Bottjer and Jablonski (1988) evaluated the origination and diversification of the Tellinacea, and found that the earliest occurrences of this veneroid superfamily were in inner-shelf and nearshore environments. Thereafter, tellinaceans expanded offshore, although most of their subsequent global radiation was apparently associated with diversification in inner and middle shelf habitats (Figure 6.2B).

Thus, there is substantial evidence that in both the Palaeozoic and post-Palaeozoic, the global diversification of bivalves was spatially dynamic on a local scale. As noted by Miller (1988), spatio-temporal patterns such as those illustrated in Figure 6.2A were probably not associated with any single evolutionary process or class of processes. Rather, they almost certainly resulted from the complex interplay of both physical and biological factors, some of which uniquely affected bivalves (see Miller, 1988, for a more detailed discussion of potential underlying mechanisms). However, the onshore–offshore environmental patterns for bivalves discussed by Jablonski and Bottjer (1983), Bottjer and Jablonski (1988), and Miller (1988) may, in part, reflect a general onshore–offshore spatial pattern in the diversification of many higher taxa: early diversification primarily in shallow, nearshore environments, followed by expansion into open shelf and deep water habitats, and, in some (but not all) instances, eventual restriction to deep settings as new groups originate closer to shore. Sepkoski and Sheehan (1983) and Sepkoski and Miller (1985) recognised this kind of pattern in the Palaeozoic diversification of global evolutionary faunas. Jablonski and Bottjer (in press), Bottjer *et al.* (1988) and Bottjer and Jablonski (1988) have now provided detailed documentation of these patterns, or close variations, in a diverse group of higher taxa including cheilostomes, isocrinids, tellinaceans, and the trace fossils *Ophiomorpha* and *Zoophycus*. (Jablonski and Bottjer, in press, found apparent hierarchical limits, from a taxonomic standpoint, to onshore origination; among investigated higher taxa that originated onshire, no such onshore bias was recognised in the originations of constituent lower taxa). All these authors, plus others, have discussed several potential mechanisms to account for the recognised patterns. McKinney (1986) suggested a possible role for heterochrony in nearshore origination of evolutionary novelties. Among fossil echinoids, he observed that in unstable environments, which are more common nearshore than offshore, there was a greater propensity towards progenetic, small-sized adults (see Chapter 9 herein). Such adults are thought to have greater 'evolutionary potential' than larger adults, because of less allometric constraint. However, Jablonski and Bottjer (in press) found little evidence for progenesis as a mechanism of onshore origination in the higher taxa that they evaluated.

Further consideration of potential evolutionary processes responsible specifically for onshore–offshore patterns is beyond the scope of this discussion. The point of bringing them up is to note that, like the patterns they are intended to explain, proposed processes generally transcend the characteristics of particular taxa. Thus, it seems likely that at least part of the

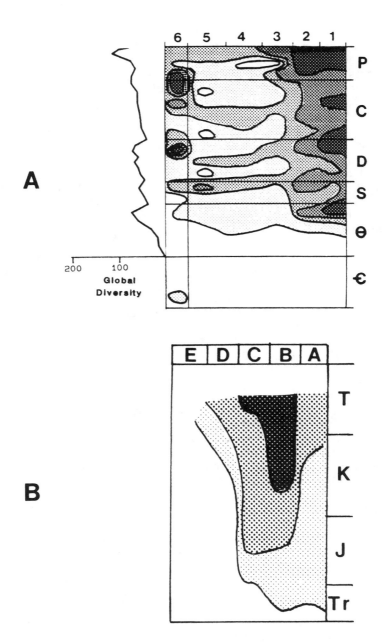

spatio-temporal history of bivalves is associated with general evolutionary processes that have little to do with innate bivalve attributes.

NET ENVIRONMENTAL TRANSITIONS: ASSEMBLY LINES FOR TRENDS?

Environmental factors are of obvious importance in evolution, but the role of environmental transition is more enigmatic. Given the complex interplay of physical and biological factors in the overall environment that an organism encounters, the expectation should be that many, if not most, environmental transitions are random with respect to time. Even in the absence of directional environmental transitions, however, evolution may proceed along rather well-defined pathways. The pervasiveness of onshore–offshore trends among higher taxa at various points throughout geologic time provides an excellent example. Although the mechanism(s) responsible for these patterns are not yet definitively understood, it is clear that they do not result from some sort of major environmental transition that is repeated again and again. Rather, whatever their cause, the pattern of initial nearshore radiation of higher taxa is almost certainly associated with differences in the characteristics of nearshore and offshore organisms that are attributable directly to the environments in which they live. Thus, differences among environments, rather than net changes within any of them, ultimately drive this evolutionary trend. Furthermore, Hoffman and Kitchell (1984) found evidence to support the contention of Van Valen (1973) that evolution, in general, may proceed in the absence of any changes, random or otherwise, to the abiotic environment.

Given that they can develop in the absence of net environmental change, it would seem logical that evolutionary trends might also result directly from clear temporal vectors in environmental conditions. To what degree have evolutionary trends among bivalves been associated with and dependent upon environmental change? Environmental changes can be grouped roughly into two categories that are not mutually exclusive: *abiotic* and *biotic*. Abiotic changes are those that can be classified as basically physical, including sea-level transitions, climatic changes, tectonic events, and large body impacts. Their greatest potential importance to evolutionary trends among bivalves would probably be manifested in such biological transitions as mass extinctions

---

Figure 6.2   A: Palaeozoic time-environment diagram for bivalve genera; bivalve global generic diversity has been plotted along the left margin for comparison. In this diagram, habitat zones 1 (nearshore) to 6 (deep water), are arrayed from right to left. For further explanation, see text (from Miller, in press). B: Generalised post-Palaeozoic time-environment diagram for tellinacean bivalves in the Euramerican region. Successively darker contours enclose zones on the diagram where tellinacean generic diversity values are, respectively, 1, 2, and 5. Habitat zones A, B, C, D, and E correspond approximately to zones 1, 2–3, 4, 5 and 6 in Figure 6.2A (after Jablonski and Bottjer, in press).

and changes in biogeographic configuration. Biotic environmental changes involve transitions in the biota with which bivalves interact directly.

The possible importance of mass extinctions among bivalves has already been touched upon with regard to the Late Permian event, which is among the most widely cited examples of the effect of mass extinction on evolutionary trends and the subsequent history of life. It was concluded earlier that the Late Permian extinction has apparently been overrated as an agent in changing the course of bivalve class-level global diversification and palaeoenvironmental distribution; this extinction also had little effect on the bivalve life-habit spectrum (Hallam and Miller, 1988; Figure 6.1C). With respect to the course of global diversification, this may also be true for most other higher taxa during most mass extinctions (Sepkoski, 1984; Gilinsky and Bambach, 1985). Several mass extinctions have substantially reduced bivalve global diversity (Hallam and Miller, 1988); in some instances these events resulted in the final demise of important constituent taxa (e.g., the Hippuritoida in the Late Cretaceous mass extinction). Still, there is little evidence to suggest that mass extinctions, in general, have ever substantially affected long-term evolutionary trends among bivalves.

Biogeographic 'events', associated with physical environmental transitions, must sometimes initiate evolutionary trends among bivalves that would not have taken place had the event not occurred. For example, the emergence of a major barrier to dispersal initially produces assemblages of cognate species. As defined by Vermeij (1978, p. 212), cognate species are 'two or more geographically isolated forms that, morphologically speaking, have diverged only slightly (and in some cases not at all) from their common . . . ancestor.' Early in the history of the barrier, these cognates might not be sufficiently distinct to represent true biological species. However, with the passage of time, no matter what the conditions on opposite sides of the barrier, it is reasonable to expect that divergence will result in the disappearance of recognisable cognates and the development of demonstrably distinct species. Thus, as a consequence of increased provinciality associated with this biogeographic event, there is an overall evolutionary trend of increased species richness. On a global scale, changes during the Phanerozoic in levels of provinciality have been invoked by Valentine and Moores (1972), Valentine (1973), and Valentine *et al.* (1978) as important arbiters in global species-diversity trends. However, there is reason to believe that, at least among bivalves, this has not been the case. The data for the time-environment diagram in Figure 6.2A were collected mainly from a single faunal province (associated with the Palaeozoic continent of Laurentia), yet diversity patterns on this figure closely parallel the global diversity curve (Miller, 1988; in press). This correspondence suggests that the global diversification of bivalves ultimately had a local, environmental basis.

Vermeij (1978) measured the degree of divergence of bivalve species associated with the emergence of the isthmus of Panama, by calculating the proportion of recognisable cognates within assemblages on opposite sides of the isthmus. The greater the proportion, the smaller the degree of divergence. He found that as a group, bivalves had not diverged as much as other taxa studied. Thus, the potential for diversification or the development of other evolutionary trends as consequences of biogeographic events may be relatively limited among the Bivalvia.

Biotic environmental transitions could have several different manifestations. With respect to bivalves, one particularly important transition may have been the development of a land-derived nutrient supply associated with the rise of land plants. In the early Palaeozoic, the general scarcity of land plants may have restricted high nutrient levels in benthic environments to shallow areas within the photic zone. However, with the mid-Palaeozoic land-plant proliferation, nutrient levels in benthic habitats beyond the photic zone may have increased substantially because of a general increase in the supply of organic detritus. Because of their nutrient requirements, this may have permitted an offshore proliferation of some bivalves (Calef and Bambach, 1973).

Vermeij (1987) argued cogently for the role of biological interaction as perhaps *the* primary agent in the development of evolutionary trends. Specifically, he suggested that, 'the history of life is characterized by two simultaneous trends: increasing risks to individuals from potential enemies and increasing incidence and expression of aptations to cope with these risks' (Vermeij, 1987, p. 4). Thus, among bivalves, evolutionary trends might commonly develop in direct response to transitions exhibited by organisms with which they interact. Indeed, Vermeij suggested that several Phanerozoic trends in bivalve shell morphology and locomotion were related to increased biotic environmental 'risk' associated with the evolution of various predators and competitors.

Given the complex hierarchy of trends recognised throughout bivalve history, it should not be surprising that the various mechanisms and processes proposed to explain them comprise a rather daunting list. The mechanisms discussed here only scratch the surface, but they suffice to illustrate the dichotomy between extrinsic and intrinsic factors. Extrinsic factors involve the potential development of a trend as a consequence of, or in parallel with, net environmental transitions. Intrinsic factors are those associated with innate characteristics of organisms in the contexts of the environments in which they live; trends resulting from intrinsic mechanisms show no obvious parallel relationship to environmental transitions. Moreover, because they are not dependent upon events that might only happen once or, at best, infrequently, it is possible for evolutionary trends associated with intrinsic mechanisms to be repeated again and again. Examples include the cases of heterochrony and onshore–offshore paleoenvironmental transitions cited earlier.

It is difficult, and perhaps unfair, to make a blanket statement about the relative roles of extrinsic and intrinsic mechanisms in the evolutionary history of the Bivalvia. Nevertheless, it can be noted that most of the trends discussed here are probably associated with intrinsic factors.

DISCUSSION: TRENDS, RANDOMNESS, CHAOS AND PROGRESS

Webster's New Collegiate Dictionary defines a 'trend' as, 'a line of general direction or movement . . . a prevailing tendency or inclination . . . the general movement in the course of time of a statistically detectable change.' Certainly, the patterns evaluated herein qualify as trends, and, needless to say, this chapter has catalogued just a minuscule fraction of the numerous evolutionary trends documented for bivalves. However, the intention here was not to provide such a catalogue, but to place some of the more pervasive patterns into a hierarchical context.

It is noteworthy that many apparent global trends may simply represent epiphenomenal accumulations of trends recognised further down the hierarchical ladder. For example, as discussed earlier, the global diversification of byssally attached bivalves actually represents the cumulative effect of a paedomorphic trend that occurred polyphyletically in several lineages, at several different points in geologic time. Recognition of this hierarchy lessens concern with respect to the possibility that global trends are actually random walks. As noted by Raup (1977a, p. 64):

> In palaeontology, many so-called evolutionary trends are established because the path followed by a morphological character through time clearly does not fit the preconceived idea of the random walk . . . In other words, the interpretation of the evolutionary trend is based on the assumption that if the observations were literally random in character, the line would have many reversals in direction . . .

Raup (1977a; 1977b) demonstrated that, in fact, random walks tend to behave in a rather counterintuitive fashion. Computer-generated random walks commonly display long intervals of monotonic change. Thus, it is essential to look beyond the simple graphical picture of an apparent trend when invoking causation. In the case of the global development of byssate forms, the most conclusive evidence that it does not represent a random walk is its hierarchical structure. Surely, the occurrence, over and over again, of the same paedomorphic pattern is suggestive of a non-random pattern. Had the global diversification been related to the proliferation of this morphology in just a single lineage, the possibility of a random walk would have been more difficult to put aside.

Earlier in this chapter, it was suggested that the synoptic, global diversification of bivalves, as a class, conformed rather closely to the expectations of a coupled logistic system. But how can such a uniform class-level pattern accommodate the complex hierarchy and variety of trends that lie beneath this umbrella? In part, an explanation is provided by Raup *et al.* (1973), who developed a stochastic, branching model of phylogeny in a system where total diversity behaved logistically. This model was subsequently criticised, most notably for problems of scaling (Stanley *et al.* 1981). Despite its shortcomings, results of simulations conducted with the model demonstrated that in a logistic system, where overall diversity is well ordered and constrained, constituent clades that comprise a logistically diversifying 'superclade' are, themselves, relatively unconstrained in the variety of trends that they exhibit or the times that they originate. There is no tendency for trends within individual clades to mirror those in the superclade. Thus, the apparently chaotic array of trends exhibited at lower taxonomic levels is by no means irreconcilable with an orderly diversification pattern at the class level.

Gould (1988) suggested that in many instances, evolutionary trends that have an apparent direction of 'increase' or 'decrease' are attributed incorrectly to anagenesis. For the group being evaluated, these 'trends' may simply result from increases or decreases in the amount of structural variance, rather than some form of progressive improvement (in the case of an increase) or honing (in the case of a decrease). Thus, Gould cautioned against ascribing trends to some form of progress.

Does the trend of continuously increasing bivalve global diversity represent progress? The number of bivalve genera, and presumably species as well, has increased fairly continuously throughout the Phanerozoic; arguably, this represents a measurement of the group's increasing success. However, as suggested by Gould, it may simply be a measure of increasing variance. In many respects the bivalve die was cast during the Ordovician, when the class underwent an initial radiation; by the end of the period, all four principal life-habit groups had undergone initial development. Arguably, many present-day bivalves exhibit the same morphologies and life habits as their Ordovician counterparts; there is little indication that they have 'progressed', by any objective measure, during the intervening time interval. Moreover, the lack of temporal constraint in the lineage-level trends that hierarchically comprise global trends suggests that progress, if measured as some form of net change in the way the system operates, has been limited. There is every reason to believe that some present-day free-burrowing lineage could, in the future, develop an adult byssus through paedomorphosis, just as several lineages have in the geologic past. Thus, for bivalves, the rules of the game remain largely unchanged.

On the other hand, many new adaptive themes have developed throughout the Phanerozoic that are difficult to attribute simply to increases in variance. For example, the advent of siphons provided a pathway for the development of a variety of unprecedented aptations for life below the sediment–water interface; a portion of the increase in bivalve global diversity is certainly associated with these new themes. Whether the genetic bases for these aptations resided untapped in Ordovician bivalves, waiting, perhaps, for the advent of predators that encouraged them (*sensu* Vermeij, 1987) to increase their variance deep into the sediment, is problematical.

The question of progress is a difficult one, and there are no objective means to measure it, let alone define it. As suggested by Gould (1988), our search for progress, and our sense of it, are certainly culturally embedded. Whatever the role of progress, the variety of evolutionary trends exhibited among bivalves in the past 500 million years leads one to believe that the next 500 million years will yield many patterns that could not be predicted on the basis of past history.

ACKNOWLEDGEMENT

I thank Ben Greenstein for providing helpful comments on an earlier draft of this chapter.

REFERENCES

Boardman, D.R., II, Yancey, T.E., Mapes, R.H. and Malinky, J.M., 1983, A new model of succession of Middle and Late Pennsylvanian fossil communities in North Texas, mid-continent, and Appalachians with implications on black shale controversy, *Okla. Geol. Notes*, **43**: 73–74.

Boardman, D.R., II, Brett, C.E., Kammer, T.W. and Mapes, R.H., 1984a, The late Palaeozoic dysaerobic biofacies, *Geol. Soc. Am. Abs. Prog*, **16**: 447.

Boardman, D.R. II, Mapes, R.H., Yancey, T.E. and Malinky, J.M., 1984b, A new model for depth-related allogenic community succession within North American Pennsylvanian cyclotherms and implications on the black shale problem. In N.J. Hyne (ed.), *Limestones of the mid-continent*, Tulsa Geological Survey Special Publication, **2**: 141–182.

Bottjer, D.J., Droser, M.C. and Jablonski, D., 1988, Palaeoenvironmental trends in the history of trace fossils, *Nature*. **333**: 252–5.

Bottjer, D.J. and Jablonski, D., 1988, Palaeoenvironmental patterns in the evolution of post-Palaeozoic benthic marine invertebrates, *Palaios*, **3**: 540–60.

Bretsky, P.W., Jr., 1970, Upper Ordovician ecology of the central Appalachians, *Bull. Peabody Mus. Nat. Hist.*, **34**: 1–150.

Calef, C.E. and Bambach, R.K., 1973, Low nutrient levels in lower Palaeozoic (Cambrian–Silurian) oceans, *Geol. Soc. Am. Abs. Prog.*, **5**: 565.

Cox, L.R., Newell, N.D., Boyd, D.W., Branson, C.C., Casey, R., Chaven, A., Coogan, A.H., Dechaseaux, C., Fleming, C.A., Haas, F., Hertlein, L.G., Kauffman, E.G., Keen, A.M., Larocque, A., McAlester, A.L., Moore, R.C., Nuttall, C.P., Perkins, B.F., Puri, H.S., Smith, L.A., Soot-Ryen, T., Stenzel, H.B., Truemen, E.R., Turner, R.D. and Weir, J., 1969, *Treatise on invertebrate paleontology, Part N: Mollusca 6 (Bivalvia)*, Geological Society of America and University of Kansas, Lawrence.

Eldredge, N., 1987, *Life Pulse*, Facts on File Publications, New York.

Frey, R.C., 1987a, The occurrence of pelecypods in early Palaeozoic epeiric-sea environments, *Palaios*, **2**: 3–23.

Frey, R.C., 1987b, The paleoecology of a Late Ordovician shale unit from southwest Ohio and southeastern Indiana, *J. Paleont.*, **61**: 242–67.

Gilinsky, N.L. and Bambach, R.K., 1985, The roots beneath patterns of taxonomic diversity: implications for extinction, *Geol. Soc. Am. Abs. Prog.*, **17**: 592.

Gould, S.J., 1985, The paradox of the first tier: an agenda for paleobiology, *Paleobiology*, **11**: 2–12.

Gould, S.J., 1988, Trends as changes in variance: a new slant on progress and directionality in evolution, *J. Paleont.*, **62**: 319–29.

Gould, S.J. and Calloway, C.B., 1980, Clams and brachiopods: ships that pass in the night, *Paleobiology*, **6**: 383–96.

Hallam, A., 1968, Morphology, palaeoecology, and evolution of the genus *Gryphaea* in the British Lias, *Phil. Trans. R. Soc.*, series B, **254**: 91–128.

Hallam, A. and Miller, A.I., 1988, Extinction and survival in the Bivalvia. In G.P. Larwood, (ed.), *Extinction and survival in the fossil record*, Systematics Association Special Volume no. 34, Clarendon Press, Oxford: 121–38.

Hoare, R.D., Sturgeon, M.T., and Kindt, E.A., 1979, Pennsylvanian marine Bivalvia and Rostroconchia of Ohio, *Bull. Geol. Surv. Ohio*, **67**: 1–77.

Hoffman, A. and Kitchell, J.A., 1984, Evolution in a pelagic ecosystem: a paleobiologic test of models of multispecies evolution, *Paleobiology*, **10**: 9–33.

Jablonski, D. and Bottjer, D.J., 1983, Soft-bottom epifaunal suspension feeding assemblages in the Late Cretaceous: implications for the evolution of benthic communities. In M.J.S. Tevesz and P.L. McCall (eds), *Biotic interactions in Recent and fossil benthic communities*, Plenum, New York: 747–812.

Jablonski, D. and Bottjer, D.J., in press, Onshore–offshore trends in marine invertebrate evolution. In R.M. Ross and W.D. Allmon (eds), *Biotic and abiotic factors in evolution: a paleontologic perspective*, University of Chicago Press, Chicago.

Jell, P.A., 1980, Earliest known pelecypod on earth—a new Early Cambrian genus from South Australia, *Alcheringa*, **4**: 233–9.

Kammer, T.W., Brett, C.E., Boardman, D.R. and Mapes, R.H., 1986, Ecologic stability of the dysaerobic biofacies during the Late Paleozoic, *Lethaia*, **19**: 109–21.

Kauffman, E.G., 1967, Coloradoan macroinvertebrate assemblages, central western Interior, United States. In E.G. Kauffman and H.C. Kent, (eds), *Paleoenvironments of the Cretaceous Seaway in the western Interior: a symposium*, Colorado School of Mines, Golden: 67–143.

Kauffman, E.G., 1978, Evolutionary rates and patterns among Cretaceous Bivalvia. In C.M. Yonge and T.E. Thompson (eds), *Evolutionary systematics of bivalve molluscs, Phil. Trans. R. Soc.*, series B, **284**: 277–304.

Kennedy, W.J., Morris, N.J. and Taylor, J.D., 1970, The shell structure, mineralogy, and relationships of the Chamacea (Bivalvia), *Palaeontology*, **13**: 379–413.

Kriz, J., 1984, Autecology of Silurian Bivalvia. In M.G. Bassett and J.D. Lawson (eds), *Autecology of Silurian organisms*, Spec. Pap. Palaeont., **32**: 183–95.

McKinney, M.L., 1986, Ecological causation of heterochrony: a test and implications for evolutionary theory, *Paleobiology*, **12**: 282–9.

Miller, A.I., 1986, The spatio-temporal development of the class Bivalvia during the Palaeozoic Era, unpublished doctoral dissertation, University of Chicago.

Miller, A.I., 1988, Spatio-temporal transitions in Palaeozoic Bivalvia: an analysis of North American fossil assemblages, *Hist. Biol.*, **1**: 251–73.

Miller, A.I., 1989, Spatio-temporal transitions in Palaeozoic Bivalvia: a field comparison of Upper Ordovician and Upper Palaeozoic bivalve-dominated fossil assemblages, *Hist. Biol.*, **2**: 227–60.

Miller, A.I., in press, The relationship between global diversification and spatio-temporal transitions in Palaeozoic Bivalvia. In W. Miller, III (ed.), *Paleocommunity temporal dynamics: processes and patterns of long-term community development*, Paleontological Society Special Publication.

Miller, A.I. and Sepkoski, J.J., Jr. 1988, Modeling bivalve diversification: the effect of interaction on a macroevolutionary system, *Paleobiology*, **14**, 364–9.

Morris, N.J., 1978, The infaunal descendants of the Cycloconchidae: an outline of the evolutionary history and taxonomy of the Heteroconchia, superfamilies Cycloconchacea to Chamacea. In C.M. Yonge and T.E. Thompson (eds), *Evolutionary systematics of bivalve molluscs, Phil. Trans. R. Soc.*, series B, **284**: 259–75.

Morton, B., 1970, The evolution of the heteromyarian condition in the Dreissenacea, *Palaeontology*, **13**: 563–72.

Nevesskaya, L.A., 1967, Problems of species differentiation in light of paleontological data, *Paleont. Jl.*, **1 (4)**: 1–17.

Newell, N.D. and Boyd, D.W., 1970, Oyster-like Permian Bivalvia, *Bull. Am. Mus. Nat. Hist.*, **143**: 217–82.

Newell, N.D. and Boyd, D.W., 1975, Parallel evolution in early Trigonacean bivalves, *Bull. Am. Mus. Nat. Hist.*, **154**: 53–162.

Pojeta, J., 1971, Review of Ordovician pelecypods, *Prof. Pap. U.S. Geol. Surv.*, **695**: 1–46.

Pojeta, J., 1975, *Fordilla troyensis* Barrande and early pelecypod phylogeny, *Bull. Am. Palaeont.*, **67**: 363–84.

Pojeta, J., 1978, The origin and early taxonomic diversification of pelecypods. In C.M. Yonge and T.E. Thompson (eds), *Evolutionary systematics of bivalve molluscs*, *Phil. Trans. R. Soc.*, series B, **284**: 225–46.

Pojeta, J., 1987a, Class Pelecypoda. In R.S. Boardman, A.H. Cheetham and A.J. Rowell (eds), *Fossil invertebrates*, Blackwell Scientific Publications, Oxford: 386–435.

Pojeta, J., 1987b, The first radiation of the Class Pelecypoda, *Geol. Soc. Am. Abs. Prog.*, **19**: 238–9.

Raup, D.M., 1977a, Stochastic models in evolutionary palaeontology. In A. Hallam (ed.), *Patterns of evolution*, Elsevier, Amsterdam: 59–78.

Raup, D.M., 1977b, Probabilistic models in evolutionary paleobiology, *Am. Sci.*, **65**: 50–57.

Raup, D.M., Gould, S.J., Schopf, T.J.M. and Simberloff, D.S., 1973, Stochastic models of phylogeny and the evolution of diversity, *J. Geol.*, **81**: 525–42.

Reid, R.G.B. and Brand, D.G., 1986, Sulfide-oxidizing symbiosis in lucinaceans: implications for bivalve evolution, *The Veliger*, **29**: 3–24.

Runnegar, B., 1974, Evolutionary history of the bivalve subclass Anomalodesmata, *J. Paleont.*, **48**: 904—39.

Runnegar, B. and Bentley, C., 1983, Anatomy, ecology, and affinities of the Australian Early Cambrian bivalve *Pojetaia runnegari* Jell, *J. Paleont.*, **57**: 73–92.

Runnegar, B. and Pojeta, J., 1974, Molluscan phylogeny: the paleontological viewpoint, *Science*, **186**: 311–17.

Seilacher, A., 1984, Constructional morphology of bivalves: evolutionary pathways in primary versus secondary soft-bottom dwellers, *Palaeontology*, **27**: 207–37.

Sepkoski, J.J., 1984, A kinetic model of Phanerozoic taxonomic diversity: III. Post-Palaeozoic families and mass extinctions, *Paleobiology*, **10**: 246–67. .

Sepkoski, J.J. and Miller, A.I., 1985, Evolutionary faunas and the distribution of Palaeozoic benthic marine communities in space and time. In J. Valentine (ed.), *Phanerozoic diversity patterns: profiles in macroevolution*, AAAS and Princeton University Press, Princeton, NJ: 153–90.

Sepkoski, J.J., and Sheehan, P.M., 1983, Diversification, faunal change, and community replacement during the Ordovician radiations. In M.J.S. Tevesz and P.L. McCall (eds), *Biotic interactions in Recent and fossil benthic communities*, Plenum, New York: 673–713.

Skelton, P.W. 1978, The evolution of functional design in rudists (Hippuritacea) and its taxonomic implications. In C.M. Yonge and T.E. Thompson (eds), *Evolutionary systematics of bivalve molluscs*, *Phil. Trans. R. Soc.*, series B, **284**: 305–18.

Stanley, S.M., 1968, Post-Palaeozoic adaptive radiation of infaunal bivalve molluscs — a consequence of mantle fusion and siphon formation, *J. Paleont.*, **42**: 214–29.

Stanley, S.M., 1970. Relation of shell form to life habits in the Bivalvia, *Mem. Geol. Soc. Am.*, **125**: 1–296.

Stanley, S.M., 1972, Functional morphology and evolution of byssally attached bivalve mollusks, *J. Paleont.*, **46**: 165–212.

Stanley, S.M., 1977, Trends, rates, and patterns of evolution in the Bivalvia. In A. Hallam, (ed.), *Patterns of evolution*, Elsevier, Amsterdam: 209–50.

Stanley, S.M., 1979, *Macroevolution: pattern and process*, Freeman, San Francisco.

Stanley, S.M., 1985, Rates of evolution, *Paleobiology*, **11**: 13–26.

Stanley, S.M., Signor, P.W., III, Lidgard, S. and Karr, A.F., 1981, Natural clades differ from 'random' clades: simulations and analyses, *Paleobiology*, **7**: 115–27.

Steele-Petrovic, H.M., 1979, The physiological differences between articulate brachiopods and filter-feeding bivalves as a factor in the evolution of marine level-bottom communities, *Palaeontology*, **22**: 101–34.

Thayer, C.W., 1979, Biological bulldozers and the evolution of marine benthic communities, *Science*, **203**: 458–61.

Thayer, C.W., 1981, Ecology of living brachiopods. In T.W. Broadhead, (ed.), *Lophophorates: notes for a short course*, University of Tennessee Department of Geological Sciences Studies in Geology, **5**: 110–26.

Thayer, C.W., 1983, Sediment-mediated biological disturbance and the evolution of the marine benthos. In M.J.S. Tevesz and P.L. McCall (eds), *Biotic interactions in Recent and fossil benthic communities*, Plenum, New York: 479–625.

Thayer, C.W., 1985, Brachiopods versus mussels: competition, predation and palatability, *Science*, **228**: 1527–8.

Thayer, C.W., 1986, Are brachiopods better than bivalves? Mechanisms of turbidity and their interaction with feeding in articulates, *Paleobiology*, **12**: 161–74.

Thomas, R.D.K., 1975, Functional morphology, ecology, and evolutionary conservatism in the Glycymerididae (Bivalvia), *Palaeontology*, **18**: 217–54.

Valentine, J.W., 1973, Phanerozoic taxonomic diversity: a test of alternate models, *Science*. **180**: 1078–9.

Valentine, J.W. and Moores, E.M., 1972, Global tectonics and the fossil record, *J. Geol.*, **80**: 167–84.

Valentine, J.W., Foin, T.C. and Peart, D., 1978, A provincial model of Phanerozoic marine diversity, *Paleobiology*, **4**: 55–66.

Van Valen, L., 1973, A new evolutionary law, *Evol. Theory*, **1**: 1–30.

Vermeij, G.J., 1978, *Biogeography and adaptation*, Harvard University Press, Cambridge, MA.

Vermeij, G.J., 1987, *Evolution and escalation*, Princeton University Press, Princeton, NJ.

Yancey, T.E. and Stevens, C.H., 1981, Early Permian fossil communities in N.E. Nevada and N.W. Utah. In J. Gray, A.J. Boucot and W.B.N. Berry (eds), *Communities of the past*, Hutchinson Ross, Stroudsburg, PA: 243–70.

Yonge, C.M., 1939, The protobranchiate mollusca: a functional interpretation of their structure and evolution, *Phil. Trans. R. Soc*, series B, **230**: 79–147.

Yonge, C.M., 1948, Formation of siphons in Lamellibranchia, *Nature*, **161**: 198.

Yonge, C.M., 1962, On the primitive significance of the byssus in the Bivalvia and its effects on evolution, *J. Mar. Biol. Ass. U.K.*, **42**: 112–25.

Yonge, C.M., 1967, Form, habit, and evolution in the Chamidae (Bivalvia) with reference to conditions in the rudists (Hippuritacea), *Phil. Trans. R. Soc.*, series B, **252**: 49–105.

Chapter 7

AMMONOIDS

Jean-Louis Dommergues

INTRODUCTION

With a combination of highly dynamic evolution, extreme morphological plasticity, flexible ontogeny and a range of about 300 million years, from the Devonian to the Late Cretaceous, ammonoids are one of the major fossil groups on which detailed palaeobiological analysis can be carried out. Partly on account of their exceptional use in biostratigraphy, ammonoids are also one of the most studied groups of fossils. The huge literature (mainly descriptive and biostratigraphic) dealing with ammonoids, offers a major challenge to any attempt to synthesise such a topic as evolutionary trends in a single chapter. One particular difficulty is the scarceness of modern palaeobiological analyses. Fortunately ammonoids constitute a rather homogenous group organised around a single *Bauplan* from which only a few groups deviate appreciably (i.e. Clymeniida and various heteromorph taxa). Moreover, it is probable that, throughout the range of the order, most ammonoids inhabited homologous or at least similar marine ecosystems. These morphofunctional and ecological continuities allow extrapolation of general palaeobiological assumptions from only a few detailed analyses of trends, particularly if they are restricted to a limited time period. To this end, this chapter concentrates in some detail on selected examples from among Jurassic ammonites. The choice of this geological system is based both on subjective and objective arguments: first, it is on Jurassic ammonoids that my research on evolutionary trends has concentrated; and second, available studies with a clear palaeobiological aim are more common for ammonoids of this age than for other time periods.

Having deliminated the range of the study it is necessary to specify what exactly a trend is. Simpson (1953) defined a trend as 'a sustained prevailing tendency in a phylogenetic progression'. He further states: 'Almost all fossil sequences long enough to be called sustained show prevailing tendencies in some characters and, over a part at least, of sequence'.

To Simpson trends were indissociable from gradualist perspectives and connected concepts. More recently, discussions about the model of punctuated equilibria have led to fundamentally different assumptions based on the externalist idea of 'species selection' (Eldredge and Gould, 1972; Stanley, 1979) and on the more internalist 'effect hypothesis' concept (Vrba, 1980). The heuristic choice from among these different concepts is critical. Indeed, if sustained morphological tendencies (trends) are real phenomena, gradualist or punctualist assumptions are often little more than *ad hoc* hypotheses dependent mainly on the author's own preconceptions or on the scale of observation and/or on the frequently irrefutable continuity of the fossil record. Thus, any such hypothesis about evolutionary tempo will be discussed here only if the data appear sufficiently accurate. The principal aim of this chapter will be the analysis and the comparison of trends in terms of morphogenesis, aptative (*sensu* Gould and Vrba, 1982) significance, evolutionary processes associated with ontogeny (e.g., heterochrony) and the relevant importance of intrinsic versus extrinsic factors. Special attention will be paid to the relationship between ontogeny and phylogeny. Indeed ammonoids, being molluscs with accretionary growth, provide an opportunity for taking ontogeny into account. Moreover, some tentative studies which propose alternative standards for biological age determinations of individuals are available for ammonoids. These constitute an effective and up-to-date method of studying the role of heterochrony in evolutionary trends.

It is not easy to estimate, in a major fossil group, the respective contributions of sustained and orientated patterns (trends) versus unsustained and unorientated patterns (e.g., the sudden appearance of a new lineage in the fossil record). Indeed, how is one to compare historical patterns of such different natures, the former apparently deterministic and the latter not. Nevertheless, trends are certainly the most frequently observed pattern throughout the fossil record of ammonoids. This is underlined by the high frequency of detailed biostratigraphical zonations which are more often than not based on trends.

At the origin of the order, the earliest Devonian coiled ammonoids arose from Bactritina nautiloids in a trend described by Erben (1964). This starts with orthoconic *Bactrites* and *Lobobactrites*, then follows through the slightly arched *Anetoceras* and *Erbenoceras* to the slightly coiled gyrocone—as, for example, in *Convoluticeras*—and ends with the true tightly coiled *Werneroceras*. If this trend is partly idealised (House, 1982) a progressive coiling by peramorphosis (peramorphocline *sensu* McNamara, 1982; see also Chapter 3 herein) is apparent. At the other end of the range, just before the extinction of the order, trends remained a common pattern, especially among Upper Cretaceous heteromorph ammonoids (Kennedy and Wright, 1985). Heterochrony, demonstrating an underlying flexibility in ontogeny, has been frequently demonstrated (for example, by Landman, 1988). Between these two temporal extremes, Jurassic ammonoids play the part of representative examples to demonstrate evolutionary trends in ammonoids.

Although only the shell is usually preserved, palaeontologists concerned with ammonoids have at their disposal several fairly different categories of features, such as the general morphology of the shell (i.e. types of coiling, shape of section), the ornamentation (i.e. ribbing, tubercles, keel), the suture line, shell variability (i.e. ornamental polymorphism, sexual dimorphism) and also characters able to yield by indirect means information on the biological age (i.e. adult size, rib or septa density). All these different kinds of features may be involved in evolutionary trends. These features will be independently analysed, taking into account their respective emphases and roles in ammonoid evolution, with special attention being paid to their intrinsic versus extrinsic significance.

TRENDS IN GENERAL MORPHOLOGY OF THE SHELL

In ammonoids the hydrodynamic characteristics of the shell (e.g., buoyancy or mobility properties) are particularly important and intimately related to shell morphology (Reyment, 1973; Swan, 1988; Tintant *et al.*, 1982; Ward, 1980; Westermann, 1989). In point of fact, among features usually available to palaeontologists, general shell morphology and suture lines can yield by far the most information on the organism's morphofunctional fitness. Nevertheless, we must not forget that hard parts can only yield a partial idea of the corresponding living animal. So although we may wish to place morphological trends within a coherent environmental framework, many such unknown factors restrict the extent to which generalisations can be made.

Trends in Psilocerataceae

The Lower Jurassic super family Psilocerataceae provides well-documented examples of evolutionary trends involving mainly general shell morphology. These ammonoids show several independent trends leading from platycone to more or less oxycone shells (Figure 7.1). Such trends were first described as far back as the nineteenth century (Hyatt, 1889). Several recent works have discussed this pattern (such as Corna, 1987; Donovan, 1987) and it seems possible to recognise among Psilocerataceae at least three trends leading to sub-oxycone or even true oxycone shells: *Coroniceras* to *Agassiceras*; *Caenisites* to *Radstockiceras* (via *Eparietites* and *Oxynoticeras*); and *Caenisites* to *Gleviceras* (via *Asteroceras*). The two latter trends are usually interpreted (Hyatt, 1889; Donovan, 1987) as a single trend, but unpublished data suggest the existence of two parallel but independent lineages.

The trend leading from *Caenisites* to *Radstockiceras* is a good example of these evolutionary trends (Figure 7.1). Its duration is about 5 million years (at least), starting in the Lower Sinemurian and extending to the Upper Pliensbachian. If, in the fossil record, the attainment of an oxycone shell

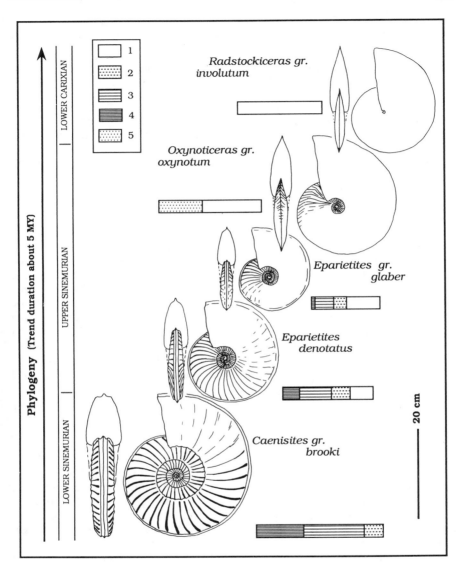

Figure 7.1. Diagrammatic illustration of the major peramorphic trend (mainly by acceleration) among Psilocerataceae leading from evolute and ribbed shell to involute and smooth shell. Only some steps of the trend are illustrated. 1–4 represent transitions from rather evolute shell with stout, simple, slightly arched ribs and 'three-keeled' ventral area to strictly involute and smooth shell with sharp ventral carina (= oxycone morphology). 5 represents bifurcated and weak ribs of *Oxynoticeras* inner and middle whorls (such complex ornamentation is not a close equivalent to that found in *Caenisites* and *Eparietites*).

appears as a consistently polarised trend, it does not correspond with gradual changes. Indeed , although the stratigraphical successions for the Sinemurian and Pliensbachian are particularly well documented, the successive species along the trend are separated by observable morphological gaps. The progressive morphological changes occurring during geological time (including decrease of umbilicus diameter and, conversely, increase of whorl height; decrease in the length of the juvenile ornamented stage; and acquisition of a sharp siphonal keel as the lateroventral keels disappear) can be attributed to peramorphosis, probably by a combination of acceleration and predisplacement (see Chapter 3 and Glossary herein). Nevertheless, as the progressive shortening of the ribbed juvenile stage is accompanied by an increase in rib density, it is possible, if a change in rib density indicates a change in growth rate (Dommergues, 1988), that a combination of hypermorphosis and dwarfism (*sensu* Gould, 1977) can also play an important part in the evolution of the lineage.

Such kinds of palaeontological data (polarised morphological drift under heterochronical control, with successive steps separated by slight but unambiguous morphological gaps) conform well with the peramorphocline model proposed by McNamara (1982) and discussed in Chapter 3 herein. Yet this model assumes that an environmental pressure directs the heterochronic trend. In the present example it is difficult to demonstrate such heuristic determinism. One can only assume, as some authors have (among them Tintant *et al.*, 1982; Westermann, 1989), that the progressive acquisition of oxycone shells would be linked with an improved fitness in a nectobenthonic environment. Tintant *et al.* (1982) considered these trends to be an integral part of the series of Lower Jurassic evolutionary radiations associated with the colonisation of the epicontinental seas that accompanied the Liassic transgression.

Trends in Cardioceratidae

This Middle and Upper Jurassic family which originated in the boreal sea (Bajocian) and ultimately spread southward (Callovian) into the European epicontinental seas, also displays interesting examples of trends leading to oxycone shell morphology (Callomon, 1985; Dommergues *et al.*, 1989). It is particularly interesting to compare these trends with those previously described for Psilocerataceae. Although, for the two taxa, trends result in convergent morphology (adult oxycone shell), the evolutionary processes that led to the convergence are clearly distinct, while the historical and palaeoecological contexts are noticeably different.

In the Cardioceratidae the tendency towards adult oxycone morphology was evident at least four times: in the trend leading from the early species of the family, *Cranocephalites borealis* to *Cadoceras nordenskjoeldi*; in the punctuated event which produced *Chamoussetia chamousseti*; in the event at the origin of '*Chamoussetia' galdrynus*; and in the trend which led from *Cadoceras sublaeve* through *Longaeviceras nikitini* to the genus *Quenstedtoceras*, then continued into *Cardioceras*. This last trend (Figure 7.2) will be

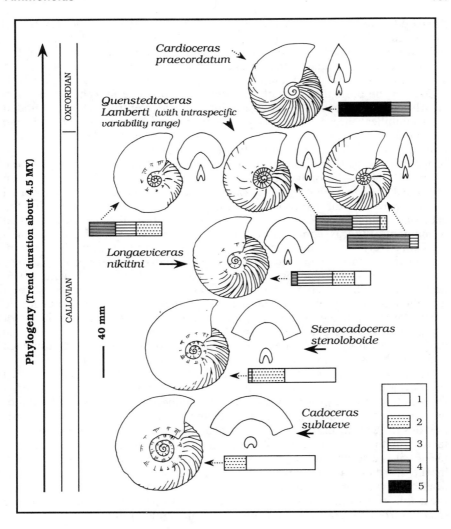

Figure 7.2. Diagrammatic illustration of the major proterogenetic trend (juvenile innovation plus neoteny) among Cardioceratidae (Stephanocerata-ceae), leading from cadicone to oxycone morphology. Only some steps of the trend are illustrated and the phylogeny must be followed from the bottom left to the top right. In order to indicate that the phylogenetic paedomorphocline combines with the intraspecific variability, this latter aspect is illustrated in *Quenstedtoceras lamberti*. This shows to the left an inflated peramorphic variant and to the right a particularly compressed paedomorphic one. 1–4 represent transitions from inflated cadicone shell with rounded ventral area and flat umbilical margins to compressed sub-oxycone shell with ogival section, sharp ventral area and rounded umbilical margins. 5 represents oxycone coiling with sagittate section and independent crenulate keel (= innovation).

discussed in detail, as it provides a good example of the evolutionary trends that occurred within Cardioceratidae. The lineage ranged for about 4.5 million years, from the Upper Callovian to the Lower Oxfordian. As for the older species in the family, the ontogeny of *Longaeviceras nikitini* starts with compressed inner whorls and ends with depressed outer whorls (Figure 7.2). Indeed in the particular case of *L. nikitini*, the juvenile morphology is soon clearly sub-oxycone with an ogival section, a keeled ventral area and rounded lateroumbilical areas. The body chamber retains the typical cadicone morphology (depressed section, an arched ventral area without a keel and angular lateroumbilical margin). In the ontogeny of *L. nikitini* there is a progressive transition between these two main kinds of morphology. In the course of the evolutionary trends from *L. nikitini* to *Cardioceras* one can observe a paedomorphic tendency (neoteny) corresponding with a centripetal extension of the sub-oxycone morphology and a progressive decay of the cadicone adult morphology. In a parallel direction there was a gradual refinement of the oxycone morphology in the inner whorls. Indeed, new and retrospectively progressive morphologies appear in the juvenile stage in conjunction with the neotenic tendency. Such an evolutionary process agrees well with the proterogenesis concept (Pavlov, 1901; Schindewolf, 1936; McNamara, 1986; Dommergues *et al.*, 1986) These changes in the general morphology of the shell are accompanied by other ornamentation changes, in particular ribs and keel (Callomon, 1985; Marchand, 1986). Both evolution of ornamentation and general morphology of the shell follow proterogenetic pathways. Such a proterogenetic model can equally apply to the other lineage which leads towards the oxycone morphology among Cardioceratidae.

If, in both Psilocerataceae and in Cardioceratidae, evolution leads toward the development of oxycone or at least sub-oxycone shells, and if in both cases heterochronic processes have a major influence, then it is interesting to note that peramorphosis is the rule for Psilocerataceae while paedomorphosis prevails for Cardioceratidae. These facts underline the importance of intrinsic factors in the determination of evolutionary strategies. As with Psilocerataceae, it is also difficult to ascertain the nature of selection pressure, if any, in the Cardioceratidae. For Callomon (1985) it remains totally obscure. Marchand (in Dommergues *et al.*, 1989), on the other hand, considered that the acquisition of the sub-oxycone or oxycone morphology in Cardioceratidae resulted from adaptations to a nectobenthonic mode of life.

Conclusions

From the previous examples two main impressions emerge: on the one hand heterochrony appears to be a common pattern of evolutionary change, and on the other hand it is often possible to interpret these changes in terms of morphofunctional adaptation. At first sight, it seems that evolutionary trends involving shell shape and size change often result from the connection between opportunistic heterochronic transformations (internal potentialities) and progressive ecological changes (external constraints), polarised in a suitable way with these internal potentialities. Yet it is difficult to extend this interpretation any further and to establish if external constraints prevail over internal potentialities, or if the contrary is true.

TRENDS IN ORNAMENTAL FEATURES

Like trends involving general shell morphologies, trends associated with ornamental features (such as ribs, tubercles, spines or keels) are among the most frequently described in the literature. In morphofunctional terms, such ornamental features are usually difficult, or even impossible, to understand with regard to their adaptive significance. This is especially true if one considers the entire ornamentation and not just single features, such as rib density or tubercle strength, which can only sometimes be associated with environmental or (and) palaeogeographical constraints (Dommergues *et al.*, 1989; Ward, 1981; Westermann, 1989). Such independence from environmental constraints makes it possible to use ornamental features as tests of internal versus external factors in evolution.

The Pliensbachian family Liparoceratidae

Among ammonites this family is arguably one of the most studied taxa in terms of evolution (Trueman, 1919; Spath, 1938; Callomon, 1963; 1980; Phelps, 1985; Marchand and Dommergues, 1988; Dommergues, 1983; 1988). This is because of an especially good fossil record (both in space and time), great variability, mainly controlled by ontogenetic factors (Dommergues *et al.*, 1986) and rapid evolutionary rates, which it is tempting to interpret in terms of heterochrony. Diagrammatically one can recognise two main morphological groups within the family: in one, sub-sphaerocone shells bear a complex ornamentation consisting of ribs divided on the ventral area and two lateral rows of tubercles; in the other, sub-platycone shells bear simple ribs; this latter group is here designated as 'capricorn' liparoceratids and the former as 'sphaerocone' liparoceratids. Moreover, there is a third group, of less importance, intermediate between sphaerocones and capricorns; this form is designated here as 'androgyn' liparoceratids. They possess a capricorn morphology in their inner and middle whorls and a sphaerocone morphology in their outer whorls (Callomon, 1963; Dommergues *et al.*, 1986). In the more recent work on the family, 'capricorn' liparoceratids are viewed as an independent lineage derived from sphaerocone liparoceratids by neoteny. In these terms androgyn and capricorn can be regarded as being, respectively, peramorphic and paedomorphic morphotypes within the variability of the capricorn lineage (*sensu lato*). A different interpretation has been proposed by Callomon (1963; 1980) who suggests that sphaerocone, androgyn and capricorn liparoceratids constitute a single, highly variable, lineage, with the sphaerocone–androgyn pair as macroconchs and the capricorn as microconchs. We can ignore this divergence of views here, because we will only analyse the evolutionary trend involving rib density in the capricorn forms (Figure 7.3).

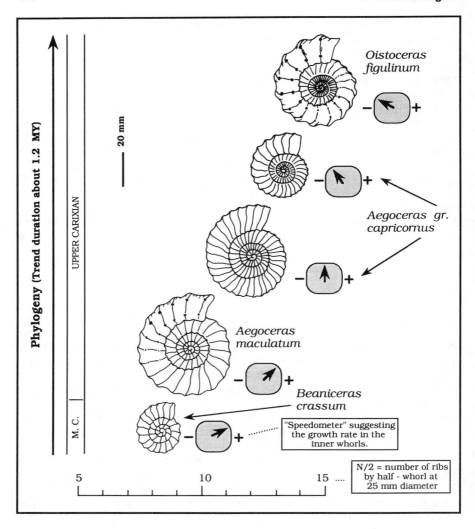

Figure 7.3. Diagrammatic illustration of the trend among some Liparocera-tidae (Eoderocerataceae) leading from coarse to close inner whorl ribbing. Only some steps of the trend are illustrated. The average rib density in the inner whorls is indicated *in abscisse*. The symbolical 'speedometers' suggest the growth rate indicated by septal density at the same ontogenetic stage. Notice the direct relationship between rib density decay and growth rate decrease.

From the earliest true capricorn (*Beaniceras crassum*) to the last (*Oisto-ceras figulinum*) a trend leads from forms with coarse-ribbed inner whorls to forms with closely ribbed ones. Slow to develop early on, the trend quickly increased. The result of these morphological changes has been to provide an efficient and accurate biostratigraphical tool for the Upper Carixian (Figure 7.3). Moreover, it is interesting to note that both rib and septal densities display strictly congruent patterns of growth and variability. This implies that rib and septal densities are directly under the control of shell growth rate (Dommergues, 1988). If one accepts this hypothesis, morphological trends involving ribs and septal density can be understood as a mechanical result of a trend at first involving the shell growth rate. Thus the development of capricorn liparoceratids, leading towards a reduction in the space between ribs or between septa, is better regarded as a trend in reduction of shell growth rate (Dommergues, 1988). Such trends involving growth rate have a direct consequence on the 'biological age' (length of life, age of sexual maturity) of extinct organisms. If one is to seek the adaptive significance of this, it is probably more fruitful to look at reproductive strategies than at assumptions concerning morphofunctional fitness.

Trends in Epipeltoceras

The Upper Oxfordian genus *Epipeltoceras*, collected in detail by Enay (1962), provides an interesting example of a trend involving both changes in body size and ornamental features. The interpretation of this trend as a process induced by heterochrony was proposed by Marchand and Dommergues (1988). The trend is spread over about 1.8 million years (Figure 7.4). It starts with a small species, *Epipeltoceras semimammatum*, about 25 mm in diameter. A typical example is characterised by simple ribs in the inner whorls and, in the body chamber, a weak tendency for two lateroventral tubercle rows to appear; furthermore, the ventral area, which is regularly rounded in the young stages, becomes increasingly flattened between these tubercles. Moreover, in later growth stages of some especially peramorphic variants of *E. semimammatum*, the rib strength declines further in the ventral area, which becomes smooth close to the aperture.

The evolutionary trends are continued by species such as *E. berrense*. In comparison with *E. semimammatum*, variability in *E. berrense* is shifted towards peramorphic morphologies. So, a body chamber with a smooth and flat, or even slightly depressed, ventral area between two rows of well-developed tubercles becomes the rule. In this species the regularly rounded ventral area and uninterrupted ribs become restricted to the earlier whorls. In a parallel direction, the body size slightly increases (to about 35 mm in diameter). The trend leads finally to the *E. bimammatum* group, the largest (with a diameter usually reaching 50 mm) and the most peramorphic species of the lineage. On average, the adult morphologies of *E. berrense* or of *E. treptense* are now limited in the medium or even in the inner whorls and the body chamber displays a stout ornamentation of strong lateral ribs terminated at the lateroventral edge by coarse and prominent tubercles which

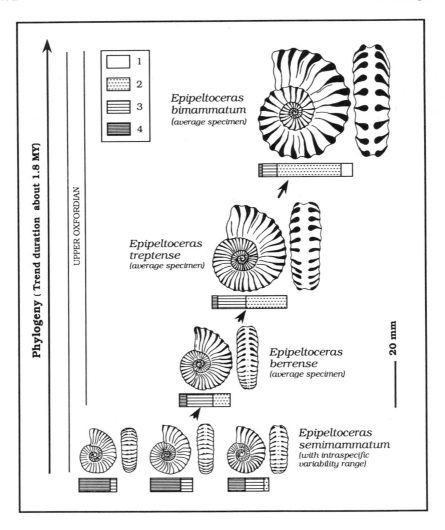

Figure 7.4. Diagrammatic illustration of the peramorphocline (by hyper-morphosis and acceleration) leading within the genus *Epipeltoceras* (Peri-sphinctaceae) from the small slenderly ribbed *E. semimammatum* to the large coarsely ornamented *E. bimammatum*. To show the heterochronic scheme both at the phylogenetic and at the intraspecific level, the variability of one species (*E. semimammatum*) is illustrated with paedomorphic variant to the left and peramorphic one to the right. 1 represents stout ornamentation of ribs termin-ated by such coarse and prominent lateroventral tubercles that the smooth ventral area seems depressed. 2 represents the same, but with less bold ornamentation. 3 represents continuous ribs that only weaken on the ventral area. 4 represents slender ornamentation comprising ribs crossing the regularly rounded ventral area without interruption or weakening.

delimit a smooth and deeply depressed ventral area.

In this example the evolutionary trend is clearly controlled by heterochronic (peramorphic) processes, combining acceleration and hypermorphosis. Thus this evolutionary lineage provides a good illustration of a peramorphocline (McNamara, 1982). However, it is difficult to demonstrate a link between the morphological changes and possible environmental constraints.

Conclusions

Although only two examples are described in this section, their selection yields a probably reasonable idea of trends in ornamental characters in ammonoids. The salient feature is the high frequency and the prevalence of heterochrony (both paedomorphosis and peramorphosis) during these trends. Another important characteristic is the frequency of the link between heterochrony involving ornamentation and affecting either the shell shape or body size. In view of such complex patterns it is difficult to interpret the possible adaptative significance of the ornamental changes. In most cases it is even probable that such a question may have little significance when ontogenetic internal constraints seem prevalent (Dommergues *et al.*, 1989), particularly so when many of the morphological features of the shell constitute an integrated system.

TRENDS IN THE SUTURE LINE

The importance of trends involving ammonoid suture lines has been stressed by a number of authors following Schindewolf's (1954; 1962) publications on suture lines and evolution. These works discuss many examples from Palaeozoic to Cretaceous ammonoids (Glenister, 1985; Gould, 1977; Kullman and Wiedmann, 1982; Miller and Furnisch, 1958; Tanabe, 1977; Wiedmann, 1970; Wright and Kennedy, 1979). In a recent review paper, Landman (1988) synthesised the main results and demonstrated that trends involving ammonoid suture lines are chiefly controlled by heterochrony from the ordinal to the intraspecific level. Moreover, Landman has stressed that most of the trends that have been analysed involve peramorphosis including, either independently or in conjunction, predisplacement, acceleration and hypermorphosis. On the other hand, paedomorphosis of suture lines is rare in the literature and Landman quotes only two examples, one by neoteny in Cretaceous ammonoids (Wright and Kennedy, 1979) and one by progenesis in Late Palaeozoic ammonoids (Glenister, 1985). If such a scarcity of paedomorphosis partly reflects reality, it is probably artificially magnified by the predominance of works on higher-level taxa; paedomorphosis would appear more frequent at the specific or generic levels. Moreover, in practically all studies, all elements of the suture lines are viewed as being subjected to single heterochronic processes. While such simplified approaches may sometimes be satisfactory, in other cases one must consider single suture-line

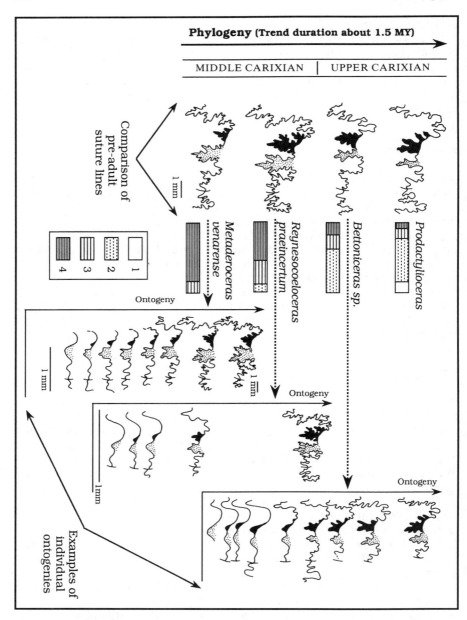

Figure 7.5. Diagrammatic illustration of a peramorphocline (mainly by acceleration) demonstrating among Eoderoceratidae (*Metaderoceras*) and Dactylioceratidae (*Reynesocoeloceras, Bettoniceras* and *Prodactylioceras*) an interesting trend in the suture line. There is a dramatic decrease of the lateral lobe (dotted area) and, conversely, an increase of the internal incision (in black) of

elements individually. Heterochronic trends between different sutural elements with strong differential allometries can induce more important structural changes than heterochronic processes affecting suture lines which show little allometric differences.

Evolution of suture lines in Metaderoceras and Bettoniceras

Between these two genera a trend leads from a rather weakly incised to a deeply incised suture line (Dommergues, 1986). Yet, unlike the most frequently observed pattern in ammonoids involving an increase of suture-line complexity, here it is the ventral and not the umbilical part of the suture which is affected. Such a pattern, extremely rare during the Mesozoic, seems more frequent among Palaeozoic ammonoids. In any case from *Metaderoceras* to the close genera *Bettoniceras* and *Prodactylioceras* (Figure 7.5) a progressive deepening of the internal 'incision' of the external saddle can be observed which rapidly becomes as deep as the lateral lobe. If one considers only the adult suture line of *Bettoniceras* or *Prodactylioceras*, it is tempting to interpret the pair as having been formed by the new incision and the lateral lobe as a primitive, widely bifid lateral lobe. Like the suture line, the shell morphology and the ornamentation in this lineage evolved by peramorphosis (Dommergues, 1986). Indeed, during the ontogeny of *Metaderoceras* the internal 'incision' of the external saddle increases proportionately faster than the lateral lobe (Figure 7.5). The dramatic product of the trend is a simple consequence of the accentuation by acceleration of this ontogenetic allometry.

Thus from *Metaderoceras* to *Bettoniceras*, ornamentation, shell morphology and suture line all evolved by the same combination of heterochronic processes: acceleration and predisplacement. It is difficult to interpret the significance of this in terms of selective pressure versus internal constraint; yet it must be stressed that the kind of shell morphology, the close ornamentation and the complex suture line of *Bettoniceras* and *Prodactylioceras* are features usually common among pelagic Tethyan ammonoids. For example, such features characterise the major part of the Lytocerataceae, a superfamily chiefly abundant in the Tethyan realm. This implies an adaptive determinism for the trend. Whenever heterochrony plays a major part in evolution one should not neglect the internal constraints, even if their role remains indirect.

Figure 7.5 – cont.
the external saddle. Only some steps of the trend are illustrated and the phylogeny must be followed in the left-hand part of the figure from bottom to top; in the right-hand part some individual ontogenies are illustrated for comparison. 1–4 represent transitions from a suture line with an incision of the external saddle deeper than the lateral lobe *sensu stricto* to a suture line with a deep lateral lobe and an external saddle hardly incised.

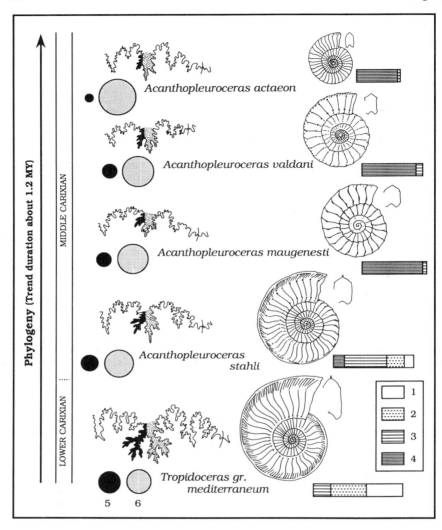

Figure 7.6.   Diagrammatic illustration of the trend leading within the Acantho-pleuroceratidae (Eoderocerataceae) from a widely bifid lateral lobe (*Tropido-ceras*) to a narrow trifid one (*Acanthopleuroceras*). These transformations are accompanied by an overall decrease in sutural complexity. Only some of the main steps of the trend are illustrated. Shell morphologies display a significant paedomorphic change which leads from sub-oxycone to platycone morpho-logies. 1–4 represent transitions from an involute coiling associated with a complex ornamentation to an evolute coiling with a simple ornamentation. 5 and 6 represent, respectively, external and internal parts of the lateral lobe, the disc diameters indicate the changes of proportion between these two parts during the course of the trend.

Suture line evolution in the Acanthopleuroceratidae

This Liassic (Pliensbachian) family exhibits an interesting example of a suture-line evolutionary trend where heterochrony is not obviously involved. This trend leads from the chiefly Tethyan genus *Tropidoceras* (Figure 7.6) to the principally north-west European genus *Acanthopleuroceras* (Dommergues and Mouterde, 1978; Dommergues, 1987). Shell morphology and ornamentation display a change along the evolutionary sequence, from sub-oxycone and close ornamentation in *Tropidoceras* to platycone and coarse-ribbed in *Acanthopleuroceras* (Figure 7.6). During the Pliensbachian, such a transformation usually developed each time a Tethyan group colonised the north-west European epicontinental seas; these transformations are frequently accompanied by sutural simplifications induced by paedomorphic (chiefly neotenic) processes.

Similar paedomorphic sutural simplifications occur frequently in Mesozoic ammonoids, as shown by Wright and Kennedy (1979) for a Cretaceous example from *Salaziceras* to *Flickia*. If in the case of Acanthopleuroceratidae the issue is also a decrease of sutural complexity, the evolutionary process appears quite different. Indeed it is progressive changes in the lobe structure which induced the simplification. From *Tropidoceras* to late *Acanthopleuroceras* (Figure 7.6), lobes, and especially the lateral lobe, shift from a primitive bifid to a derived narrow trifid structure; these changes do not occur as a result of simple shifts of ontogenetic timing (e.g., neoteny). For each step in the trend, novelties become visible in the earliest ontogenetic stages. Nevertheless, it will be difficult to exclude the effects of cryptic predisplacements so long as accurate analyses of suture-line ontogeny within the lineage are not available. In addition for the shell morphology and the ornamentation, the trend in suture-line development seems to point towards an increased fitness for the north-west European seas environments during the Pliensbachian.

Conclusions

Any attempt to synthesise data in the literature and the examples analysed in this chapter is not easy. Indeed, if one considers evolutionary trends involving suture lines at higher taxonomic levels, it appears that heterochrony, usually peramorphosis, plays an important role. Nevertheless, if one studies suture-line evolutionary trends in detail at lower taxonomic levels, within genera or at most within subfamilies, patterns appear more complex and heterochrony seems less important or their role appears less obvious. For example, it appears that the different sutural elements are able to evolve independently even if changes are under heterochronic control. It appears that global suture-line changes usually have morphofunctional significance: simplifed and highly dissected sutures correspond, respectively, with shallow and pelagic environments.

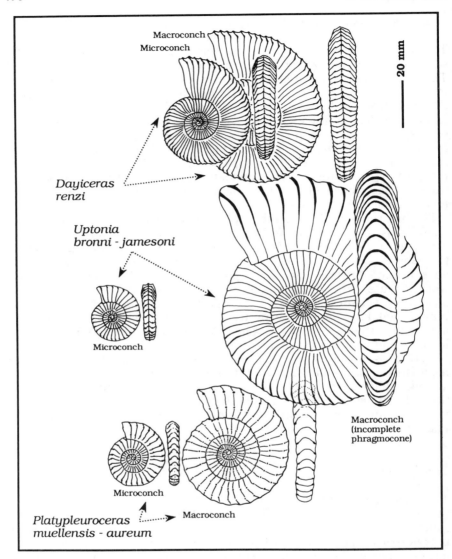

Figure 7.7.   Illustration of three dimorphic pairs (*Platypleuroceras, Uptonia* and *Dayiceras*) selected among those analysed in Figure 7.8. All the shells are adult and entire except the macroconch of *Uptonia jamesoni*, which is an incomplete phragmocone.

TRENDS IN POLYMORPHISM

The most striking expression of morphological variability in ammonoids is polymorphism. This, too, can display evolutionary trends. Kennedy and Cobban (1976) have given an example of such a trend for the Albian genus *Neogastroplites*. They have shown how progressively during geological time the ratio of one morphotype (the so-called 'stout nodose' type), declines within a basically trimorphic variability. Since this work was published, few studies have dealt with similar problems. Moreover almost all of these studies show restricted fluctuations of the microconch/macroconch ratio (= sex ratio?). It is important to look beyond the morphotype-ratio analysis, which usually reveals only simple ecological control without any true evolutionary connotations. A promising line of research is to search for evidence to show the extent that polymorphism is able to change along evolutionary trends.

Trends in polymorphism in Polymorphitidae

This is one of the major Middle Liassic (Carixian) families and is chiefly found in north-western Europe. It provides an interesting example of an evolutionary trend showing the morphological expression of sexual polymorphism (Dommergues, 1987). While this family was widespread and a common element of ammonoid faunas during the Early Carixian, it became almost completely restricted to the Lusitanian basin (north-western Portugal) in the later Carixian. So it is only in this area that it is possible to obtain a reasonably complete idea of the family history. As Donovan *et al.* (1981) have suggested, it is possible to recognise two main lineages in the Polymorphitidae. The first one, with sub-sphaerocone macroconch (*Parinodiceras*) and small sub-platycone microconch (*Polymorphites* s.s.), is only common in middle-western Europe (e.g., south-west Germany) and did not persist beyond the Lower Carixian. The second lineage (*Platypleuroceras*, *Uptonia* and *Dayiceras*), with both platycone microconchs and macroconchs (Figure 7.7), is more widely spread and persisted until the Middle Carixian in the Lusitanian basin. Only this lineage will be analysed here.

Seen as a whole, the evolution of sexual dimorphism in platycone Polymorphitidae, which at first appears to be a complex process, can be reduced to three successive episodes (Figure 7.8): a gradual and well-sustained trend of increased dimorphism, persisting for about 1 million years; a faster, and perhaps partly punctuated reverse trend, which led over about 0.75 million years to an almost complete disappearance of dimorphism; a succession of two punctuated events inducing both a dramatic morphological change and the reappearance of sexual dimorphism.

The sequence of the two first episodes appears to be a simple increase then decrease of dimorphism. However, a more elaborate analysis shows that the second episode is not a simple reversal of the former (Figure 7.8). The first episode (increased dimorphism) starts with a small platycone species, *Platypleuroceras* aff. *rutilans*, without dimorphism. The largest known specimen does not exceed 3 cm in diameter. The trend ends with the particularly large species, *Uptonia lata*, in which the largest macroconchs exceed 30 cm in diameter, whereas the microconch retains the diameter of the ancestral species, about 3 cm (Figure 7.7). The trend can be explained as a simple peramorphocline produced by hypermorphosis, involving only the macroconch.

The following trend which leads towards the disappearance of dimorphism is more complex, resulting from the operation of several heterochronic processes. It is not only a case of simple progenesis (the opposite process to hypermorphosis—see Chapter 3 herein) involving the macroconch which has occurred. Indeed even if the macroconch size decrease resulted from pro-genesis, the contribution of this process is of insufficient importance to compensate for the size increase which has occurred by hypermorphosis during the previous episode. The almost virtual disappearance of dimorphism in *Dayiceras renzi* implies a simultaneous microconch size increase, probably induced by hypermorphosis. If the trend is analysed further, it appears that it is not possible to reduce the evolutionary sequence to a simple case of microconch and macroconch size convergence by, respectively, hyper-morphosis and progenesis. If morphotype adult sizes are important in dimorphism the other morphological and especially ornamental features are also important. Thus, if one considers ornamentation, it is striking to observe from the late *Uptonia* to the *Dayiceras* an obvious and progressive expansion by neoteny of previously juvenile ornamentation towards the adult stages (e.g., keeled ventral area and ribs interrupted at the lateroventral edge) for both the microconch and the macroconch.

It is important to stress that all these diagnoses of heterochronic processes are based on the assumption that adult sizes yield close estimations of biological age. Recent work on other Liassic ammonoids (Dommergues, 1988) suggests that this assumption may be made only with prudence. For instance, the neotenic hypermorphosis pattern that is suspected for the microconch during the second trend could be a simple case of giantism (*sensu* Gould, 1977) if size and age are not coupled (Dommergues *et al.*, 1986).

This kind of uncertainty is common in palaeontology when it is impossible to obtain independent data on biological age. Such unreliability in hetero-chronic pattern diagnosis makes all further assumptions on the selection pressure hypothetical. For instance, if giantism and neotenic hypermorphosis can induce in some cases close morphological results, their consequences on demographic strategies are very different: giantism does not imply change in life duration and generation turnover while hypermorphosis does.

For the example considered here, if it is possible to assert that trends involving sexual dimorphism have important morphological consequences and that these are chiefly under heterochronic constraint, it is difficult to propose heuristic assumptions about their adaptive (*sensu lato*) signifi-cance (sexual selection, strategies of trophic resources exploitation, internal constraint).

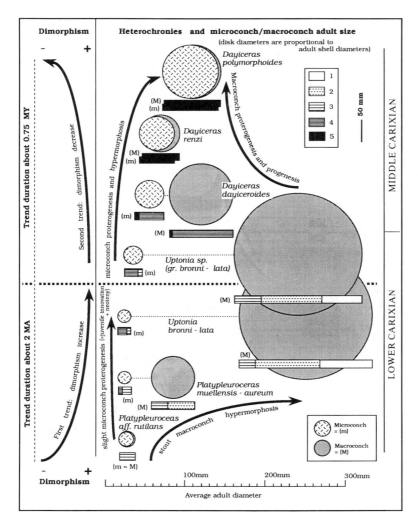

Figure 7.8. Schematic representation of two successive trends inducing an increase in sexual dimorphism, then a reverse tendency among the platycone Polymorphitidae (Eoderocerataceae). The disc diameters are proportional to the adult shell diameters (see Figure 7.7 for illustration of some species morphologies); for each step microconchs are on the left and macroconchs on the right. 1 represents a compressed and eliptic section with simple stout ribs crossing the regularly rounded ventral area without interruption. 2 represents the transitional morphology between 1 and 3. 3 represents a thicker section than in 1, with a subtectiform ventral area crossed by ribs clearly projected forward (possible lateroventral and lateroumbilical tubercles). 4 represents the same but with a keel borne on the ventral area. 5 represents a similar morphology to 4, but with an increase in rib density and appearance (innovation) of a crenulate keel.

Conclusions

Studies of evolutionary changes in polymorphism remain so scarce that it is not yet possible to draw any firm conclusions about their frequency and (or) their importance in the fossil record. Nevertheless, these first observations suggest that trends involving morphological polymorphism can be seen as direct reflections of the changes in mode of life to exploit environmental resources; demographic stategies are also probably involved. Thus analysis of such evolutionary trends will probably produce interesting results about strategies used by ammonoids to improve their adaptive fitness. Yet these studies will depend largely on the ability to estimate and compare the biological age of the different morphotypes.

CONCLUSION

Trends (sustained tendencies) are the most common evolutionary pattern in the phylogenetic progressions throughout the range of ammonoids from Devonian to Late Cretaceous; yet it is impossible to state that unorientated and ephemeral evolutionary patterns (e.g., the punctuated appearance of a new form, often at the beginning of a future trend) play only a minor part in ammonoid history, taken as a whole. Indeed, deterministic trends and factual events are complementary and indissociable aspects of historical processes.

In the absence of sufficient data throughout the entire range of ammonoids it is difficult to speak about trend durations. For the Jurassic, to which this chapter has been confined, it appears that durations range from about one to three biochronological zones (about 1–5 million years). Yet, if one takes into account ammonoid history as a whole, Jurassic durations are probably rather short, because evolutionary rates appear to be particularly rapid at this time. Moreover, durations of Jurassic trends are frequently bounded by extinction events that are associated with faunal substitutions, without environmental changes being recorded in the rocks.

The common link between trends and heterochrony is one of the most striking features to emerge from this analysis. Indeed, only the case of the evolution of suture lines in Pliensbachian Acanthopleuroceratidae cannot be easily assigned to a heterochronic cause. If one considers the examples analysed in this chapter, it appears that peramorphoclines occur almost as often as paedomorphoclines. This result is different from that obtained by Hallam (1989) from data compiled from various recent palaeontological publications, which suggests a predominance of paedomorphoclines. On the one hand, Landman (1988), in a review of ammonoids, recorded more peramorphoclines than paedomorphoclines, whereas Swan's (1988) results agree with Hallam's, indicating a prevalence of paedomorphoclines by neoteny for Namurian ammonoids. Yet it seems that if one considers true sustained and uninterrupted trends at a rather lower taxonomic level (e.g., genus or family) then either peramorphic or paedomorphic processes can occur. But at higher taxonomic levels, trends are essentially peramorphic processes modifying chiefly the sutural pattern, as discussed by Wiedmann and Kullman (1980) and by Landman (1988).

Special mention must be made of progenesis (involving early maturity at small size). While in the literature this type of heterochrony is almost always associated with punctuated events, it is shown here to have been involved in a trend (e.g., in Pliensbachian Polymorphitidae). In this example progenesis is associated with neoteny leading to 'hyper-paedomorphic' results (*sensu* Dommergues *et al.*, 1986). However, in this example this complex heterochronic pattern must be understood in a context of declining sexual dimorphism. Nevertheless, if this example suggests that progenesis can sometimes occur in trends, it is probable that it is associated much more frequently with punctuated events (see Chapter 3 herein), as shown by the abundance in the fossil record of trends starting from small species. It is quite evident that such smaller species have often evolved from larger ones (see Chapter 4 herein).

It now remains to discuss the question of the 'aptative' significance (*sensu* Gould and Vrba, 1982) of trends in ammonoid evolution. In fact the relationship between trends and heterochrony is so close that such a question is almost synonymous with what is the 'aptative' significance of heterochrony in ammonite evolution. This problem has been discussed in two recent papers (Swan, 1988; Dommergues *et al.*, 1989). These two works stress that any heterochronic modes imply an evolutionary option restricted by specific features of the organism's ontogeny (see Chapter 3 herein). Although heterochrony offers only a restricted choice of possible morphological changes, it does allow, on the other hand, important morphological transformations in mature morphology to be induced by isolated mutations in regulatory genes (Raff and Kaufman, 1983; McKinney and McNamara, 1990). Such genetic alteration does not imply important risks of declining viability because large-scale structural mutations are not required. Moreover, heterochrony, especially those processes related to changes in life durations and in timing of reproduction (progenesis and hypermorphosis), have direct and immediate results on ecological strategies (Gould, 1977; Stearns, 1976; Swan, 1988; Dommergues *et al.*, 1989; McKinney and McNamara, 1990).

For Swan (1988) any heterochronic modes observed among Namurian ammonoids implied a close balance between the intrinsic potentialities of heterochronic change and the extrinsic constraints imposed by the environment. For Jurassic ammonoids discussed in this chapter and in Dommergues *et al.* (1989) the relationships between intrinsic potentialities and environmental constraints (*sensu lato*) are usually rather obvious, even if they often remain hardly demonstrable by heuristic methods. Nevertheless the assumption that such relationships exist may sometimes be unjustified. This is particularly so because hypotheses concerning ecological strategies must often be cautiously proposed in order to avoid confusions between processes such as progenesis and hypermorphosis, on the one hand, and proportional dwarfism and giantism, on the other. This presupposes an understanding of the age at maturity which is generally difficult to obtain from fossil material, and especially from ammonoids which are characterised by possible important distortions between 'age' and 'size' (Dommergues, 1988).

Whatever difficulties are encountered when trying to test assumptions about the 'aptative' or 'non-aptative' significance of trends, it appears that intrinsic potentialities, mainly ontogenetic constraints, make an important contribution to evolution. Correlations between morphological trends and changes of habitat are often tenuous, particularly if one compares trends above the genus or sub-family level (Dommergues *et al.*, 1989). If environmental conditions can further enhance one particular option among a restricted set of potential heterochronic processes, they are usually unable to deflect noticeably the channelling once initiated and enforced by the ontogenetic constraints (Dommergues *et al.*, 1989). How, in view of such strong dynamic equilibrium, is it possible to suggest a hierarchy among the extrinsic and intrinsic factors controlling trends? Are not these two basic and complementary elements by their very nature impossible to compare? Yet, even if it sometimes seems possible to propose an evolutionary scenario and to suggest whether extrinsic or intrinsic factors are responsible for the initiation and direction of the trend, such a priority appears to depend more on particular circumstances than on any fundamental predominance of either extrinsic or intrinsic factors.

ACKNOWLEDGEMENT

I would like to thank K.J. McNamara for his help with the English language and for offering valued suggestions for improving the manuscript.

REFERENCES

Callomon, J.H., 1963, Sexual dimorphism in Jurassic ammonites. *Trans. Leicester Lit. Phil. Soc.*, **57**: 21–56.

Callomon, J.H., 1980, Dimorphism in ammonoids. In M.R. House and J.R. Senior (eds). *The Ammonoidea*, Systematics Association Special Volume no. 18, Academic Press, London: 257–73.

Callomon, J.H., 1985, The evolution of the Jurassic ammonite family Cardioceratidae, *Spec. Pap. Palaeont.*, **33**: 49–90.

Corna, M., 1987, Eléments de phylogénie des Ariétitidés d'après les données du Jura méridional, *Les Cahiers de l'Institut Catholique de Lyon, Sciences*, **1**: 93–104.

Dommergues, J.-L., 1983, L'Évolution des Liparoceratidae capricornes (Ammonites, Jurassique, Lias moyen); diversité des rythmes évolutifs. In *Modalités, rythmes, mécanismes de l'évolution biologique*, Colloques internationaux du CNRS no. 330., Paris: 105–13.

Dommergues, J.-L., 1986, Les Dactylioceratidae du Carixien et du Domérien basal, un groupe monophylétique. Les Reynesocoeloceratinae nov. subfam, *Bull. Sci. Bourg.*, **1**: 1–26.

Dommergues, J.-L., 1987, L'Évolution chez les Ammonitina du Lias moyen (Carixien, Domérien basal) en Europe occidentale, *Docum. Lab. Géol. Lyon*, **98**: 1–297.

Dommergues, J.-L., 1988, Can ribs and septa provide an alternative standard for age in ammonite ontogenetic studies?, *Lethaia*, **21**: 243–56.

Dommergues, J.-L. and Mouterde, R., 1978, Les Faunes d'ammonites du Carixien inférieur et moyen des Cottard (Cher), *Geobios*, **11**: 345–65.

Dommergues, J.-L., Cariou, E., Contini, D,. Hantzpergue, P., Marchand, D., Meister, C. and Thierry, J., 1989, Homéomorphies et canalisations évolutives: le rôle de l'ontogenèse. Quelques exemples pris chez les ammonites du Jurassique, *Geobios*, **22**: 5–48.

Dommergues, J.-L., David, B. and Marchand, D., 1986, Les relations ontogenèse-phylogenèse: applications paléontologiques, *Geobios*, **19**: 335–356.

Donovan, D.T., 1987, Evolution of the Arietitidae and their descendants. *Les Cahiers de l'Institut Catholique de Lyon, Sciences*, **1**: 123–38.

Donovan, D.T., Callomon, J.H. and Howarth, M.K., 1981, Classification of the Jurassic Ammonitina. In M.R. House and J.R. Senior (eds), *The Ammonoidea*, Systematics Association Special Volume no. 18, Academic Press, London: 101–55.

Eldredge, N. and Gould, S.J., 1972, Punctuated equilibria: an alternative to phyletic gradualism. In T.J.M. Schopf (ed.), *Models in paleobiology*, Freeman, San Francisco: 82–115.

Enay, R., 1962, Contribution à l'étude paléontologique de l'Oxfordien supérieur de Trept (Isere). Stratigraphie et ammonites, *Trav. Lab. Géol. Fac. Sc. Lyon*, NS, **8**: 9–81.

Erben, H.K., 1964, Die Evolution der ältesten Ammonoidea (Leif.I), *N. Jb. Geol. Pal. Abh*, **120**: 107–212.

Glenister, B.F., 1985, Terminal progenesis in Late Paleozoic ammonoid families. In *Abstract, 2nd International Cephalopods, Present and Past, Tübingen, 1985*, (unpublished), Tübingen: 9.

Gould, S.J., 1977, *Ontogeny and phylogeny*, Belknap, Cambridge, MA.

Gould, S.J. and Vrba, E.S., 1982, Exaptation – a missing term in the science of form, *Paleobiology*, **8**: 4–15.

Hallam, A., 1989, Heterochrony as an alternative to species selection in the generation of phyletic trends, *Geobios, mém. spec.*, **12**: 193–8.

House, M.R., 1982, On the origin, classification and evolution of the Early Ammonoidea. In M.R. House and J.R. Senior (eds), *The Ammonoidea*, Systematics Association Special Volume no. 18, Academic Press, London: 3–36.

Hyatt, A., 1889, Genesis of the Arietidae, *Bull. Mus. Comp. Zool. Harvard*, **16**: 1–238.

Kennedy, W.J. and Cobban, W.A., 1976, Aspects of ammonite biology, biogeography and biostratigraphy, *Spec. Pap. Palaeont.*, **17**: 1–94.

Kennedy, W.J. and Wright, C.W., 1985, Evolutionary patterns in Late Cretaceous ammonites, *Spec. Pap. Palaeont.*, **33**: 131–43.

Kullman, J. and Wiedmann, J., 1982. Bedeutung der Rekapitulationsentwicklung in der Paläontologie, *Verh. Naturwiss. Ver. Hamburg, Hamburg*, N.F., **25**: 71–92.

Landman, N.H., 1988, Heterochrony in ammonites. In M.L. McKinney (ed.), *Heterochrony in evolution: a multidisciplinary approach*, Plenum, New York: 159–82.

Marchand, D., 1986, L'Évolution des Cardioceratinae d'Europe occidentale dans leur contexte paléobiogéographique (Callovien supérieur, Oxfordien moyen), unpublished doctoral dissertation, Univ. Dijon.

Marchand, D. and Dommergues, J.-L., 1988, Rythmes évolutifs et hétérochronies du développement: Exemples pris parmi les Ammonites Jurassiques. In J. Wiedmann and J. Kullmann (eds), *Cephalopods present and past*, Schweizerbart'sche Verlag, Stuttgart: 67–78.

McKinney, M.L. and McNamara, K.J., 1990, *Heterochrony: the evolution of ontogeny*,

Plenum, New York.

McNamara, K.J., 1982, Heterochrony and phylogenetic trends, *Paleobiology*, **8**: 130–42.

McNamara, K.J., 1986, A guide to the nomenclature of heterochrony, *J. Paleont.*, **60**: 4–13.

Miller, A.K. and Furnisch, W.M., 1958, Middle Pennsylvanian Schistoceratidae (Ammonoidea), *J. Paleontol.*, **32**: 253–68.

Pavlov, A.P., 1901, Le Crétacé inférieur de la Russie et sa faune, *Nouv. Mém. Soc. Imp. Nat.*, Moscow, N.S., **16**: 1–87.

Phelps, M.C., 1985, A facies and faunal analysis of the Carixian, Domerian boundary beds in north-west Europe (2 vols), unpublished Ph.D. Thesis, Birmingham University.

Raff, R.A. and Kaufman, C., 1983, *Embryos, genes and evolution*, Macmillan, New York.

Reyment, R.A., 1973, Factors in the distribution of fossil cephalopods. 3: Experiments with exact models of certain shell types, *Bull. Geol. Inst. Univ. Uppsala*, N.S., **4**: 7–41.

Schindewolf, O.H., 1936, *Paläontologie, Entwicklungslehre und Genetik: Kritik und Synthese*, Bornträger, Berlin.

Schindewolf, O.H., 1954, On development, evolution and terminology of ammonoid suture line, *Bull. Mus. Comp. Zool. Harvard*, **112**: 217–37.

Schindewolf, O.H., 1962, Studien zur Stammesgeschichte der Ammoniten. II Psilocerataceae, Eoderocerataceae, *Akad. Wiss. Lit. Mainz, Abh. math.-nat.*, Mainz, **8**: 425–572.

Simpson, G.G., 1953, *The major features of evolution*, Columbia University Press, New York.

Spath, L.F., 1938, *A catalogue of the ammonites of the Liassic family Liparoceratidae in the British Museum (Nat. Hist.)*, Brit. Mus. (Nat. Hist.), London.

Stanley, S.M., 1979, *Macroevolution: pattern and process*, Freeman, San Francisco.

Stearns, S.C., 1976, Life-history tactics: a review of an idea, *Q. Rev. Biol.*, **51**: 3–47.

Swan, R.H., 1988, Heterochronic trends in Namurian ammonoid evolution, *Palaeontology*, **31**: 1033–51.

Tanabe, K., 1977, Functional evolution of *Otoscaphites puerculus* (Jimbo) and *Scaphites planus* (Yabe). Upper Cretaceous ammonites, *Mem. Fac. Sci. Kyushu Univ., Geol.*, **23**: 367–407.

Tintant, H., Marchand, D. and Mouterde, R., 1982, Relations entre les milieux marins et l'évolution de Ammonoïdés: les radiations adaptatives du Lias, *Bull. Soc. Géol. France*, ser. 7. **24**: 951–61.

Trueman, A.E., 1919, The evolution of the Liparoceratidae, *Q. Jl. Geol. Soc. Lond.*, **74**: 247–98.

Vrba, E.S., 1980, Evolution, species and fossils: how does life evolve?, *S.A. J. Sci.*, **76**: 61–84.

Ward, P., 1980, Comparative shell shape distribution in Jurassic–Cretaceous ammonites and Jurassic–Tertiary nautilids, *Paleobiology*, **6**: 32–43.

Ward, P., 1981, Shell sculpture as a defensive adaptation in ammonoids, *Paleobiology*, **7**: 96–100.

Westermann, G.E.G., 1989, New developments in ecology of Jurassic, Cretaceous ammonoids, Preprint from *Proceedings of 2nd Pergola symposium*, 1987, Pergola: 1–21.

Wiedmann, J., 1970, Probleme der Lobenterminologie, *Eclogae Geol. Helv.*, **63**: 909–22.

Wiedmann, J. and Kullman, J., 1980, Ammonoid sutures in ontogeny and phylogeny. In M.R. House and J.R. Senior (eds), *The Ammonoidea*, Systematics Association Special Volume no. 18, Academic Press, London: 215–55.

Wright, C.W. and Kennedy, W.J., 1979, Origin and evolution of the Cretaceous micromorph ammonite family Flickiidae, *Paleontology*, **22**: 685–704.

Chapter 8

CRINOIDS

Michael J. Simms

INTRODUCTION

Palaeontologists investigating the evolution of particular groups have often sought to identify morphological trends within that group. Crinoids, with their complex multielement skeletons and growth patterns, are ideal subjects for the documentation of such evolutionary trends. These trends can only be documented unequivocally where ancestor–descendant relationships are known at the species level. Apparent trends between related genera or higher taxa cannot be identified with the same degree of confidence unless the transitions between genera are also documented as an ancestor–descendant sequence at the species level. Seldom is documentation thorough enough for this. Since genera and higher taxa are essentially artificial, any supposed trends documented above species level may be equally artificial, perhaps excluding lower taxa which show morphological changes contrary to those of the 'trends' in question. Such exclusion of non-congruent taxa may occur through non-preservation or collection failure, or through artificial exclusion. The latter may be due to inadequate taxonomic procedure failing to recognise the true affinities of non-congruent taxa, or by a subconscious tendency to assume that taxa which do not follow the trend being documented must represent offshoots from the main lineage. Furthermore, there is often an inherent tendency among palaeontologists to arrange taxa in order of increasing complexity or to assume that evolution proceeds inexorably from simple to more complex forms, thereby generating artificial evolutionary trends. These factors do not mean that all evolutionary trends documented at generic or higher level must be dismissed as based on inadequate data, though they should be examined with this in mind. Gould's (1988) idea of trends as changes in variance is often applicable to branching clades, particularly at higher taxonomic levels. Many represent increases in variance. Others, in which primitive morphotypes are entirely replaced by more advanced ones, represent subsequent decreases in variance through loss of the primitive morphotypes. To some extent these may be considered to represent real trends, inasmuch as there is a real shift through time in the modal value for that particular character.

Real evolutionary trends, with no suggestion of changes in variance, can best be identified in unbranched lineages at the species level. Their recognition requires at least three consecutive species to exhibit unidirectional morphological change in at least one character. In this simplest example there is, however, an equal likelihood that the supposed 'trend' is an essentially random shift in morphology. The probability of random morphological shifts decreases as the number of consecutive species through which the 'trend' can be traced increases. Similarly, an increase in the number of morphological characters showing unidirectional change produces a corresponding decrease in the probability of these changes being random.

In the following account I shall discuss the processes which appear to underlie the diversification of crinoids and the generation of evolutionary trends. I shall consider a number of supposed evolutionary trends at various taxonomic levels and attempt to evaluate whether they are real trends. Finally, I shall describe the evolution of the Pentacrinitidae, the only well-documented lineage of obligate pseudoplankton, and compare this with the evolution of contemporaneous benthic isocrinids to assess the role of extrinsic factors in generating evolutionary trends.

PROCESSES OF DIVERSIFICATION AND THE GENERATION OF REAL EVOLUTIONARY TRENDS

Among crinoids diversity appears to increase through the operation of two main processes: paedomorphosis and the optimisation of adult morphology (Simms, in press). Expansion of a clade's adult morphological diversity into that occupied by the juveniles, through heterochrony, seems almost inevitable since the very existence of adult crinoids implies that the juvenile morphology is viable. On the other hand, natural selection will tend to optimise existing morphology and feeding strategies since this will allow a greater proportion of the energy budget to be invested in reproductive effort. This process of morphological optimisation may also involve heterochrony, but operating with differing polarities and strategies on particular structures rather than operating relatively homogeneously on the whole individual.

Diversification into juvenile morphologies through paedomorphosis may be triggered by extrinsic factors, although chance may also play a significant role. Paedomorphic forms often pursue an opportunistic lifestyle and their advantage over the ancestral morphotypes may lie in this or in the attainment of greater evolutionary plasticity in the juvenile morphotype. In contrast, optimisation of the existing adult morphology may be a direct result of selection pressure for mechanically and energetically superior morphotypes.

If this same selection pressure persists over a long time interval it might be anticipated that any clade subject to its influence will exhibit true morphological trends through time associated with the 'fine-tuning' of the existing strategies of feeding, reproduction, attachment, protection against predators, and so on.

PUNCTUATED EQUILIBRIA, GRADUALISM AND EVOLUTIONARY TRENDS

Prior to the publication of Eldredge and Gould's (1972) article on punctuated equilibria, the evolution of many lineages was interpreted as essentially gradualistic. Subsequently many of these lineages were reinterpreted as punctuational, to the virtual exclusion of gradualistic hypotheses. It is now generally accepted that both modes of evolution have equal validity but, as discussed by Fortey (1985), three asymmetries between the gradualistic and punctuational theories lead to a strong bias towards identification of stasis and punctuational change in fossil lineages. This bias is further compounded by artefacts of perception. Fortey (1985) suggested that the supposed evolution of lineages by punctuated equilibria could be verified if an example of an unbroken gradualistic lineage occurred in the same strata. Conversely, examples of 'perceptual stasis' could be recognised by the characteristic pattern of a series of stratigraphically overlapping taxa pursuing a morphological trend through time. The implication of this is that evolutionary trends are characteristic of gradualistically evolving lineages and that lineages which apparently evolved in a punctuational fashion yet display morphological trends actually represent examples of 'perceptual stasis'.

The limited evidence from crinoids, however, suggests that morphological trends do occur in lineages which, as far as can be judged using the criteria of Fortey (1985), were evolving by punctuated equilibria.

EVOLUTIONARY TRENDS IN CRINOIDS

General 'trends': real or artefact?

Much has been written concerning overall trends in the evolution of crinoids. Many of these 'trends' can almost certainly be attributed to changes in variance, as discussed by Gould (1988). Where such an increase in variance is followed by a subsequent decrease through loss of the primitive morphotypes, there will be a real shift through time in the modal value for that character. Such cases may be regarded as examples of either an evolutionary trend or a change in variance, depending on how strictly one defines the limits of these two processes. Evolutionary trends have been noted in many crinoid groups, though there are few detailed accounts (Broadhead, 1988; Brower, 1973; 1976; 1982; Lane, 1978; Lane and Strimple, 1978; Lane and Webster, 1966; Moore and Laudon, 1943; Ubaghs, 1953; 1978b; Webster and Lane, 1967). Only a brief summary can be given here.

One of the most frequently noted 'trends' is that of phyletic size increase (see Chapter 4 herein), often referred to as 'Cope's Rule' (Raup and Stanley, 1978). Phyletic size decrease, as noted by Gould (1983), is much less prevalent, at least as far as documented examples are concerned. Moore and Laudon (1943, p. 20), in their discussion on general evolutionary trends in crinoids, commented that there was 'an increase in the average size of specimens and, later on, a decrease'. However, any suggestion that these represented real trends was discredited by them, unknowingly, in the following sentence, where they remarked that 'the geologic age of the most robust representatives of any large group of Palaeozoic crinoids coincides approximately with the time of maximum differentiation and profusion of the group'. It would be difficult to find a clearer and more concise example of changing variance.

Apparently genuine examples of phyletic size increase have been noted in the camerate Melocrinitidae (Brower, 1976), in the early part of the evolution of the Jurassic genera *Isocrinus* and *Balanocrinus* (Simms, 1985; 1988a) and in Palaeozoic microcrinoids. The latter show a gradual, but apparently real, phyletic size increase over well-documented parts of their known range (G.D. Sevastopulo, pers. comm.). By Late Permian times many crinoid taxa with a microcrinoid-type morphology (i.e. showing extreme progenesis) were far larger than the 1–2 mm thecal size typical of the Carboniferous forms. These might be interpreted as the culmination of this trend, though lack of documentation of true microcrinoids from the Late Permian leaves open the distinct possibility that these 'giant microcrinoids' represent merely one extreme of an (unevenly sampled) overall increase in size variance rather than any genuine trend towards larger size.

General 'trends' have been noted many times in features of the stem, cup and arms. Although detailed documentation might reveal real trends in some lineages, on the coarse scale of existing documentation most show clear evidence of changing variance.

The stem
The earliest known crinoid, the Middle Cambrian *Echmatocrinus*, and a few other early Palaeozoic forms, had a stem consisting of a fairly spacious cavity surrounded by an integument of numerous irregularly arranged plates. In all extant crinoids and the vast majority of fossil ones the columnals are holomeric, each consisting of a single ossicle. However, throughout the Palaeozoic a small minority of crinoids had stems constructed of meric columnals, composed of more than one ossicle and apparently representing an intermediate form between the primitive poly-plated stem and the typical holomeric stem. The transition from meric columnals to holomeric ones has been discussed by Donovan (1985) and Ubaghs (1978a). Meric columnals, which typically are pentameric, decreased in abundance through the Palaeozoic (Sieverts-Doreck, 1957) and are unknown after the Permian. The basic morphological sequence leading from the polyplated stem to holomeric columnals is a good example of changing variance. Initially there must have been an increase in variance as the various stem types appeared. However, this phase of morphological diversification must have occurred prior to the first

record of a true holomeric columnal near the base of the Ordovician. The fossil record of meric columnals preserves only the subsequent decline of a morphological group which, even in early Ordovician times, were of comparatively minor importance. The apparent trend towards the dominance of holomeric stems therefore represents a decrease in variance as the other stem types disappeared.

The other major 'trend' seen in the evolution of crinoid stems is towards increasing stem length. Crinoids, as stemmed organisms able to elevate their feeding structures well above the sea-floor, have been much discussed in the context of tiering in suspension-feeding communities. Ausich and Bottjer (1985) consider that there was a definite trend during the Palaeozoic for crinoids to move into higher tiering levels through increasing the length of the stem. However, this was achieved only by certain crinoid taxa, with many others remaining at lower, previously attained, tiering levels. The development and subsequent decline in late Palaeozoic times of the highest tiering levels correlates fairly closely with total crinoid diversity levels through the Palaeozoic. As such it is a clear example of changing variance rather than a real trend. The development of successively higher tiering levels was probably driven by interspecific competition associated with high taxonomic diversity. The disappearance of these highest tiers once diversity began to decline in the late Palaeozoic suggests that they were viable only when conditions were particularly favourable and the competition particularly severe.

The cup
Among Palaeozoic crinoids the cup has generally been accorded high taxonomic value. Several general trends have been noted in aspects of the cup. Both Lane (1978), looking at the Flexibilia, and Lane and Strimple (1978), who considered the inadunates, noted a general tendency for cup shape to change through time from the predominant cone shape among earlier Palaeozoic taxa to low bowl-shaped cups, often with an invaginated base, in Late Palaeozoic taxa. Almost certainly there is an increase in variance producing a modal shift in cup shape through time but the only quantitative analysis of trends in cup shape (Lane and Webster, 1966; Webster and Lane, 1967) indicated that invaginated cup bases decreased in relative abundance through the late Palaeozoic.

A striking and frequently discussed aspect of crinoid evolution is the trend towards a simpler cup, with fewer plates and perfect pentamerism. Simplification of the cup has two main components: a reduction in number or size of the infrabasals, the lowest of the three principal circlets of plates in the cup, and their eventual elimination; and the upward displacement or resorption of various plates found in the cup in addition to these three primary circlets.

No clear trend is discernible towards predominantly dicyclic cups (with infrabasals) or monocyclic cups (without infrabasals). Among the camerates, the monocyclic monobathrids survived to the end of the Permian whereas the dicyclic diplobathrids were extinct by Late Carboniferous times. In contrast, the Late Palaeozoic dicyclic cladids were much more abundant than the monocyclic disparids, which instead had their greatest diversity earlier in the Palaeozoic. In some apparently monocyclic groups the lowermost, infrabasal, circlet is vestigal or it can be otherwise demonstrated that the monocyclic condition was derived from an originally dicyclic cup: such forms are known as 'cryptodicyclic'. Lane (1978) noted a general trend among the Flexibilia for the cup to change from being dicyclic, with large exposed infrabasals, to cryptodicyclic, with the infrabasals entirely concealed beneath the base of the cup. Similarly, all but the most primitive post-Palaeozoic articulates, and a handful of aberrant post-Triassic taxa, are cryptodicyclic.

Concerning the various other plates which are commonly found in the cups of Palaeozoic crinoids, Moore and Laudon (1943, p. 20) stated that:

a universal evolutionary trend in the life history of Paleozoic crinoid stocks is the upward displacement and ultimate elimination from the dorsal cup of all plates except the radials and the one or two circlets of plates below them. Interbrachial and anal plates are thus reduced in number and they are finally absent. Pentamerous symmetry and maximum simplicity of structure are attained.

The cup of Palaeozoic crinoids may contain a varying number of plates in addition to the three primary circlets: radials, basals and infrabasals. These include anal plates, interradials, interbrachials, fixed brachials and a number of other minor plate types. Broadly speaking, earlier Palaeozoic crinoids have relatively large numbers of extra plates while later in the Palaeozoic many groups had far fewer, often having only a single anal plate or even none at all (Simms, 1988b). Ubaghs (1978b) noted a trend among camerate crinoids towards the elimination of fixed brachials, interbrachials and anal plates, though this was by no means ubiquitous since the reverse is true of the Melocrinitidae, where some of the later taxa had more of these extra plates than the earlier ones (Figure 8.1). This 'trend' towards perfect pentamerous symmetry of the cup culminated in the post-Palaeozoic articulates, where an anal plate is found only in the larval stages and its absence is a diagnostic character of adult articulates. The precise adaptive significance of this reduction in cup complexity is unclear other than in reducing the amount of calcite stereom which the animal had to produce. A significant component of this supposed trend is a change in variance. In Late Palaeozoic times there was a wide range of cup morphotypes: from those with many anal plates to those with only one or none at all. Considering the broad range of morphotypes present, the probability that only one lineage lying at one extreme of variation (i.e. with no anal plates in the cup) should survive the end-Permian crisis seems remarkable and suggests that the reduction and loss of anal plates from the cup was by no means adaptively neutral. Another interesting, but as yet unexplained, feature of these changes in number of anal plates in the cup, is an apparent reversal of heterochronous polarity associated with the presence of anal plates in the cup. Among Palaeozoic crinoids the reduction or loss of anal plates is often associated with neoteny, as exemplified by many of the highly aberrant Late Permian taxa. Anal plates are lacking from the cups of all articulate crinoids, except for a handful of teratological specimens from the Triassic, yet consistently are present in the late larval stages where these have been documented. Hence the absence of anal plates in the cup changes from being a paedomorphic trait in Palaeozoic crinoids to being a peramorphic one in the post-Palaeozoic articulates.

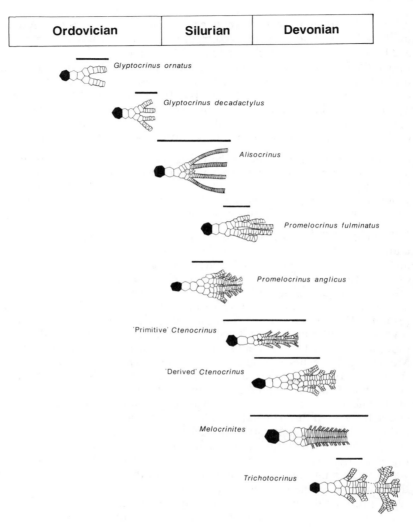

Ordovician	Silurian	Devonian

Figure 8.1.   Morphology and stratigraphic distribution of the camerate family Melocrinitidae to show changes in arm-branching complexity and number of interbrachial plates. Radial plates are shaded black; interbrachial plates are stippled. Based on Ubaghs (1953) and Brower (1976).

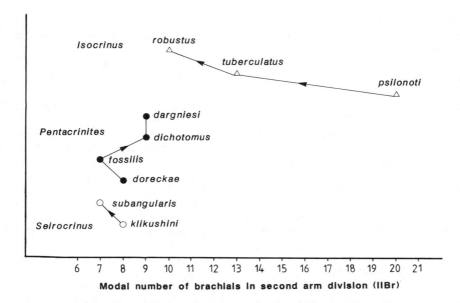

Figure 8.2. Changes in the modal number of secundibrachials in three Late Triassic–Early Jurassic crinoid genera. Note the remarkable parallelism in development of shorter brachitaxes in *Seirocrinus* and the first two species of *Pentacrinites*. Direction of arrows indicates sequence of species.

Arms

The pattern of arm branching in crinoids generally shows such variety at any one time that it is difficult to identify any real trends or even changes in variance. Within particular crinoid groups some trends have been noted, however. The most prevalent of these, noted by Moore and Laudon (1943) in several groups, was the migration of isotomous arm divisions towards the base of the cup. This trend has also been observed among several Early Jurassic clades (Figure 8.2; see also Simms, 1988a) where it is considered to represent an improvement in the effectiveness of the filtration fan by bringing large numbers of pinnules nearer to the cup, thus shortening the travel distance for food particles.

Probably the most pervasive general 'trend' in the evolution of crinoid arms is the increasing prevalence of pinnulate forms through time. However, again this 'trend' appears to represent an initial increase in variance with the appearance of pinnulate arms, later followed by a decrease in variance caused by the extinction of all non-pinnulate groups. Pinnules probably evolved independently several times, in camerates, cladids and disparids, early in the Ordovician but were rare prior to the Devonian. Pinnulate forms became increasingly dominant through the late Palaeozoic, though, with one possible exception (Broadhead, 1988), the pinnulate condition was never attained by flexible crinoids. Pinnulation is ubiquitous among the post-Palaeozoic articulates but we can only speculate about whether this would be the case if all but one, pinnulate, lineage had not become extinct at the end of the Permian.

At lower taxonomic levels a few groups may show trends in the evolution of the arms. Ausich and Lane (1982) record an apparent increase in the arm branching of Mississippian species of the cladid *Cydrocrinus* but admit that the sequence of species is beyond biostratigraphic resolution and is arranged by them in order of increasing complexity, thereby artificially generating an evolutionary trend. Among inadunates Lane and Strimple (1978) noted a frequent tendency towards the development of simple atomous arms rather than the isotomous branching of earlier taxa. One group, the Allagecrinacea, developed multiple atomous arms on broad radial facets. Commencing with the Devonian Anamesocrinidae, this trend reached its acme in the Late Palaeozoic Allagecrinidae and Catillocrinidae, where up to 34 simple atomous arms were developed from each radial facet. Ubaghs (1978b) noted a number of trends in camerate arms. These included a trend from uniserial arms to biserial arms and an increase in the number of arms. In many cases the latter appears to have developed through pinnule differentiation (Broadhead, 1987). These two trends are exemplified by the evolution of the melocrinitids. The camerate family Melocrinitidae have what is often held to be one of the best-documented evolutionary lineages among Palaeozoic crinoids (Brower, 1976; Ubaghs, 1953). However, there is some indication that Brower assumed that the lineage could be identified simply on the basis of increasing morphological complexity and, furthermore, he admitted that some of the genera included in his reconstructed lineage were probably polyphyletic. Brower (1976) considered the two main trends to be an increase in calyx size and the development of hypertrophied arms or ray trunks. Other trends include the development of biserial arms from uniserial ones and, initially, an increase in the number of interbrachial plates incorporated in the cup (Figure 8.1). From the simple uniserial dichotomous arms of the ancestral form, *Glyptocrinus*, there followed a sequence of taxa with increasingly complex patterns of arm branching culminating, through the loss of certain arm branches and the fusion of others into ray trunks, in the exceptionally complete, biserial, endotomous filtration fan of *Melocrinites* and *Trichotocrinus* (Figure 8.1). Cowen (1981) drew an analogy between the arrangement of arms in *Melocrinites* and the ideal arrangement of roads on a banana plantation. He viewed both as representing the optimum arrangement for an efficient harvesting network with the minimum expenditure of energy and materials in construction. Brower (1976) considered the trends seen in the arms to represent a classic case of size-related allometry. However, Cowen (1981) pointed out the rarity of endotomous arm branching among crinoids despite the obvious advantages of this morphology. The most advanced pattern and largest size is found in the Devonian *Melocrinites*, which both Wells (1941) and McIntosh (1978) have suggested was pseudoplanktonic. Significantly, the only other pseudoplanktonic crinoids known, the Early Jurassic *Pentacrinites* and *Seirocrinus*, show remarkable convergence with *Melocrinites* in attaining very large size and highly developed endotomy.

These traits probably reflect the severe biological constraints imposed by a pseudoplanktonic lifestyle, as discussed later in this chapter and elsewhere (Simms, 1986; 1988a; Wignall and Simms, 1990).

Evolutionary trends in benthic isocrinids

The evolution of two early Jurassic genera, *Isocrinus* and *Balanocrinus*, support the hypothesis of initial diversification through paedomorphosis followed by evolution through a sequence of progressively more 'optimal' forms. Other aspects of the evolution of these genera suggest that unidirectional morphological change is not restricted to lineages evolving gradualistically but can also occur in lineages which, on the criteria of Fortey (1985), evolved by punctuated equilibria. *Balanocrinus* shows clear trends in a number of characters despite showing evidence for evolution by punctuated equilibria. The 'control' here is provided by the contemporaneous genus *Isocrinus* which shows clear evidence of episodes of gradualistic change, also unidirectional, between longer periods of stasis (Simms, 1988a). These trends appear, to some extent, to have been driven by extrinsic factors. In the origin and early evolution of the genus *Balanocrinus* (Simms, 1985; 1988a), the earliest species, *B. quiaiosensis*, appears to have arisen from the much larger *Isocrinus* through progenesis (precocious sexual maturation—see Chapter 3 herein; see also McNamara, 1986). Five species are known between the Sinemurian and Domerian. The first three show clear trends in increasing maximum size, increasingly robust cirri (Figure 8.3), increased height of nodals relative to internodals and decreasing nodal frequency. Of the remaining two species, from the upper Domerian, one, *B. solenotis*, shows a clear continuation of the trends already observed in the first three. In contrast the other species, *B. donovani*, is very much smaller. The evolution of *Balanocrinus* in the early Jurassic therefore shows aspects both of real evolutionary trends and, in the Domerian, of an increase in variance. The first three species show unidirectional changes in several characters, a strong indication that the trends are real rather than a random walk. In the Domerian, *B. solenotis* continues these trends but the simultaneous appearance of *B. donovani* demonstrates an increase in variance. If we consider the minimum and maximum morphological limits for several isocrinid characters and compare these values with those of early Jurassic isocrinids, we have the possibility of assessing the relative influence of chance and specific selection pressures in generating morphological trends or increases in variance. With few exceptions the maximum adult stem diameter for isocrinids is about 15 mm; the theoretical minimum is about 0.4 mm (Simms, 1989), but in practice few, if any, isocrinid species have a maximum adult stem diameter of much less than 3 mm. Cirral scar width may be less than 10 per cent of nodal diameter at the lower limit of its range but the maximum is geometrically constrained to about 65 per cent of nodal diameter in examples with five cirri per nodal. The earliest species of *Balanocrinus*, *B. quiaiosensis*, lies close to the minimum size of isocrinids, whereas the contemporaneous *Isocrinus*, with a stem diameter of more than 10 mm, lies towards the upper size limit. In the

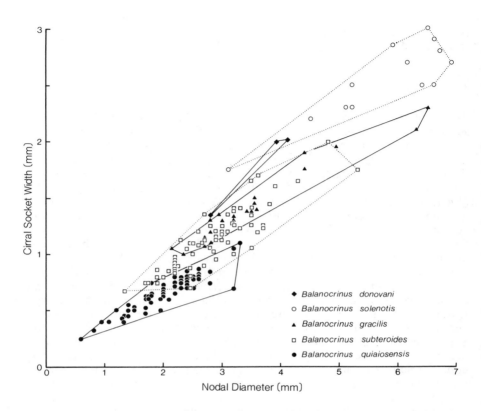

Figure 8.3. Bivariate plot for five species of *Balanocrinus* in the Early Jurassic, showing trends towards larger overall size and increase in relative size of the cirral scars through time. The stratigraphic sequence proceeds: from *quiaio-sensis* through *subteroides* through *gracilis* to *solenotis* and *donovani*.

evolution of *Balanocrinus* there is a progressive increase in maximum adult stem diameter, from less than 3 mm in the first species to more than 6 mm in the third. From the latter species, *B. gracilis*, arose *B. solenotis*, with a stem diameter of up to 6.5 mm, and *B. donovani*, reaching less than 4 mm in diameter. It should perhaps be stressed here that these relatively small differences in stem diameter reflect very considerable differences in the overall size of species and, perhaps more significantly, differences in tiering level (Ausich and Bottjer, 1985). The progressive increase in size of the first three species might be viewed as a non-random trend, though it might equally well be considered to represent essentially random shifts away from the ancestral form. Since the ancestor lies virtually at the minimum size limit, the observed trend may reflect selection pressure not so much for increasing size but just for different size. The appearance of the small *B. donovani* alongside the large *B. solenotis* perhaps indicates that a point had been reached at which the gap between the minimum size and the size of the ancestor was sufficient to allow random shifts towards a smaller size. *B. donovani* may have reoccupied the tiering level formerly occupied by small species of *Balanocrinus* but vacated as successively larger species evolved to occupy higher tiering levels.

Although the size increase observed in *Balanocrinus* may be due to random change, this does not seem to be the case for some of the other characters. The cirral scars of *B. quiaiosensis* are about 33 per cent of nodal diameter, similar to those of *Isocrinus* and significantly larger than in the contemporaneous genus *Hispidocrinus*. Although there is clearly scope for a decrease in cirral scar width, subsequent species show a progressive increase in width to 45 per cent in *B. solenotis* (Figure 8.3). They are even wider (49 per cent of nodal diameter) in *B. donovani*, probably reflecting the paedomorphic origin of this species. This trend towards larger cirral scars (Figure 8.3) may be viewed as a paedomorphocline resulting from selection pressure for more robust cirri to anchor successively larger species in the progressively higher-energy environments which they inhabited, the latter associated with an overall shallowing of the Liassic sea from the Sinemurian to the earliest Toarcian.

Similar unidirectional morphological trends have been observed in the contemporaneous genus *Isocrinus*. Some, such as the progressive decrease in length of brachitaxes in consecutive species from the Hettangian to the Domerian (Figure 8.2), are most readily interpreted as optimisation of the existing feeding strategy associated with constant selection pressure. Increasing the frequency of branching increases the efficiency of the filtration fan. Others, such as the progressive, though slight, decrease in width of cirral scars, are peramorphic traits contrary to those observed in *Balanocrinus* and have unclear evolutionary significance.

Evolutionary trends in obligate pseudoplankton: the role of extrinsic factors

The Pentacrinitidae were a low-diversity group of articulate (isocrinid) crinoids. At their peak of abundance, in the early Jurassic, they were almost exclusively pseudoplanktonic in habit and hence, with the possible exception of some Late Devonian melocrinitids attached to driftwood (McIntosh, 1978), are ecologically unique among crinoids. Furthermore, they are one of the few well-documented examples of obligate pseudoplankton known from the fossil record and the only group of pseudoplankton for which an evolutionary lineage has been reconstructed (Simms, 1988a; Wignall and Simms, 1990).

A pseudoplanktonic lifestyle imposes constraints on an organism's biology which differ significantly from those operating on either benthic or truly planktonic forms. Such constraints arise from the fact that the ultimate fate of the floating attachment site is independent of the life history of the pseudoplanktonic organism. Hence pseudoplankton must attain reproductive maturity rapidly before the attachment site sinks due to waterlogging and/or the increasing weight of the attached organisms. This problem is exacerbated by the relative scarcity of suitable floating substrata, necessitating production of large numbers of offspring to ensure the colonisation of new attachment sites (Wignall and Simms, 1990). Such constraints must have influenced the evolution of pseudoplanktonic taxa, necessitating the rapid attainment of sexual maturity and large size. Hence the pentacrinitid lineage not only is important for the study of crinoid evolution but also has implications for all pseudoplanktonic taxa. They provide the opportunity to compare the evolution of a pseudoplanktonic taxon with contemporaneous benthic ones and thereby assess the role of biotic and abiotic factors in bringing about evolutionary change.

The contemporaneous benthic crinoid clade, the Isocrinidae, diversified rapidly in the Early Jurassic from a single species (Simms, 1988a). Both *Isocrinus* and *Balanocrinus* show clear trends towards changing cirral scar size, nodal frequency, frequency of arm branching and overall size (Simms, 1988a). In most cases these changes are sufficiently great that species are easily distinguished from one another. Morphological change in these benthic taxa appears to correlate with changes in the benthic environment. In particular, the widespread development of benthic anoxia in the Early Toarcian caused a major turnover in benthic crinoid taxa and other groups (Hallam, 1986).

The earliest definite pentacrinitids are from the Upper Norian (uppermost Triassic). Their precise ancestry among earlier Isocrinida is unclear, but on morphological grounds *Pentacrinites* is considered the more primitive, later giving rise to *Seirocrinus* through peramorphosis (Simms, 1988a). The subsequent history of these two lineages is of low diversity, with only six described species (two spp. of *Seirocrinus* and four spp. of *Pentacrinites*), and extreme evolutionary conservatism. Morphologically, the earliest species differ very little from the latest. Both show a disjunct stratigraphic distribution with considerable gaps both within and between the known ranges of constituent species, which are generally longer than the ranges typical of Early Jurassic benthic isocrinids (Simms, 1988a). There is no evidence for more than a single species of each genus at any one time.

All pentacrinitids have an endotomous pattern of arm branching, unusual among Mesozoic crinoids. Such an arrangement is considered optimal for an efficient yet materially economical filtration fan (Cowen, 1981) and seems to have been essential to attain the rapid growth rates required by pseudo-plankton (Simms, 1986). The various species of pentacrinitid are distinguished largely on minor, though consistent, differences in features of the arms. The Late Triassic *Seirocrinus klikushini* Simms and early Jurassic *S. subangularis* (Miller) differ only in the slightly higher frequency of arm divisions and a slightly more complex pattern of branching in the latter species (Simms, 1988a). The first two species of *Pentacrinites*, *P. doreckae* Simms (Hettangian to Lower Sinemurian) and *P. fossilis* Blumenbach (Upper Sinemurian) are also virtually indistinguishable from each other but, in the slight increase in arm divisions, show remarkable parallelism with the changes seen in *Seirocrinus* (Figure 8.2). The most parsimonious explanation of these increases in branching frequency is as adaptations towards increased efficiency of the filtration fan, for which selection pressure may have been quite intense considering the biological constraints on pseudoplankton (Simms, 1986; Wignall and Simms, 1990).

The morphology of the two later species of *Pentacrinites*, *P. dichotomus* M'Coy (Carixian to Toarcian) and *P. dargniesi* (Terquem and Jourdy) (Aalenian to Bathonian) departs significantly from that of earlier species of *Pentacrinites*, and also from *Seirocrinus*. Both have a shorter stem, rarely more than twice the length of the arms, bristling with long, closely spaced cirri. Furthermore, branching frequency of the arms decreases (Figure 8.2) while syzygial articula, absent from the other four pentacrinitid species, are present at one or two points in the arms and thereby interrupt the otherwise complete pinnulation (Simms, 1986).

This apparent reversal of the trend towards optimum morphology between *P. fossilis* and *P. dichotomus* (Figure 8.2) suggests a fundamental change in the mode of life of pentacrinitids, more specifically a shift away from the obligate pseudoplanktonic habit which so constrained the earlier species. Limited support for this comes from taphonomic evidence. *Pentacrinites doreckae*, *P. fossilis* and *Seirocrinus subangularis* are characteristically found in black shale facies in close association with fossil driftwood (data for *S. klikushini* is not available) and fulfil all the criteria of obligate pseudo-plankton. In contrast *Pentacrinites dichotomus* and *P. dargniesi* have, to my knowledge, never been found with driftwood or any other floating object, though the possibility still remains that they exploited a substratum which is not preserved, such as floating algae. *P. dargniesi* and an unidentified Corallian form are found in high-energy carbonate or clastic facies, often as isolated ossicles in bioclastic debris beds, but are unknown from black shale facies such as occur widely in the Callovian. Such observations suggest a benthic mode of life for Aalenian to Corallian pentacrinitids. The ecological position of *P. dichotomus* remains enigmatic, however. Other than the absence of associated driftwood, *P. dichotomus* fulfils all of the taphonomic paradigms for a pseudoplanktonic crinoid, yet its morphology seems in-compatible with this habit. This species must therefore have become either truly planktonic, which seems unlikely, or else it utilised an unusually stable floating substrate, perhaps a *Sargassum*-like alga, which released it from some of the biological constraints suffered by earlier species.

The influence of abiotic factors on crinoid evolution is most clearly demonstrated by comparing the fate of the pentacrinitids, *Seirocrinus sub-angularis* and *Pentacrinites dichotomus*, with contemporaneous benthic iso-crinids between the Domerian and Toarcian. A major anoxic event in the Early Toarcian (tenuicostatum–falciferum Zone boundary) caused signifi-cant biotic turnover among a wide range of benthic and nektobenthic organisms but left the plankton virtually unaffected (Hallam, 1986). Benthic isocrinids showed a major change across this event whereas the two pseudo-planktonic taxa were unaffected, as might be expected from their position high in the water column (Simms, 1988a).

In conclusion, the evolution of obligate pseudoplankton appears to be characterised by long species ranges and only very minor changes between successive species. Any evolutionary changes which do occur appear to be confined to the fine-tuning of existing morphology. This contrasts with the evolution of benthic isocrinids, in which species ranges are generally shorter and morphological changes more pronounced and apparently influenced by extrinsic factors. The extreme evolutionary stability of pseudoplanktonic pentacrinitids indicates that selection pressure for the optimum morphology is much stronger than in benthic taxa. Since a variety of different selection pressures may operate on a benthic organism, evolutionary changes in morphology are more variable and significantly less conservative than those seen in pseudoplankton.

CONCLUSIONS

Crinoids, with their complex multielement skeletons, are an ideal group in which to document evolutionary trends. Many general evolutionary trends in particular crinoid characters have been identified but most, if not all, of these supposed trends appear to represent changes in variance. Real evolutionary changes can only be confidently identified at low taxonomic levels where there is thorough documentation. Consequently they are rare in the literature.

Initial diversification of clades occurs through expansion into juvenile morphotypes (paedomorphosis), often followed by optimisation of adult morphology. The latter appears often responsible for the generation of evolutionary trends, presumably under the influence of a continuing specific selection pressure, which can occur even within lineages where evolutionary change is punctuational and separated by long periods of evolutionary stasis. Extrinsic factors appear to play a major role in the generation of many evolutionary trends in benthic crinoids. In pseudoplanktonic crinoids evolutionary trends, if they occur at all, are driven by selection pressures common to all crinoids, such as improvement of the filtration fan.

ACKNOWLEDGEMENTS

I thank George Sevastopulo for useful discussion and for reading through the original manuscript.

REFERENCES

Ausich, W. and Bottjer, D., 1985, Phanerozoic tiering in suspension-feeding communities on soft substrata: implications for diversity. In J.W. Valentine (ed.), *Phanerozoic diversity patterns: Profiles in macroevolution*, Princeton University Press, Princeton, NJ: 255–74.

Ausich, W. and Lane, N.G., 1982, Evolution of arm structure in the early Mississippian crinoid *Cydrocrinus*. In J.M. Lawrence (ed.), *International Echinoderm Conference, Tampa Bay*, Balkema, Rotterdam: 139–43.

Broadhead, T.W., 1987, Heterochrony and the achievement of the multibrachiate grade in camerate crinoids, *Paleobiology*, **13**: 177–86.

Broadhead, T.W., 1988, The evolution of feeding structures in Palaeozoic crinoids. In C.R.C. Paul and A.B. Smith (eds), *Echinoderm phylogeny and evolutionary biology*, Oxford University Press, Oxford: 256–68.

Brower, J.C., 1973, Crinoids from the Girardeau Limestone (Ordovician), *Paleontographica Americana*, **7**: 263–499.

Brower, J.C., 1976, Evolution of the Melocrinitidae, *Thalassia Jugoslavica*, **12**: 41–9.

Brower, J.C., 1982, Phylogeny of primitive calceocrinids. In J. Sprinkle (ed.), Echinoderm faunas from the Bromide Formation (Middle Ordovician) of Oklahoma, *University of Kansas Paleontological Contributions, Monograph*, **1**: 90–110.

Cowen, R., 1981, Crinoid arms and banana plantations: an economic harvesting analogy, *Paleobiology*, **7**: 332–43.

Donovan, S.K., 1985, Biostratigraphy and evolution of crinoid columnals from the Ordovician of Britain. In B.F.Keegan and B.D.S. O'Connor (eds), *Proceedings of the 5th International Echinoderm Conference, Galway*, Balkema, Rotterdam: 19–24.

Eldredge, N. and Gould, S.J., 1972, Punctuated equilibria: an alternative to phyletic gradualism. In T.J.M. Schopf (ed.), *Models in Paleobiology*, Freeman, San Francisco: 85–115.

Fortey, R.A., 1985, Gradualism and punctuated equilibria as competing and complementary theories. In J.C.W. Cope and P.W. Skelton (eds), *Evolutionary case histories from the fossil record, Spec. Pap. in Palaeont.*, **33**: 17–28.

Gould, S.J., 1983, Phyletic size decrease in Hershey Bars. In S.J. Gould, *Hen's teeth and horse's toes*, Penguin, London.

Gould, S.J., 1988, Trends as changes in variance: A new slant on progress and directionality in evolution, *J. of Paleont.*, **62**: 319–29.

Hallam, A., 1986, The Pliensbachian and Tithonian extinction events, *Nature*, **319**: 765–8.

Lane, N.G., 1978, Evolution of flexible crinoids. In R.C. Moore and C. Teichert (eds), *Treatise on Invertebrate Paleontology, Part T, Echinodermata 2 (Crinoidea)*, Vol. 1, T301–2, Geological Society of America and University of Kansas Press, Lawrence.

Lane, N.G. and Strimple, H.L., 1978, Evolution of inadunate crinoids. In R.C. Moore and C. Teichert (eds), *Treatise on Invertebrate Paleontology, Part T, Echinodermata 2 (Crinoidea)*, Vol. 1, T292–301, Geological Society of America and University of Kansas, Lawrence.

Lane, N.G. and Webster, G.D., 1966, New Permian crinoid fauna from southern Nevada, *Univ. of Calif. Pub. Geo. Sc.*, **63**: 1–58.

McIntosh, G.C., 1978, Pseudoplanktonic crinoid colonies attached to Upper Devonian logs, *Geo. Soc. Am., Abst. Prog.*, **10** (7): 453.

McNamara, K.J., 1986, A guide to the nomenclature of heterochrony, *J. Paleont.*, **60**: 4–13.

Moore, R.C. and Laudon, L.R., 1943, Evolution and classification of Palaeozoic crinoids, *Geological Society of America Special Paper*, **46**: 1–153.

Raup, D.M. and Stanley, S.M., 1978, *Principles of paleontology*, Freeman, San Francisco.

Sieverts-Doreck, H., 1957, Bemerkungen über altpaläozoische Crinoiden aus Argentinien, *Neues Jahrbuch für Geologie, Paläontologie, no. 4, Monatschrift, Abt. B*, **1957**: 151–6.

Simms, M.J., 1985, The origin and early evolution of *Balanocrinus* (Crinoidea:

Articulata). In B.F. Keegan and B.D.S. O'Connor (eds), *Proceedings of the 5th International Echinoderm Conference, Galway*, Balkema, Rotterdam: 169.

Simms, M.J., 1986, Contrasting lifestyles in Lower Jurassic crinoids: A comparison of benthic and pseudopelagic Isocrinida, *Palaeontology*, **29**: 475–93.

Simms, M.J., 1988a, Patterns of evolution among Lower Jurassic crinoids, *Historical Biology*, **1**: 17–44.

Simms, M.J., 1988b, The phylogeny of post-Palaeozoic crinoids. In C.R.C. Paul and A.B. Smith (eds), *Echinoderm phylogeny and evolutionary biology*, Oxford University Press, Oxford: 269–84.

Simms, M.J., 1989, Columnal ontogeny in articulate crinoids and its implications for their phylogeny, *Lethaia*, **22**: 61–8.

Simms, M.J., (in press), The radiation of post-Palaeozoic echinoderms. In P.D. Taylor and G. Larwood (eds), *Major Evolutionary Radiations*, Systematics Association.

Ubaghs, G., 1953, Classe des Crinoides. In J. Piveteau (ed), *Traité de Paléontologie*, **3**: 658–773.

Ubaghs, G., 1978a, Skeletal morphology of fossil crinoids. In R.C. Moore and C. Teichert (eds), *Treatise on Invertebrate Paleontology, Part T, Echinodermata 2 (Crinoidea)*, Vol. 1, T58–216, Geological Society of America and University of Kansas, Lawrence.

Ubaghs, G., 1978b, Evolution of camerate crinoids. In R.C. Moore and C. Teichert (eds), *Treatise on Invertebrate Paleontology, Part T, Echinodermata 2 (Crinoidea)*, Vol. 1, T281–92, Geological Society of America and University of Kansas, Lawrence.

Webster, G.D. and Lane, N.G., 1967, Additional Permian crinoids from southern Nevada, *University of Kansas Paleontological Contributions*, **27**: 1–32.

Wells, J.W., 1941, Crinoids and *Callixylon*, *Am. J. Sc.*, **239**: 454–6.

Wignall, P.B. and Simms, M.J., 1990, Pseudoplankton, *Palaeontology*: **33**: 359–78.

Chapter 9

ECHINOIDS

K.J. McNamara

INTRODUCTION

Of all the groups of marine invertebrates, echinoids provide one of the best opportunities for investigating evolutionary trends. They have a rich fossil record, particularly in post-Palaeozoic strata, largely because of their rigid (in most cases) calcite endoskeleton and their mode of life. While the fossil record of regular echinoids (which are epifaunal) is, by and large, poor (Kier, 1977), irregular echinoids (which are often infaunal) are ideal candidates for fossilisation. Moreover, because of their use of external appendages for mobility and feeding, echinoid morphology is closely related to function and thus the environment (McKinney, 1988). Consequently, quite plausible interpretations of the functional significance of evolutionary trends can be made. This is further enhanced by their high speciation rates. For example, as of 1970, 3672 Mesozoic and 3250 Caenozoic species have been described (Smith, 1984).

Echinoids can also provide us with the opportunity of examining the nature and causes of evolutionary trends at two separate periods in the history of the group. It is generally accepted (Durham and Melville, 1957; Kier, 1974) that the mass extinction at the end of the Permian accounted for all but one lineage of echinoids, the sole survivor in the Early Triassic being *Miocidaris*. From this genus it is likely that all subsequent species evolved (Kier, 1974). We therefore have the intriguing situation of being able to compare and contrast the nature of evolutionary trends in two quite separate groups of echinoids at two distinct periods.

Any review of evolutionary trends in echinoids cannot ignore two major works that were published in the 1960s and 1970s: Porter Kier's analyses of evolutionary trends and their functional significance in Palaeozoic (Kier, 1965) and post-Palaeozoic (Kier, 1974) echinoids. But this chapter aims to achieve more than merely synthesise Kier's findings. He concentrated on the overall patterns of trends, and tried to explain these changes in purely functional terms. Since 1974 there has been an upsurge in interest in evolutionary trends in echinoids which has focused on lower taxonomic levels. Most of these studies have attempted to examine the nature of both inter- and intraspecific evolutionary trends (see, for example, McNamara, 1985; 1987a; 1988a; 1989; McNamara and Philip, 1980; 1984; Smith and Paul, 1985; McKinney, 1984). Moreover, in addition to explaining the changes in purely functional terms, other aspects have been examined, in particular the relationship between trends and life-history strategies (McKinney, 1986; Jablonski and Bottjer, 1988). The other emphasis in studies of evolutionary trends in the last decade has been the examination of the intrinsic factors that may have contributed towards the development of trends, in particular heterochrony (reviewed in McNamara, 1988a; 1989). So, in this chapter the broad-scale trends that Kier recorded will be reviewed in the context of small-scale trends, with the particular aim of trying to ascertain to what extent small-scale trends have influenced large-scale trends. Following an examination of the patterns of these trends, their functional significance will be reviewed, followed by the role that heterochrony has played in trend generation. Finally, a new factor in the equation will be explored: the process, or processes, that actually drive these trends. While there might be the intrinsic 'potentiality' (in the form of heterochrony) and the extrinsic facility (in the form of environmental polarities) to generate trends, what are the driving forces that initiate these trends and keep them going in particular directions?

PATTERNS OF EVOLUTIONARY TRENDS

Four aspects will be examined: trends in ambulacra; interambulacra; test shape; and test size. Evolutionary changes in the ambulacra reflect changes in the functioning of the water vascular system, because the external expression of this system is revealed by the pore pairs which pierce the ambulacral plates. Changes in the numbers of plates and in the numbers, distribution and form of pore pairs are often unidirectional within lineages and are likely to be attributable to selection pressure and consequent adaptation to changing environmental conditions. Evolutionary trends in the interambulacra involve not only the number of plates but also the structures that they carry, namely spines. Test shape is often constrained by extrinsic factors and provides a good indicator of changing functions and changing habitats. Studies of changing test size along lineages provide the opportunity of assessing the role of changing life-history strategies in determining the directionality of trends. It should not be thought, however, that selection pressures operate independently on these four major features of echinoid morphology. Far from it. There is often strong covariation between parts, and it is changes in degree of covariation that provide an insight into the intrinsic factors controlling morphological changes along the trends.

Evolutionary trends in ambulacra

In both Palaeozoic and post-Palaeozoic echinoids there are large-scale trends towards increasing both the numbers of ambulacral plates and the numbers of pore pairs. However, these two groups achieved this by quite different means. Palaeozoic echinoids did so principally by increasing the numbers of ambulacral columns. Rather surprisingly in some ways, the earliest echinoids, which occurred in the Late Ordovician, are closer to post-Palaeozoic echinoids in this regard than they are to other Palaeozoic echinoids. The Ordovician echinoids possessed just two columns of plates within each ambulacrum. Likewise, no post-Palaeozoic echinoid had more than two columns. Throughout the rest of the Palaeozoic there was a trend of increase in the number of columns within each ambulacrum. Thus, the Late Silurian *Echinocystites* and the Early Devonian *Rhenechinus* have four columns, while Carboniferous genera have no fewer than six columns, with a maximum reached by *Proterocidaris*, with 20! This trend does not only occur at higher taxonomic levels, but can be traced within single clades. The Devonian to Carboniferous *Lepidesthes* show that the earliest species have the least number of columns (eight), while the youngest attain 20. This trend arose from an increase in variance (Figure 9.1), as some of the later species also possess 'only' eight columns. The great plasticity in ambulacral column number in Palaeozoic echinoids can be interpreted as arising from an inherent flexibility in the developmental system. In echinoids, plate number increases through ontogeny. The generation of multiple columns arose by peramorphosis, and the great variability is indicative of poor developmental regulation (see below). Kier (1965) proposed an adaptive explanation for the trends of increase in numbers of ambulacra and, *ipso facto*, increase in number of tube feet. He interpreted these increases as reflecting advantages to descendant species in locomotion, food gathering and respiration.

The acquisition of more pore pairs in post-Palaeozoic echinoids was achieved by increasing the number of plates, rather than the number of columns. In order to accommodate large numbers of plates in the columns compound plates formed. These first evolved in the Late Triassic (Kier, 1974) when two elemental plates were covered by a single tubercle. The number of elemental plates within a compound plate increased through the Jurassic from three in the Early Jurassic to four by the Middle Jurassic. The mechanism that allowed this to occur is discussed below in the section on intrinsic mechanisms.

Interspecific changes in ambulacra have been commonly recorded, particularly in some irregular echinoids, such as spatangoids and clypeasteroids, where pore pairs and tube feet are differentiated morphologically and functionally in different areas of the test. Thus in spatangoids tube feet may be used solely for feeding (phyllodal), respiration (petaliferous) or sensory purposes (in ambulacrum III). Directional selection has, in many lineages,

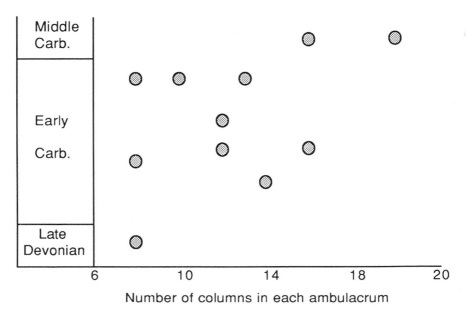

Figure 9.1. Increase in variance in number of ambulacral columns in Late Devonian to Middle Carboniferous species of *Lepidesthes* (data from Kier, 1965).

acted specifically on particular types of pore pairs—in other words, on only specific parts of the ambulacral column. For instance, the overall trend of reduction in pores in each pair from two to one, reported in the phyllodal pore pairs of some spatangoids by Kier (1974) and McNamara and Philip (1980), can be traced in single lineages. McNamara (1985) has shown how in the southern Australian spatangoid *Protenaster* the phyllodal pores change from paired in the Late Eocene *P. preaustralis* to single in later forms. This anagenetic trend involves the raised interporal partition on each ambulacral plate around the mouth in this species being breached in the Late Oligocene *P. philipi*. The Lower Miocene *P. antiaustralis* then shows the development of a curved ridge, where the two breached halves have regrown on one side of the pore. The living species, *P. australis*, continues this trend with this periporal ridge swelling. Such specific adaptations to feeding structures can be attributed to changing sources of food (see below). This reduction in number of pores on each ambulacral plate, albeit on localised areas of the test, is an interesting throwback to the ancestral echinoid condition. Just a single pore is present on each ambulacral plate in the Late Ordovocian *Aulechinus*. Kier (1965) considers this to be the ancestral condition. In the contemporaneous *Ectinechinus* the pores are double. In *Eothuria* multiple pores occur on each ambulacral plate.

Spatangoids also show changes in the number of respiratory pores in the petals, as well as changes in the depth of the petals. Numbers of pores and

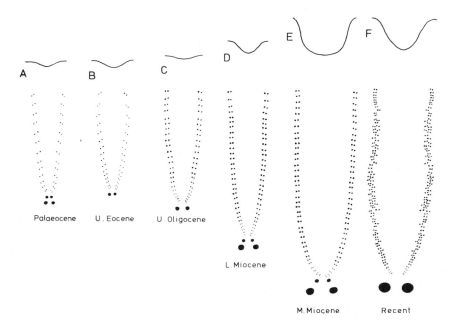

Figure 9.2. Evolutionary trend of increase in number of pore pairs in the anterior ambulacrum III in Australian Tertiary species of *Schizaster*. A: *S. (Paraster) carinatus* McNamara and Philip. B: *S. (Paraster) tatei* McNamara and Philip. C: *S. (Schizaster) halli* McNamara and Philip. D: *S. (Schizaster) abductus* Tate. E: *S. (Schizaster) sphenoides* Hall. F: *S. (Ova) portjacksonensis* McNamara and Philip. Reproduced from McNamara and Philip (1980), with permission of the Association of Australasian Palaeontologists.

lengths of petals can show either increases along lineages, such as in the *Paraster–Schizaster* lineage (Figures 9.2 and 9.7) described from the Tertiary of southern Australia (McNamara and Philip, 1980), or a decline, as occurs in a *Hemiaster* lineage (see Figure 9.9 below) described from the same strata (McNamara, 1987a). Being well adapted to inhabiting deep burrows in the sediment (Nichols, 1959), a number of spatangoid lineages show trends of increasing the length of aboral ambulacrum III. This is perhaps best exemplified in the *Paraster–Schizaster* lineage. From the Late Eocene *Paraster tatei* to some of the living forms of *Schizaster (Ova)*, ambulacrum III not only progressively lengthens (and the anterior notch deepens), but the density of pore pairs increases (Figure 9.2). With trends such as these, we are not seeing the anagenetic replacement of one morphotype by another. On the contrary, there is an overall cladogenetic increase in variance (see Chapter 2 herein), for the ancestral morphotypes that evolved in the Early Tertiary are still present today. However, Kier (1974) has noted how in Early Jurassic to Early Cretaceous holasteroids the trend of increased depth of the anterior

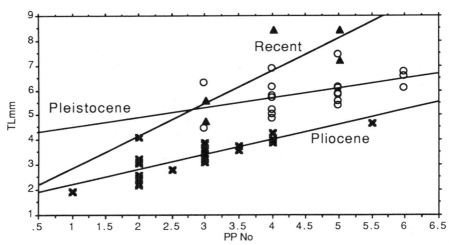

Figure 9.3. Intraspecific evolutionary trend showing increase in numbers of pore pairs in aboral ambulacra from the Pliocene to Recent in the fibulariid *Echinocyamus planissimus* from the Perth Basin, Western Australia. Late Pliocene form represented by crosses; Pleistocene form by open circles; and Recent form by triangles.

groove was not an increase in variance but an actual shift in morphology from 'no groove' through 'slight groove' to 'deep groove', the successive ancestral morphotypes disappearing.

Evolutionary trends in ambulacral characters can also be recognised intraspecifically. In a study of the tiny fibulariid *Echinocyamus planissimus* from Pliocene to Recent sediments in the Perth Basin in Western Australia, it has become apparent that a steady decrease in pore pair number can be traced even along a single species lineage. Pore pairs in the five dorsal petals increase in number during ontogeny (Figures 9.3 and 9.4). The earliest, Pliocene, members of the species have on average two pore pairs at a test length of 2.5 mm, increasing to six pore pairs at a test length of just over 4 mm. In Early Pleistocene forms only two pore pairs are present at 4 mm length, increasing to six at 6.5 mm. In living specimens only three are present at 5 mm, and a maximum of five at 8.5 mm. As the species shows a progressive increase in size along the lineage (maximum lengths: Pliocene, 5.5 mm; Pleistocene, 7 mm; Recent 10 mm), and as many other characters show no apparent morphological change along the lineage, the relative decrease in number of pore pairs along the lineage is most likely to be due to either a reduction in rate of pore-pair production (neoteny) or delayed onset of pore-pair development (postdisplacement).

Evolutionary trends in interambulacra

As with the ambulacral columns, there was a somewhat similar large-scale trend of increase in column number in the interambulacra of Palaeozoic echinoids, in all but the Cidaroida, where this trend was reversed. In the Echinocystitoida the numbers of columns produced were far fewer than the ambulacral columns. In the echinocystitoid *Proterocidaris*, where the test is little more than a mass of ambulacral plates (see Kier, 1965, Plate 60, Figure 1), the interambulacra comprise just a single, irregular row. Kier (1965) considered that the cidaroid *Polytaxicidaris*, with most interambulacral columns, was the most 'primitive', while *Miocidaris*, with just two columns per interambulacrum, was the most 'advanced'. The only consistent trend that Kier suggested could be determined in the interambulacra of Palaeozoic echinoids was an increase in regularity. The only significant trend in inter-ambulacrum number in post-Palaeozoic echinoids occurred in the Atelo-stomata. In spatangoids this is expressed primarily in shape changes, princi-pally involving an increase in the size of plates near the peristome (mouth), resulting in the formation of the plastron. This pair of large interambulacral plates carried spines adapted for burrowing and locomotion. Combined with other morphological changes this facilitated the adaptation of an infaunal habitat. In holasteroids plating of the plastron changed from protosternous, where the plates were paired, to meridosternous, shown, in genera such as *Echinocorys*, by a single row of interambulacral plates (see Smith, 1984, Figure 3.25).

More significant than plate size or number were changes in the spines that were attached to the interambulacral plates. While post-Palaeozoic cidaroids show little change in spine number, they did undergo an increase in diversity of shape. At the other extreme was the great reduction in spine size, combined with increased density of spines, in irregular echinoids. Moreover, as with the tube feet, in some groups of echinoids, most notably spatangoids and clypeasteroids, spines became differentiated both morphologically and functionally. In spatangoids some became adapted for defensive purposes, while others were used for locomotion or digging. One of the most notable developments in spine evolution was the reappearance of major defensive dorsal spines in spatangoids, such as brissids. In clypeasteroids spines differentiated into eight types (Ghiold, 1984, Table 2) that were used for defence, locomotion and feeding. The major trend apparent in spine differen-tiation in clypeasteroids was an increase in variance. Early forms, such as clypeasterids, fibulariids and laganids had only four types, while forms that evolved later, such as astriclypeids and mellitids, developed seven.

Selection pressure seems to have targeted spine density in many lineages and has formed characteristic evolutionary trends. At the interspecific level this can be shown in the Tertiary echinoid *Lovenia*. Three southern Australian species form an evolutionary lineage, from the Late Oligocene–Early Miocene *Lovenia forbesi*, through an undescribed Middle Miocene species, to the Late Miocene *Lovenia woodsi* (see Figure 9.9). Changes affecting the spines involve the number of dorsal interambulacral columns that actually carry the spines; the number of spines upon each of these

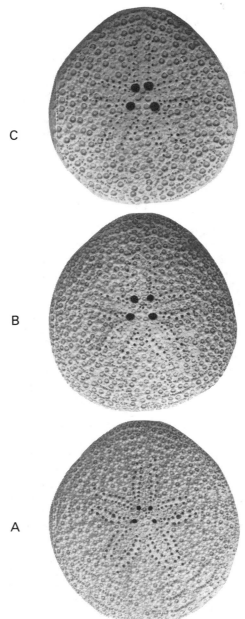

Figure 9.4. *Echinocyamus planissimus*. A: Specimen from the Pliocene Ascot beds. B: Specimen from the Early Pleistocene Jandakot beds. C: Recent specimen. All × 8. Evolutionary trends exist in decline in number of pore pairs, and in increase in tubercle size.

columns; and the number of spines carried on the ventrolateral inter-ambulacra. The numbers of spinose interambulacra decreases along the lineage, from eight columns in the oldest, through six in the intermediate, to only four in the youngest species (McNamara, 1989). The spine density within each of the ambulacra, however, varies quite independently, increasing between the first two species in the lineage, but then decreasing in the last. A clearer trend is apparent in the ventrolateral spines, which show a progressive increase in density along the lineage. Few other characters undergo much evolutionary change, suggesting that strong unidirectional selection pressure acted on two of the three factors (number of spinose interambulacra columns and density of ventrolateral spines). The functional significance and the nature of the selection pressure are discussed below.

In addition to unidirectional selection pressure producing evolutionary trends in spine density and distribution at the interspecific and suprageneric levels, it can also be detected intraspecifically. For instance, analysis of the Pliocene–Recent *Echinocyamus planissimus* lineage has revealed that the tubercles which supported the spines increased in width (the average width of the Pleistocene form being two-fifths as large again as the width of the Pliocene form; whereas the living form is half as wide again as the Pliocene form). As a consequence, tubercle concentration declined by almost half along the lineage (Figure 9.4).

Evolutionary trends in test shape

Both the earliest Palaeozoic and earliest post-Palaeozoic echinoids were essentially spherical in shape. As with other test parameters, large-scale evolutionary trends in test shape involve an increase in variance. This was greater in post-Palaeozoic echinoids than in Palaeozoic forms. Kier (1965) considered that Carboniferous and Permian genera such as *Proterocidaris*, *Pholidocidaris*, *Meekechinus* and *Pronechinus* developed very flattened tests, not unlike the clypeasteroids that evolved in the Tertiary. The evolution of irregular echinoids during the Jurassic saw a great increase in diversity of shapes, particularly the development of a flattened test (exemplified by the clypeasteroids—see Seilacher, 1979) and the well-known attainment of bilateral symmetry, as many taxa evolved elongate tests. The extreme development of this is seen in the extant deep sea pourtalesiid holasteroids, such as *Echinosigra* (David, 1989).

Tracing evolutionary trends in ambital outline of the test can also be accomplished at the specific level. For example, in the *Paraster–Schizaster* lineage the large-scale trend of test elongation has been demonstrated (McNamara and Philip, 1980). However, increased variance is shown by the reverse trend, documented in the lineage of *Hemiaster* in the Tertiary of southern Australia (McNamara, 1987a). Here the earliest species, the Late Eocene *H. subidus*, possessed a test width 90–96 per cent of test length. Through three intermediate Oligocene and Early Miocene species, the test width broadened to 97–103 per cent of test length in the Late Miocene *H. callidus*. The test margin can also show trends of increasing complexity.

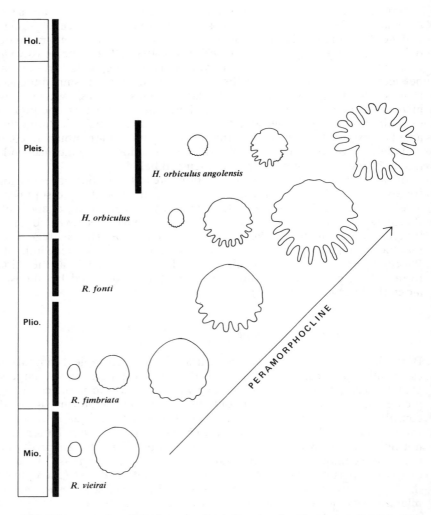

Figure 9.5. Peramorphocline in marginal lunule development in rotulids. Reproduced from McNamara (1988a).

In rotulid clypeasteroids the test evolved a denticulate posterior margin (Figure 9.5). This development can be traced in single lineages (McNamara, 1988a). For instance, in the Late Miocene *Rotuloidea vierirai* slight indentations are present in the posterior margin of adult tests (marginal lunules). In the succeeding Early Pliocene *R. fimbriata* nine more prominent indentations are present, whereas in the later Pliocene *R. fonti* the 11 deeper lunules occur. The descendant *Heliophora orbiculus* continues the trend and develops 13 even deeper lunules. One subspecies, *H. orbiculus angolensis*, carried this to an even greater extreme and evolved deep lunules around the entire margin of the test (McNamara, 1988a, Figure 12.4).

Test-height changes saw not only the clypeasteroid type of flattening, but also, in a number of lineages of cassiduloids, holasteroids and spatangoids, *increases* in test height. This is shown in the archiaciid cassiduloids *Archiacia* and *Claviaster* (Smith and Zaghbib-Turki, 1985; Zaghbib-Turki, 1989) whereby species such as *A. palmata* and *A. acuta*, which have slight raised apical areas, gave rise to *A. sandalina* and then *Claviaster libycus*. The most striking example in holasteroids is the evolution of species of *Hagenowia*, which possess an extremely elongate rostrum, from species of *Infulaster* (Gale and Smith, 1982). Not only did the rostrum progressively elongate along the lineage, but it also became more vertical. As is discussed below, such changes are thought to be intimately related to changes in feeding habits. A frequent change that occurred in spatangoid lineages was an increase in test height. This occurred either posteriorly or anteriorly. McKinney (1988) has noted how the increase in posterior test height in lineages such as *Paraster–Schizaster* resulted in the test attaining a wedge shape. This, he believes, resulted in improved mobility through the sediment, and was particularly useful in moving through fine-grained sediments. Progressive increase in anterior test height has been documented within a lineage of the spatangoid *Pericosmus* from the Tertiary of southern Australia (McNamara and Philip, 1984). This was achieved by increased swelling of the interambulacra close to the apical system. In the earliest species, the Late Oligocene *P. maccoyi*, there is no swelling and the test is low. In the earliest Miocene species, *P. compressus*, the posterior interambulacrum is swollen. In later Miocene species, such as *P. torus* and *P. quasimodo*, it is the anterior interambulacra that are swollen. Although there is the development of a similar wedge shape to that attained in the *Paraster–Schizaster* lineage (Figure 9.7), the wedge narrows posteriorly in the *Pericosmus* lineage, not anteriorly.

Evolutionary trends in test height can also be recognised intraspecifically. Smith and Paul (1985) have shown how the Early Cenomanian *Discoides subucula* varied in test height over an 11 m section. Early forms were conical, gradually being replaced by flatter forms mid-way up the section, then reverting to conical in the upper part of the section. Smith and Paul showed how these changes correlated directly with sediment grain size, conical forms dominating in finer-grained sediments.

Evolutionary trends in body size

The last group of evolutionary trends that I wish to illustrate are those involving changing size. In Chapter 4 herein it has been shown how in many groups of organisms there are temporal trends towards both increases and decreases in body size. In Palaeozoic echinoids Kier (1965) noted the trend for increased size, from the Ordovician genera that ranged between 25 and 40 mm in diameter, to forms such as the Carboniferous *Proterocidaris* which attained a diameter of 360 mm. While many lineages of post-Palaeozoic echinoids show similar trends, with some Tertiary genera, such as *Victoriaster*

(McNamara and Philip, 1984), attaining a length of 220 mm, there is again a trend of increased variance, with the evolution of a number of minute taxa. Notable among these are species of *Fibularia*, some of which are only 2–3 mm in length.

McKinney (1986) has documented a pattern of consistent trends of increase in size in 15 out of 17 species pairs of fossil echinoids from the Tertiary of south-eastern USA. This led McKinney to suggest that selection was acting not so much on morphological traits as on size, or life-history strategies related to larger size including aspects such as reproductive timing. For instance, delayed onset of maturity (hypermorphosis) will result in prolonged juvenile growth rate, and hence attainment of larger size. Because echinoid tests usually grow allometrically, increases in size will produce changes in morphology. McKinney has argued that such morphological changes are of secondary consequence to the increase in size, which itself is an indirect selection on particular life-history strategies.

A similar analysis carried out on Tertiary echinoids from southern Australia, however, shows a different picture. Of 32 species pairs, 16 show trends of size increase, 11 show size decrease, while five show no change at all. Where the two suites of echinoids are comparable is in the evolutionary trends in ecological strategies. In both cases there was a consistent trend for evolution to have been from shallow water to deep water habitats (as deduced from sedimentary characteristics). As discussed below, this widespread pattern of evolution from shallow to deep water is a major trend in echinoid evolution.

As with other evolutionary trends, large-scale patterns of evolutionary trends are mirrored at the intraspecific level. Although few studies at this level have been carried out, the *Echinocyamus planissimus* lineage shows a trend of increasing test size from the Pliocene to the present day (the living form attaining a test length twice that of the Pliocene form).

FUNCTIONAL SIGNIFICANCE OF EVOLUTIONARY TRENDS

Like all organisms, echinoids are fundamentally concerned with staying alive. Thus their prime concerns are with eating, reproducing and avoiding being eaten. For a given set of environmental conditions, those individuals that are better at any or all of these factors are the most likely to be selected for. To obtain food in particular habitats, morphological and size attributes that enable the organism to obtain food more efficiently, or that protect it from being eaten, are likely to be strong targets of selection, and so feature prominently in evolutionary trends. I would argue that most, if not all, of the evolutionary trends that I have documented in the preceeding section, and which occur in many other examples, are constrained by these factors. I will therefore outline evolutionary trends in echinoid ecological strategies and show how these were facilitated by the morphological changes.

The echinoid test is a structure that is built on a ground plan that, by modification, has allowed a wide range of marine habitats to be occupied. Because many of the physiological aspects of the organism are directly expressed in the external features of the test, the functional significance of evolutionary trends in these traits can be interpreted with reasonable confidence. Whether or not these functional changes were the prime targets of selection or were covarying with other changes, such as size or life-history strategies, is another question. There has been a tacit assumption in the past that morphological features were the prime targets of selection. This attitude has been brought into question in recent years (McKinney, 1986). The adaptive or non-adaptive nature of these changes will be discussed in the following section.

Evolution of epifaunal sediment feeding strategies

The most dramatic change in ecological strategies was the adaptation to feeding on and within unconsolidated sediment. Here is a major dichotomy between Palaeozoic and post-Palaeozoic echinoids, for it is most likely that all Palaeozoic echinoids were epibenthic, living mainly on hard substrates, and feeding by 'scraping'. The attainment of the ability to feed by sediment ingestion allowed the subsequent adoption of an infaunal habitat. This

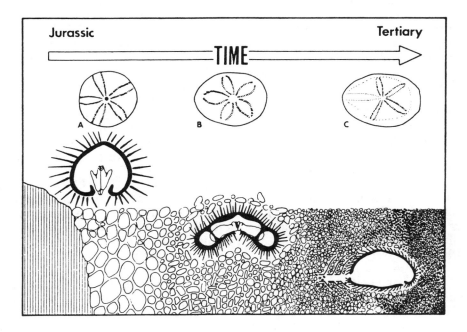

Figure 9.6. Diagrammatic illustration of the large-scale evolutionary trend from epifaunal, browsing regular echinoids, through shallow-burrowing, sand-dwelling echinoids to deeper-burrowing, mud-dwelling echinoids during the Tertiary. Persistence of ancestral ecomorphs illustrates increase in variance. Diagram reproduced by courtesy of Dr J. Ghiold.

crossing of a major adaptive threshold was the prime factor that stimulated the explosive radiation in post-Palaeozoic echinoids. For a group of marine organisms that for constructional and developmental reasons are not able to be pelagic, the only way to go was 'out' onto soft sediment, then 'down' by burrowing into the substrate (Figure 9.6). The adoption of shallow and deep burrowing habits in soft substrates resulted in a change in feeding habit to 'sediment swallowing' (Smith, 1984). This was accompanied by a change in test shape from circular to elongate, a shift in position of the periproct out of the apical system to the posterior, and anterior movement of the peristome. The differentiation of spines and tube feet also accompanied this transition, and many of the more subtle changes in these characters that can be traced as evolutionary trends in particular lineages, reflecting 'fine-tuning' to changing sediment grain sizes. The evolution of the irregular echinoids was poly-phyletic, occurring three times (Smith, 1984): in the Atelostomata (the orders Disasteroida, Holasteroida and Spatangoida), which lost the jaw apparatus and burrowed deeply in fine sediments; in the Eognathostomata (the orders Pygasteroida and Holectypoida); and the Neognathostomata (the orders Cassiduloida, Oligopygoida, Clypeasteroida and Neolampadoida), the last two of which retained the jaw apparatus and burrowed shallowly in sand.

Broad-scale trends of increasing variance in feeding strategies can be seen in clypeasteroids, with the development of food grooves. Branched food grooves evolved in at least three different lineages (Smith, 1984). In early forms, such as the Clypeasterinae, food grooves are straight and unbranched. They lack tube feet and pores as well as arched spines. With the evolution of branching, more tube feet were brought within the range of the food grooves, thus allowing more sediment to be processed. Smith (1984) suggested that this might have allowed the occupation of sediments of lower nutrient value. Within single interspecific lineages food-groove elongation can be traced. In a lineage of *Peronella* that ranges from the Pliocene to Recent in southern Australia, there was a progressive increase in food-groove length from the Pliocene form, where it is barely perceptible, through two inter-mediate species, to the living *P. tuberculata*, where the food groove extends more than half way to the ambitus (McNamara, 1988a). Likewise, in cassiduloids, changes in feeding strategies can be interpreted from changes to the bourrelets. These are swollen interambulacral areas surrounding the peristome that carry small spines used in feeding. A southern Australian fossil lineage extending from the Late Eocene *Echinolampas posterocrassus*, through two intermediate species, to the Early Miocene species *E. ovulum* shows a progressive swelling of the bourrelets.

A similar explanation can be invoked for the increased development of marginal lunules in the *Rotuloidea–Heliophora* lineage (see above). However, rather than just collecting food from the oral surface, this group and the arachnoidids also collected food from the aboral (ventral) surface (Ghiold, 1984). Small clypeasteroids, such as the fibulariid *Echinocyamus*, are epipsammic browsers. Using special buccal tube feet, the fragments are brought to the mouth, where the teeth scrape microorganisms off the surface (Ghiold, 1982). In the case of the *Protenaster* lineage, described in the preceeding section, evolutionary trends largely affected the phyllodal pore pairs, associated with tube feet used for feeding. These changes were brought about by the need to be able to feed from finer-grained sediments. The mucus-secreting phyllodal tube feet are known to collect sediment particles selectively from the substrate and pass them into the mouth (McNamara, 1985).

Evolution of infaunal sediment feeding strategies in spatangoids

Changing feeding strategies in burrowing spatangoids are inseparable from morphological changes that allowed the occupation of sediments of varying grain size. Thus in addition to trends in changing feeding strategies *per se*, many of the morphological changes can be interpreted in terms of coping with deep burrowing in sediments of different grain size. In particular, there is the need to bathe the test adequately in water for efficient respiration. While this is little problem in coarse-grained, permeable sediments, in finer-grained sediments, specific morphological innovations were necessary to allow this niche to be occuppied.

Table 9.1. *Morphological changes in some spatangoid lineages accompanying decreasing sediment grain size. Data are from McNamara and Philip (1980; 1984), McNamara (1985; 1987a; 1989) and Smith (1984)*

Genus	Test size	Test shape	Test height	Anterior groove depth	Anterior groove length	Petals	Peri-stome move	Fascioles
Micraster	increases	broadens	increases	increases	increases	deepen	anterior	broaden
Schizaster	increases	lengthens	increases	increases	increases	deepen	anterior	broaden
Hemiaster	no change	broadens	increases	increases	decreases	deepen	posterior	broaden
Psephoaster	no change	broadens	increases	no change	decreases	deepen	–	broaden
Pericosmus	increases	no change	increases	increases	increases	deepen	anterior	broaden
Protenaster	increases	no change	no change	no change	no change	variable	no change	broaden
Lovenia	no change	broadens	no change	increases	no change	no change	no change	broaden

No discussion of evolutionary trends in echinoids can ignore one of the most famous examples of an evolutionary trend ever described from the fossil record: the evolution of *Micraster* in the Upper Cretaceous of northern Europe. In Table 9.1 the major evolutionary trends in this lineage are summarised. These changes, involving increases in test size and height, anterior groove length and depth, broadening of the test and fascioles, deepening of the petals and anterior movement of the peristome, are mirrored in a number of spatangoid lineages that have been described from the Tertiary of southern Australia. While the morphological changes in the *Micraster* lineage have long been held to reflect increased depth of burrowing in the chalk sediment (Nichols, 1959), the same, or similar, changes seen in the *Schizaster–Paraster*, *Hemiaster*, *Psephoaster* and *Periscosmus* lineages

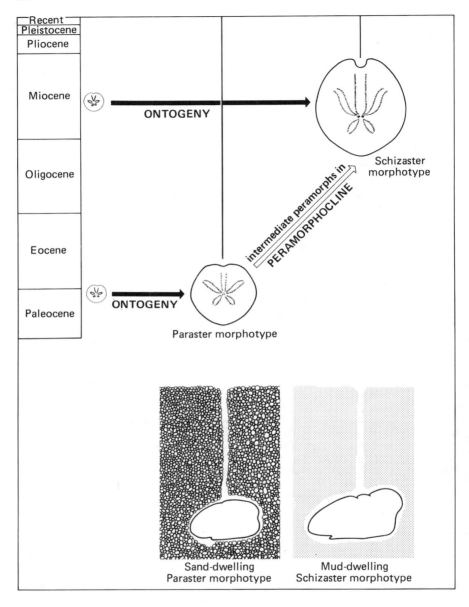

Figure 9.7. Evolutionary trends induced by peramorphosis in the Australian Tertiary heart urchin *Schizaster*, from the sand-dwelling *S. (Paraster)* to the mud-dwelling *S. (Schizaster)*. Note in particular the attainment of the wedge-shaped test profile; deep anterior notch; and longer respiratory tracts (the petals). Reproduced from McNamara (1982).

have all been attributed to adaptations to inhabiting progressively finer-grained sediments (McNamara and Philip, 1980; 1984; McNamara, 1985; 1989; see also Figures 9.7 and 9.8 herein). In a recent reappraisal of the *Micraster* lineage, Smith (1984) has shown that here also later species occupied finer-grained sediments than their progenitors, and that the morphological changes can likewise be interpreted as adaptations to inhabiting finer-grained sediments.

The evolution of deeper petals, lengthened and deeper anterior groove, higher test, anterior movement of the peristome (accompanied by an elongation of the labrum) all reflect adaptations to more efficient utilisation of a localised water source, drawn down a funnel through the otherwise increasingly poorly permeable sediment. An increase in tubercle density which also occurs in all these lineages resulted in the growth of more locomotory and digging spines, required in the finer-grained sediments. These evolutionary trends therefore reflect morphological gradients that parallel the environmental gradient of coarse- to fine-grained sediment. The difference between evolutionary trends in these genera and the *Protenaster* and *Lovenia* lineages is that these two genera are much shallower burrowers and were subjected to different selection pressures. In the case of *Lovenia* the dominant selection pressure was to avoid becoming subjected to excessive levels of predation. This is discussed further in the last section of this chapter.

Evolutionary trends in bathymetric distribution

As with many of the lineages that McKinney (1986) documented, these trends of adaptation to decreasing sediment grain size can also be interpreted as the occupation of deeper water habitats. These small-scale shallow to deeper water evolutionary trends reflect, to some degree, large-scale evolutionary trends in echinoid evolution. However, there is also evidence that there may be some discrepancies between perceived large-scale and small-scale patterns. Jablonski and Bottjer (1988) have analysed the onshore to offshore patterns in 13 orders of post-Palaeozoic echinoids (Table 9.2). They found that seven of these show a trend of migration from onshore to offshore environments. Five showed no pattern, while one, the Disasteroida, appears to show the reverse trend of offshore to onshore. Jablonski and Bottjer (1988) demonstrated that the migration offshore could happen two ways: by an 'expansion' of some forms into deeper water, but with others remaining in shallow water, so increasing the degree of ecological variance of the order; or by what they termed 'retreat' (but in the present work termed 'displacement'), whereby there was a loss of onshore representatives.

Even though these trends show a dominance of movement from onshore to offshore environments, there are cases where although at the ordinal level there is expansion, at the generic level there are many examples of displacement. Indeed, the situation could theoretically arise where 99 per cent of genera could show displacement, but the continued existence of 1 per cent of genera in the onshore environment would produce a pattern of expansion at the ordinal level. In the case of spatangoids, which at the ordinal level show

Table 9.2 *Onshore–offshore evolutionary trends in post-Palaeozoic echinoids. Data are from Jablonski and Bottjer (1988)*

Order	Direction	Method
Cidaroida	onshore to offshore	expansion
Echinothurioida	onshore to offshore	displacement*
Pedinoida	no pattern	
Hemicidaroida	no pattern	
Phymosomatoida	onshore to offshore	expansion
Temnopleuroida	no pattern	
Echinoida	onshore to offshore	probably expansion
Pygasteroida	no pattern	
Cassiduloida	no pattern	
Clypeasteroida	onshore to offshore	expansion
Disasteroida	offshore to onshore	expansion
Holasteroida	onshore to offshore	displacement
Spatangoida	onshore to offshore	expansion**

* Jablonski and Bottjer (1988) use the term 'retreat', but this is misleading as it implies a reversion to an environment previously occupied, which is not the case. The term 'displacement' is suggested as being more appropriate.
** As discussed in the text, at the generic level some genera show displacement.

expansion, there are examples of displacement at the generic level. For instance, *Hemiaster*, which in the Early Tertiary occurs in nearshore clastic environments and which in southern Australia shows a trend towards migration into finer sediments in deeper water (McNamara, 1987a), nowadays only occurs in depths of 450–3200 m, apart from a single specimen collected from 140–145 m (Mortensen, 1950). Likewise, species of *Pericosmus* were common in shallow water sediments off the coast of Australia in Miocene times (McNamara and Philip, 1984), but today are only found in water depths of 300–400 m (McNamara, 1984). It is clear that much more analysis needs to be undertaken to understand fully the nature of evolutionary trends in the bathymetric distribution of echinoids.

INTRINSIC FACTORS CONSTRAINING EVOLUTIONARY TRENDS

A common thread that runs through this volume is the part played by heterochrony in the generation of evolutionary trends. It has played a crucial role in trend formation in many groups of organisms, such as trilobites (Chapter 5), ammonoids (Chapter 7), crinoids (Chapter 8), fishes (Chapter 11) and amphibians (Chapter 12). Echinoids are no exception. Indeed, in recent years most studies of evolutionary trends in echinoids have focused on the role that heterochrony has played as a causal mechanism in trend generation (see McNamara, 1988b, for a review). In this section I will outline some of the main patterns that have been recorded and show how the correspondence between, on the one hand, morphological and size changes induced by heterochrony and, on the other, environmental gradients, has produced many of the evolutionary trends apparent in the fossil record. It is the juxtaposition of these two polarities, intrinsic trends in morphology and size and extrinsic trends in ecological strategies developed along environmental gradients, that has greatly influenced the patterns of evolution of echinoids at all taxonomic levels.

Plate growth and translocation

To understand fully the importance of heterochrony in the generation of evolutionary trends it is necessary first to analyse briefly the mechanism of plate growth in echinoids. Post-metamorphic growth of the echinoid test is two-tiered. New plates form at the apical system and are added throughout life, except in most spatangoids (Kier, 1974). Once incorporated into coronal columns, the plates undergo subsequent growth by peripheral accretion. In terms of an individual echinoid, the plates that comprise the test are of different relative ages — those closest to the apical system being the youngest, those closest to the peristome the oldest. Comparison of growth rates between plates, by the analysis of growth rings (Deutler, 1926), has shown that adjacent plates may have different growth rates. Thus on a single individual plate, size and shape may be very variable, depending on the relative age of the plate and the relative rates of plate growth. Variations to either of these parameters by heterochrony (i.e. changing onset, offset or rate of growth) can profoundly influence test shape. As each plate may be independently influenced by heterochrony, the potential variety of test shapes that could evolve is enormous. Futhermore, structures produced within plates, such as pore pairs, or upon the plates, such as tubercles and spines, are likewise affected by these changes. Even these structures may undergo heterochronic change. Furthermore, like structures on different parts of the test may be subjected to different heterochronic processes.

Although it has long been considered that ambulacral and interambulacral plates always remain in the same position relative to one another during ontogeny, I have argued elsewhere (McNamara, 1987b) that there is ample evidence, particularly in spatangoids and holasteroids (David, 1989), that growth of plates in these columns is dissociated, resulting in effective migration of plates in adjacent columns. This has been termed 'plate translocation' (McNamara, 1987b) and results not only in columns of plates 'sliding' past one another during growth, but also in plates from one column growing between plates of adjacent columns. Plate translocation in the apical system has been considered (McNamara, 1987b) to have been the principal mechanism that allowed the periproct to migrate out of the apical system in some lineages of Jurassic echinoids, and led to the evolution of irregular echinoids. Heterochronic changes to plate production and growth, facilitated by plate translocation, combined with the availability of suitable niches, were major factors in the evolution of a wide diversity of morphotypes in post-Palaeozoic echinoids.

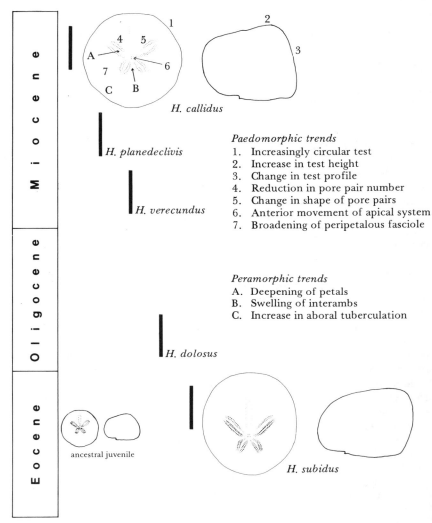

Figure 9.8. Paedomorphic and peramorphic evolutionary trends in a lineage of the heart urchin *Hemiaster* from the Tertiary of southern Australia. Successive species occupied progressively finer-grained sediments.

Effect of heterochrony on trends in test shape

The generally spherical nature of the tests of Palaeozoic echinoids and of many post-Palaeozoic regular echinoids has been attributed to the high rate of plate production, compared with individual plate growth. Consequently the growth gradient in each column was high (Raup, 1968), so that plates reached their maximum meridional size very early. The consequent high ambitus contributed to a near-spherical test shape (McNamara, 1988a). Trends towards a flattened test in some Palaeozoic and post-Palaeozoic lineages can be attributed to changes in plate growth rates. The evolution of complex intracolumnal plate allometries by heterochrony, with negative allometry in ambital plates and positive in many adoral and some aboral plates, produced a flattened test shape. Domed tests, however, evolved because of the evolution of different patterns of incremental growth. Smith and Paul (1985) suggested that the evolution of a more conical test in *Discoides* occurred in response to selection pressure for forms that were better able to maintain burrow walls in finer-grained sediments. A similar correlation between increased test height and the occupation of fine-grained sediments has also been documented in *Schizaster* (McNamara and Philip, 1980) (Figure 9.7), *Pericosmus* (McNamara and Philip, 1984), *Hemiaster* (Figure 9.8) and *Psephoaster* (McNamara, 1987a).

Many spatangoids evolved relatively large adoral plates (the plastron) because of the retention of high juvenile growth rates by these plates. In most echinoids this high juvenile rate diminished rapidly with age of the individual. Not so in spatangoids, where this high rate continued throughout growth. The importance of plate translocation was that it permitted such large plates to grow and 'slide' past adjacent smaller ambulacral plates (e.g., in *Breynia* — see McNamara, 1987b). The functional significance of such increases in plate allometry were that more burrowing and locomotory spines could be generated, so allowing such forms to burrow in the sediment more efficiently than forms with relatively smaller plastrons.

The frequent occurrence of lineages showing evolutionary trends towards increasing test height (e.g. *Hagenowia*, *Pericosmus*, *Schizaster*, *Discoides*) can be explained by peramorphosis of plate growth. Thus, for instance, juvenile *Pericosmus* have relatively low tests (McNamara and Philip, 1984), but along the lineage there is a progressive increase in adult height attained. In this character, therefore, a peramorphocline exists. The evolution of *Hagenowia* from *Infulaster*, involving elongation of the rostrum (Gale and Smith, 1982), occurred by an increase in positive allometric meridional growth of plates (acceleration), combined with changes to the timing of plate translocation (by predisplacement) in the apical system (McNamara, 1987b). Changing plate allometries by peramorphosis has also been the factor that resulted in the evolution of marginal lunules in rotulids (McNamara, 1988a). Once again, selection along an environmental gradient produced a trend in ecological strategies favouring peramorphic morphotypes. Whether lunule development improved feeding (Smith and Ghiold, 1982) or acted as an improved stabilising device (Telford, 1983) is the subject of some debate.

Effect of heterochrony on trends in test size

As discussed earlier, McKinney's (1986) observations on a suite of echinoid trends from the Tertiary of south-eastern USA indicated that selection may have been favouring aspects other than morphological characters. This he ascribed to the predominance of heterochronic processes, in particular hypermorphosis that favoured increased size. Because of the allometric nature of test growth, size increase was accompanied by morphological change. In another article, McKinney (1984) analysed a lineage of three species of the oligopygoid echinoid *Oligopygus* from the Late Eocene of Florida. Each succeeding species reached a larger maximum size than the preceeding species. Morphological differences between the three species could be interpreted as byproducts of the size increases arising from extension of the complex allometries by hypermorphosis. These morphological changes included a relative increase in peristome size, which McKinney interpreted as being necessary to allow the larger descendant forms to increase their food intake. However, the heterochronic changes are seen as selection not acting directly on such morphological traits, but acting on a range of characters and behaviours. As with many of the southern Australian lineages (see above), this lineage evolved into a deeper water (more stable) environment. Such environments are generally associated with *K*-type life-history strategies. This includes delayed maturity (produced by hypermorphosis) which results in larger body size. While larger body size may be of some selective advantage, particularly in regimes of high predation pressure (see next section), the dominant selective pressure may have been on reproductive timing, associated with the level of environmental stability. McKinney (1986) has suggested that similar selection pressures were operating on many of the other lineages of Tertiary echinoids from these same beds, which show increased size into deeper water environments.

While size increase (Cope's Rule) is a common trend in many lineages, progressive size decrease (induced by progenesis) along a paedomorphocline is rarely seen in echinoid lineages. This is not to say, however, that progenesis has not been a potent factor in echinoid evolution. It has probably been more significant at higher taxonomic levels. For instance, it has been suggested that the earliest clypeasteroids evolved by progenesis (Phelan, 1977). Likewise neolampadoids probably arose by progenesis. They therefore retained a small size, subspherical shape, and non-petaloid ambulacra carrying single rows of pores (Philip, 1963). Some of the tiny fibulariids may also be secondarily derived progenetic forms. Thus while post-Palaeozoic echinoids show a general increase in variance in test size, the dominant trend in most lineages that show any sort of size change is for test size to increase.

Dissociated heterochronoclines and the evolution of burrowing

As I have discussed, the ability of echinoids to burrow deeply, particularly into fine-grained sediments, necessitated the evolution of complex morphologies. Analysis of lineages such as the *Schizaster*, *Hemiaster* and *Pericosmus* lineages shows that similar morphologies that allowed the occupation of fine-grained sediments could be achieved in different ways. For *Schizaster* (Figure 9.7), all the morphological changes were caused by the operation of peramorphic processes: these include increasing petal and anterior furrow depth; increase in number of pore pairs in the anterior furrow; increase in posterior test height to produce a wedge-shaped test; increase in fasciole width; and increase in number of burrowing spines. The size increase along the peramorphocline indicates that hypermorphosis may have been one factor, but the proliferation of pore pairs in the anterior furrow indicates that acceleration also occurred. The pattern of this particular peramorphocline is a cladogenetic one, for ancestral morphologies still persist. The *Micraster* lineage also shows the development of a number of peramorphoclines, involving similar structural changes.

While the *Pericosmus* lineage likewise shows the development of peramorphic features, the peramorphoclines are anagenetic, rather than cladogenetic (see Chapter 2 herein). The *Hemiaster* lineage (Figure 9.8) also achieved the same end result of a morphology adapted to a fine-grained sediment, and like the *Pericosmus* lineage shows an anagenetic trend. However, it differs from the *Schizaster* and *Pericosmus* lineages in showing dissociated heterochrony: some characters evolved by peramorphosis, while others evolved by paedomorphosis. Most structures form paedomorphoclines involving, *inter alia*, changes in test outline, test height, petal length, number of pore pairs, pore shape, tuberculation within the petals and position of the apical system (McNamara, 1987a; 1989).

The morphologically plastic nature of the echinoid test has allowed suitable morphologies to evolve to cope with the occupation of an environment inimical to many organisms. While we talk of developmental 'constraint' as the intrinsic factor that channels evolution along certain environmental gradients, this provides a misleading picture of the importance of perturbations to the developmental system in generating such trends. Rather than constraining, in the negative sense of the word (see Gould, 1988), perturbations to the developmental system have opened up myriad opportunities for echinoid evolution. Although there is a degree of covariation of parts during development, it is the breakdown of this covariation in many groups of irregular echinoids that facilitated their evolution. Breakdown of the covariation laid the organisms open to dissociated heterochrony, with the consequent capacity to produce an almost endless variety of morphotypes. With such a wide diversity to choose from, it is little wonder that echinoids have adapted to such a wide diversity of marine environments.

FACTORS DIRECTING EVOLUTIONARY TRENDS

I have mentioned earlier how selection in echinoids appears to have focused on factors that enhanced the organism's capacity to feed or avoid being eaten. It is this latter aspect which is considered in this section, in particular the role of predation in directing evolutionary trends in echinoids. Recent work on the *Lovenia* lineage (McNamara, in prep.) indicates that gastropod predation

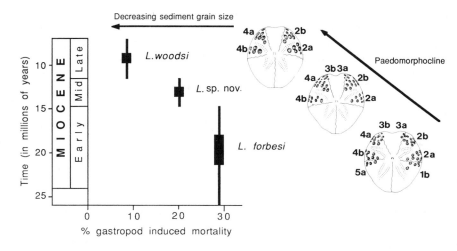

Figure 9.9. Percentage gastropod-induced mortality in the three Australian species of *Lovenia*, illustrating decrease through time, and with decreasing sediment grain size. Reconstructions illustrate the paedomorphic reduction in number of tuberculate interambulacral columns from eight to four.

pressure was the major factor directing the evolution of the lineage into deeper-water, fine-grained sediments through the Miocene. Predation pressure operates on individuals and is compounded to affect all taxonomic levels. I consider that it has been the prime agent determining the direction of many of the evolutionary trends documented in this chapter.

It has been well documented that living echinoids are preyed upon by a wide variety of predators, including other echinoids, asteroids, fishes, gastropods, crustaceans, birds, sea otters and even arctic foxes (Fell and Pawson, 1966). Only gastropods, which cut or drill holes in the test, leave a recognisable trace in the fossil record. Living echinoids are reported to suffer high levels of predation from cassid gastropods (Hughes and Hughes, 1981), which characteristically cut a disc from the echinoid test. The southern Australia *Lovenia* lineage shows a trend of reduction in levels of predation, attributed to cassids. Gastropod-induced mortality averaged 28.8 per cent in the Early Miocene *L. forbesi*, 20.2 per cent in the Middle Miocene *L.* sp. nov. and only 8.3 per cent in the Late Miocene *L. woodsi* (Figure 9.9). The influence of the primary spines in protecting the echinoid against gastropod predation is shown by the low frequency of predation holes in the region of the test covered by tubercles, and particularly immediately posterior in the area covered by the canopy of extended spines. These spines are known to serve a defensive function in living species of *Lovenia* (Ferber and Lawrence, 1976).

The oldest cassids occur in Late Eocene strata in the Northern Hemisphere. The earliest evidence for cassid predation on echinoids is on a specimen of this age form Cuba (Sohl, 1969). There is a strong correlation between the arrival of cassids in Australia in the Late Oligocene (Darragh, 1985), and the first evidence of such gastropod predation in this region. Many Miocene echinoid populations from the Australian Caenozoic show evidence of high levels of cassid predation and a rich cassid fauna occurs with these echinoids.

Predation is known to be a major agent of selection (Vermeij, 1982), which induces a wide variety of anti-predation responses in the prey. The concentration of defensive aboral spines is highest in the intermediate, Middle Miocene species of *Lovenia*, indicating that high predation pressure favoured the selection of morphotypes with greater concentrations of defensive spines. Intensification of predation pressure favoured selection of *L. woodsi*, a species capable of inhabiting a region of lower predation pressure. Species richness of cassids declines into deeper water (Abbott, 1968). The success of inhabiting a regime of lower level of predation pressure rather than increasing spine density is also reflected in the reduction in degree of successful predation, from 95 per cent in *L.* sp. nov. to only 70 per cent in *L. woodsi*.

A common pattern in many fossil groups, including some of the echinoids, such as this lineage of *Lovenia*, is one of anagenetic speciation: ancestral species in monophyletic groups failing to persist following the evolution of a descendant, even though their ancestral habitat remains. Heterochrony provides the 'internal' component in the generation of anagenetic patterns (see Chapter 3 herein). The driving force behind such trends may, as in the case of *Lovenia*, be predation pressure. Ancestral species incapable of withstanding strong predation pressure become extinct. Descendant species persist only as long as predation remains below a critical threshold level. With increased predation pressure only peripheral isolates with more effective antipredation strategies persist.

It has been suggested (Stanley, 1979) that the disappearance of Palaeozoic echinoids with flexible tests, and their replacement by forms with rigid tests, may have been partly influenced by predation pressure. Likewise, the great increase in diversity of infaunal echinoids, gastropods and bivalves since post-Palaeozoic times has been shown to correspond with an increase in diversity of predatory gastropods (Stanley, 1977). The adoption of a cryptic habit, by deep burrowing, and by displacement into deeper water niches in a number of Cretaceous echinoids (e.g., *Micraster*) may have arisen as a direct response to the adaptive radiation of predatory gastropods in the Cretaceous. Similarly, increased variance in size may also have been predation-driven—trends to both small and large size being anti-predation strategies, arising from positive selection for size classes outside of the predator's optimum prey-size range. It may be that the trends in onshore–offshore migration seen at the ordinal level merely reflect feedback from lower hierarchical levels of predation-driven trends into deeper water regimes of lower predation pressure.

The high morphological diversity of irregular echinoids during the Caenozoic has been attributed to the widespread occurrence of heterochrony. But extrinsic, as well as intrinsic, factors must have played a key part in evolutionary trends. Thus, while heterochrony provided the fuel for morphological, and thus ecological, change, the direction of the evolutionary trends was strongly influenced by predation pressure.

ACKNOWLEDGEMENT

I am grateful to Dr Joe Ghiold for kindly reading the manuscript of this chapter and offering useful suggestions for its improvement, and for allowing me to use his original of Figure 9.6.

REFERENCES

Abbott, R.T., 1968, *Indo-Pacific Mollusca*, Academy of Natural Sciences, Philadelphia.

Darragh, T.A., 1985, Molluscan biogeography and biostratigraphy of the Tertiary of southeastern Australia, *Alcheringa*, **9**: 83–116.

David, B., 1989, Jeu en mosaique des hétérochronies: variation et diversité chez les Pourtalesiidae (echinides abyssaux), *Geobios, mémoire spécial*, **12**: 115–31.

Deutler, F., 1926, Über das Wachstum des Seeigelskeletts, *Zool. Jb.*, **48**: 120–200.

Durham, J.W. and Melville, R.V., 1957, A classification of echinoids, *J. Paleont.*, **31**: 242–72.

Fell, H.B. and Pawson, D.L., 1966, General biology of echinoderms. In R.A. Boolootian (ed.), *Physiology of Echinodermata*, Wiley Interscience, New York.

Ferber, I. and Lawrence, J.M., 1976, Distribution, substratum preference and burrowing behaviour of *Lovenia elongata* (Gray) (Echinoidea: Spatangoida) in the Gulf of Elat (Aqaba), Red Sea, *J. exp. mar. Biol. Ecol.*, **22**: 207–25.

Gale, A.S. and Smith, A.B., 1982, The paleobiology of the Cretaceous irregular echinoids *Infulaster* and *Hagenowia*, *Palaeontology*, **25**: 11–42.

Ghiold, J., 1982, Observations on the clypeasteroid *Echinocyamus pusillus* (O.F. Müller), *J. exp. mar. Biol. Ecol.*, **61**: 57–74.

Ghiold, J., 1984, Adaptive shifts in clypeasteroid evolution—feeding strategies in the soft bottom realm, *N. Jb. Geol. Paläont. Abh.*, **169**: 41–73.

Gould, S.J., 1988, Trends as change in variance: a new slant on progress and directionality in evolution, *J. Paleont.*, **62**: 319–29.

Hughes, R.N. and Hughes, P.I., 1981, Morphological and behavioural aspects of feeding in the cassidae (Tonnacea, Mesogastropoda), *Malacologia*, **22**: 385–402.

Jablonski, D. and Bottjer, D.J., 1988, Onshore–offshore evolutionary patterns in post-Palaeozoic echinoderms: a preliminary analysis. In R.D. Burke, P.V. Mladenov, P. Lambert and R.L. Parsley (eds), *Echinoderm biology*, Balkema, Rotterdam: 81–90.

Kier, P.M., 1965, Evolutionary trends in Paleozoic echinoids, *J. Paleont.*, **39**: 436–65.

Kier, P.M., 1974, Evolutionary trends and their functional significance in the post-Paleozoic echinoids, *J. Paleont.*, **48** (suppl.): *Paleont. Soc. Mem.*, **5**: 1–95.

Kier, P.M., 1977, The poor fossil record of the regular echinoid, *J. Paleont.*, **3**: 168–74.

McKinney, M.L., 1984, Allometry and heterochrony in an Eocene echinoid lineage: morphological change as a byproduct of size selection, *Paleobiology*, **10**: 407–19.

McKinney, M.L., 1986, Ecological causation of heterochrony: a test and implications for evolutionary theory, *Paleobiology*, **12**: 282–9.

McKinney, M.L., 1988, Roles of allometry and ecology in echinoid evolution. In C.R.C. Paul and A.B. Smith (eds), *Echinoderm phylogeny and evolutionary biology*, Oxford University Press, Oxford: 165–73.

McNamara, K.J., 1982, Heterochrony and phylogenetic trends, *Paleobiology*, **8**: 130–42.

McNamara, K.J., 1984, Living Australian species of the echinoid *Pericosmus* (Spatangoida: Pericosmidae), *Rec. West. Aust. Mus.*, **11**: 87–100.

McNamara, K.J., 1985, Taxonomy and evolution of the Cainozoic spatangoid echinoid *Protenaster*, *Palaeontology*, **28**: 311–30.

McNamara, K.J., 1987a, Taxonomy, evolution, and functional morphology of southern Australian Tertiary hemiasterid echinoids, *Palaeontology*, **30**: 319–52.

McNamara, K.J., 1987b, Plate translocation in spatangoid echinoids: its morphological, functional and phylogenetic significance, *Paleobiology*, **13**: 312–25.

McNamara, K.J., 1988a, Heterochrony and the evolution of echinoids. In C.R.C. Paul and A.B. Smith (eds), *Echinoderm phylogeny and evolutionary biology*, Oxford University Press, Oxford: 149–63.

McNamara, K.J., 1988b, The abundance of heterochrony in the fossil record. In M.L. McKinney (ed.), *Heterochrony in evolution: a multidisciplinary approach*, Plenum, New York: 287–325.

McNamara, K.J., 1989, The role of heterochrony in the evolution of spatangoid echinoids, *Geobios, mémoire spécial*, **12**: 283–95.

McNamara, K.J. and Philip, G.M., 1980, Australian Tertiary schizasterid echinoids, *Alcheringa*, **4**: 47–65.

McNamara, K.J. and Philip, G.M., 1984, A revision of the spatangoid echinoid *Pericosmus* from the Tertiary of Australia, *Rec. West. Aust. Mus.*, **11**: 319–56.

Mortensen, T., 1950, *A monograph of the Echinoidea, Vol. 1, Spatangoida I*, Reitzel, Copenhagen.

Nichols, D., 1959, Changes in the chalk heart-urchin *Micraster* interpreted in relation to living forms, *Phil. Trans. R. Soc. Lond.*, Series B, **242**: 347–437.

Phelan, T.F., 1977, Comments on the water vascular system, food grooves, and ancestry of the clypeasteroid echinoids, *Bull. Mar. Sci.*, **27**: 400–22.

Philip, G.M., 1963, Two Australian Tertiary neolampadids, and the classification of cassiduloid echinoids, *Palaeontology*, **6**: 718–26.

Raup, D., 1968, Theoretical morphology of echinoid growth. In D.B. Macurda (ed.), *Paleobiological aspects of growth and development—a symposium*, Paleont. Soc. *Mem.*, **2**: 50–63.

Seilacher, A., 1979, Constructional morphology of sand dollars, *Paleobiology*, **5**: 191–221.

Smith, A.B., 1984, *Echinoid palaeobiology*, Allen & Unwin, London.

Smith, A.B. and Ghiold, J., 1982, Roles for holes in sand dollars (Echinoidea): a review of lunule function and evolution, *Paleobiology*, **8**: 242–53.

Smith, A.B. and Paul, C.R.C., 1985, Variation in the irregular echinoid *Discoides* during the early Cenomanian, *Spec. Pap. Palaeont.*, **33**: 29–37.

Smith, A.B. and Zaghbib-Turki, D., 1985, Les Archiaciidae (Cassiduloida Echinoidea) du Crétacé supérieur de Tunisie et leur mode de vie, *Ann. Paleont.*, **71**: 1–33.

Sohl, N.F., 1969, The fossil record of shell boring by snails, *Am. Zool.*, **9**: 725–34.

Stanley, S.M., 1977, Trends, rates, and patterns of evolution in the Bivalvia. In A. Hallam (ed.), *Patterns of evolution, as illustrated by the fossil record*, Elsevier, Amsterdam: 209–50.

Stanley, S.M., 1979, *Macroevolution: pattern and process*, Freeman, San Francisco.

Telford, M., 1983, An experimental analysis of lunule function in the sand dollar *Mellita quinquiesperforata*, *Mar. Biol.*, **76**: 125–34.

Vermeij, G.J., 1982, Unsuccessful predation and evolution, *Am. Nat.*, **120**: 701–20.

Zaghbib-Turki, D., 1989, Les Échinides indicateurs des paléoenvironnements: un exemple dans le Cénomanien de Tunisie, *Ann. Paleont.*, **75**: 63–81.

Chapter 10

BRYOZOANS

Robert L. Anstey

INTRODUCTION

The definition of 'evolutionary trend' is critical to a review of trends in bryozoan evolution. An evolutionary trend is a long-term evolutionary change in a given direction (Stanley, 1979). Therefore, it must not only be directional but also to be evolutionary, involve only homologous morphology within a lineage. Trends are usually analysed in the context of a phylogenetic tree, with morphological changes plotted against the phylogenetic pattern. Most of the classic trends in palaeontology involve sustained directional net change within a well-defined lineage in graded multistate characters, or morphoclines. Trends have been recognised in size, shape and complexity, in which many graded intermediate states exist between the extremes. The fewer the intermediates, the more difficult it is to recognise a trend. Finally, trends must be recognisable in the stratigraphic record, so that they can be observed over a well-defined temporal sequence.

Applying this definition strictly means that most of the reported 'trends' in bryozoan evolution fail as trends. Instead, most prove to be various evolutionary or ecological patterns within or among bryozoan clades, and are not gradients in homologous morphology. The definition used here excludes taxonomic trends that are not constrained within a single, evolving lineage. Non-trend patterns include: clade replacement (in which one group gains diversity as another loses it); delayed replacement, or radiation into an empty adaptive zone long after it has been vacated by an older clade; appearance or increase of character states within a lineage that are not part of a multistate morphocline; morphological grades that form a morphocline, but are polyphyletic, and hence convergent; displacements in habitat for taxa sharing

232

some uniform trait, such as a particular colony form, analogous to coelocanths 'moving' to deeper water (in a trend, the trait itself must change); and changes in relative abundance of taxa having a particular uniform trait (which may be caused by selection, and might even be related to the onset of a trend).

Other patterns known in bryozoans, however, fit the definition given above. They include gradients in biometric characters within lineages (usually species within genera), increasing convergence within a lineage in characters resembling other groups (such as generic characters within a suborder), and changes in colony form, colonial integration, and patterns of water exchange within lineages (usually species or genera within families). Some of these phylogenetic gradients represent modifications of parallel gradients in ancestral astogeny, and constitute colonial heterochrony (Anstey, 1987a). The types of trend clearly differ across levels of the taxonomic hierarchy.

MAJOR FEATURES OF BRYOZOAN EVOLUTION

Phylogenetic uncertainties plague the study of evolution in bryozoans, and account for substantial disagreement among bryozoologists. Conventional and phenetic systematics have produced at least four major phylogenies (McKinney and Jackson, 1989, Figure 2.11; Boardman *et al.* 1983, Figure 5), and molecular phylogenies have not yet been attempted. Even the latter may not satisfactorily resolve bryozoan relationships, because of homoplasy (Erwin, 1989), and absence of molecular information in 11 extinct suborders. No cladistic analysis of the major bryozoan groups has previously been published, and an alpha-level cladogram (Figure 10.1) is presented in this chapter to provide a tentative basis for discussing evolutionary trends in bryozoans. Trends cannot be analysed in a phylogenetic vacuum, and Figure 10.1 represents nothing more than a hypothesis of bryozoan relationships based on 54 two-state or additively coded multistate characters (Appendix 10.1). A wide variety of cladistic techniques, including several that assess levels of homoplasy, as well as alternative characters, await use by bryozoologists; it is clearly premature to state that bryozoan phylogeny is unresolvable using morphological characters (McKinney and Jackson, 1989, p. 51). In fact, the difficulties in bryozoan phylogeny suggested by Boardman (1984) present an excellent case for cladistic analysis.

The critical features of Figure 10.1 include the monophyletic character of the class Stenolaemata, the Palaeozoic free-walled Stenolaemata, the order Trepostomata, and the order Cheilostomata. Second, it suggests that the class Gymnolaemata and the orders Ctenostomata, Tubuliporata, Cryptostomata, and Cystoporata are paraphyletic. Third, it places the class Phylactolaemata (freshwater bryozoans) in between the gymnolaemates and stenolaemates (both marine). Fourth, it identifies most groups of extant bryozoans (except cheilostomes) as more plesiomorphous than the Palaeozoic free-walled stenolaemates, which form a cohesive, derived group. The cladogram does not 'prove' anything about the origin of the Mesozoic tubuliporates, but simply suggests that deriving them from any group other than the

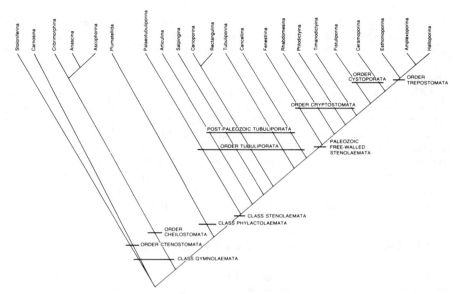

Figure 10.1.   Cladogram of bryozoan suborders, based on Wagner parsimony analysis of the characters in Appendix 10.1. The cladogram is a consensus of the two shortest trees, and was prepared by D.L. Swiderski. The length of the tree is 160 steps, and the consistency index is 0.438. The large number of reversals indicates considerably homoplasy in bryozoan evolution, and a high probability of altering this tree with the introduction of new characters.

palaeotubuliporines would be unparsimonious (see Boardman, 1984). Clade-diversity diagrams of family diversity within suborders have been plotted in cladogram order (Figure 10.2) with inferred patterns of descent, which provide an overview of bryozoan biostratigraphy, diversity and phylogeny. The reader is advised that numerous alternative views of bryozoan phylogeny exist, but no others have so far been based on a cladistic approach.

The major events in bryozoan evolution consist of two great radiations: the first in the Early Ordovician, establishing nine suborders of stenolaemates; the second in the Jurassic–Cretaceous, producing three suborders of cheilo-stomes (a new skeletal grade) and six of tubuliporates, although over a more protracted time interval. The stoloniferine ctenostomes, which appear to be the most primitive bryozoans, are unskeletonised, and appear in the fossil record only as pits or borings in calcareous shells. The bryozoan skeleton evolved twice, first in the stenolaemates during the Ordovician, and again in the cheilostomes in the Late Jurassic. No skeletal homologies exist, there-fore, between the two classes. Trends in the fossil record representing gradients in homologous skeletal morphology must be analysed separately within each class. In addition to major radiations, major extinctions periodic-ally removed bryozoan groups: four suborders during the Devonian to

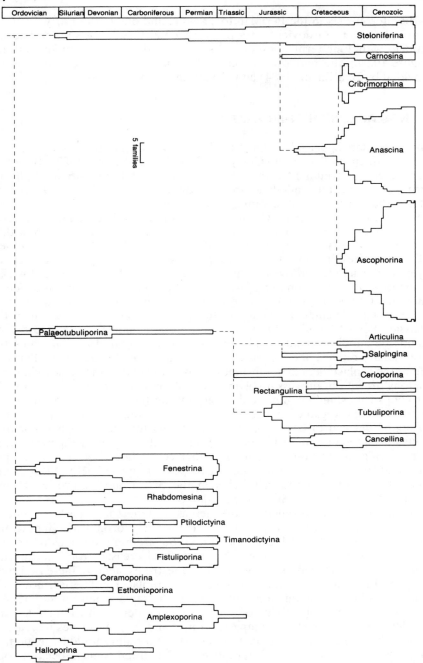

Figure 10.2. Clade diversity diagrams of bryozoan suborders, plotted in cladogram (Figure 10.1) order.

Carboniferous, six in the Permian to Triassic, and one in the Early Tertiary. During the Late Ordovician, 93 per cent of all bryozoan species became extinct, with long-term consequences in subordinal replacement patterns (Figure 10.3). Alternative views on the Permian extinction of bryozoans are provided by Boardman (1984) and Taylor and Larwood (1988).

TRENDS ACROSS HIGHER TAXA?

Nine suborders appear during the Early Ordovician (Arenig), but not in any known sequence (Taylor and Curry, 1985: Taylor and Cope, 1987). Based on morphological arguments, the fixed-walled Palaeotubuliporina might be ancestral to the free-walled suborders (Larwood and Taylor, 1979), but no intermediates are present in stratigraphic order.

Factor analysis of genera within suborders indicates that the ten Palaeozoic stenolaemate suborders belong to two temporal associations, each with a common diversity history (Figure 10.3). Factor 1 (palaeotubuliporines, esthonioporines, halloporines, ceramoporines and ptilodictyines) has its peak diversity during the Ordovician, and declines markedly after the Ordovician extinctions. Factor 2 (amplexoporines, fistuliporines, rhabdomesines, fenestrines and timanodictyines) peaks repeatedly in the Devonian, Carboniferous and Permian, declines drastically in the Late Permian, and dies out in the Triassic. Post-Palaeozoic bryozoans represent a single third factor, and cannot be subdivided into replacement groups. Figure 10.3 illustrates the evolutionary replacement of the Ordovician fauna by the Devonian–Permian fauna: the replacement is not instantaneous at the end of the Ordovician, but is a long drawn-out process involving most of the Palaeozoic. Factor 1 lost 28 per cent of its standing generic diversity during the Late Ordovician, while Factor 2 actually gained 3 per cent. Factor 1 groups declined even more (losing 62 per cent of standing diversity) during the Silurian while Factor 2 groups expanded (gaining 19 per cent), and Factor 2 groups expanded even more during the Devonian (gaining 24 per cent while Factor 1 lost 57 per cent) and Early Carboniferous (gaining 35 per cent while Factor 1 lost 25 per cent), The change in fortunes between the two groups, however, clearly began at the Ordovician–Silurian boundary.

The decline of trepostomes and the increase of fenestrines has been cited as a continuous replacement pattern over the Palaeozoic (McKinney and Jackson, 1989; Taylor and Larwood, 1988). Separating the trepostomes into suborders shows that the halloporines steadily decreased from the Silurian through to the Carboniferous, while the fenestrines steadily increased over the same interval. The halloporine–fenestrine replacement series, however, is not an isolated phenomenon, but parallels the replacement of all Factor 1 suborders by the collectivity of Factor 2 suborders, which includes the Amplexoporina. In fact, the amplexoporines and the fenestrines increase in tandem during the Mississippian and Permian. No claim should be made that long-term selection favoured fenestrines over trepostomes, because one trepostome suborder displays the same pattern of diversity increase as the fenestrines. The bad luck of the halloporines (and all of Factor 1) stems from

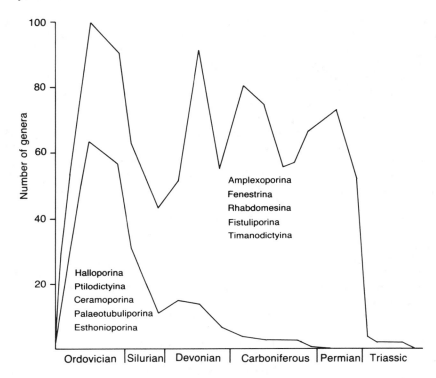

Figure 10.3. Diversity patterns of Palaeozoic bryozoan genera belonging to two temporal associations determined by cosine-theta factor analysis.

the low speciation rates of their post-Ordovician survivors, the differential survival of Factor 2 suborders into the Silurian, and their subsequent maintenance of higher speciation rates. Each free-walled order has suborders in both Factor 1 and Factor 2, and no differences can be claimed for the adaptive superiority of any ordinal *Bauplan*. Eight of the nine free-walled suborders, in fact, are present in the Early Ordovician, with no available evidence of trends in their pattern of origin. Replacement patterns are easily confused with trends,because one type of morphology is being ecologically replaced by another. For the replacement of halloporines by amplexoporines to be a trend, it would have to have involved the evolution of amplexoporine character states within the halloporines. Instead, it is a between-clade interaction in two suborders that have been separate since the Arenig. Even if there had been competitive displacement between two subclades of trepo-stomes, it would need to have been accompanied by sequential morphological changes within each group as the displacement escalated to fit the concept of trend.

A classic evolutionary trend in bryozoans concerns the degree of calcification of the frontal wall in the three cheilostome 'suborders' (Figure 10.1).

Ryland (1970, Figure 4) illustrates four evolutionary trends in the frontal wall evolution of cheilostomes. All originate from simple anascans with an uncalcified frontal wall. One leads to more complex anascans with an internal wall termed a 'cryptocyst', one to cribrimorphs with overarching spines, one to gymnocystidean ascophorans, with a vaulted roof over the frontal membrane, and, one to 'true' ascophorans, with a completely calcified frontal wall and internal ascus, or compensation sac, that is needed to extrude the lophophore. These different types of frontal shield have appeared many times in cheilostome evolution, and represent evolutionary grades (McKinney and Jackson, 1989). The 'suborders' are polyphyletic groups that taxonomically unite species with the same kind of frontal shield, because analysis of lower-level lineages has shown that closely related species and genera may have different types of frontal shield. Phylogenetic analysis of individual lineages is required to determine the existence of probably evolutionary trends from grade to grade. One trend may exist within the cribrimorphs, for example, because the oldest cribrimorphs, the myagromorphs, have only a few blunt spines around the edge of the frontal wall; younger cribrimorphs have fully extended and even laterally fused spines (Larwood, 1985).

Fully calcified frontal walls in ascophorans are supposed to provide increased protection from bryozoan predators. Diversity patterns, however, indicate no replacement of anascans by ascophorans (Figure 10.2). Instead, both appear to have diversified in tandem through the Late Cretaceous and Tertiary. This implies either that there was no great difference in adaptive 'success' (at least at the level of family diversity) or that 'success' is not simply equated with predation resistance.

Both 'classic' trends are between morphological grades which either appeared or diversified at nearly the same time. The apparent absence of trends across subordinal taxa may be related to polyphyly, or to their punctuated pattern of origins (Figures 10.2). If the suborders themselves are polyphyletic, then real trends may be buried within and between them. If they are monophyletic, and if their first appearances had been sequential or gradual in time, then character traits might have been strung out across taxa in some kind of directional pattern. Both polyphyly and penecontemporaneous origins make trends across suborders unobservable. Trends are more easily observed within lineages of lower-rank taxa.

Convergences in subordinal characters are common, and several suborders, in fact, are polyphyletic. Where convergences have been recognised, trends may exist within the lineage experiencing convergence. Examples include the introduction of trepostome-like characters into the rhabdomesines (Blake, 1980; 1983), cheilostome-like characters into the salpingines (Taylor, 1985), and fenestrate zoaria in both the salpingines and the cheilostomes (Taylor, 1987). Unrecognised polyphyly, however, prevents the recognition of trends, because such taxa have been artificially constructed and homogenised.

NON-TRENDS IN BRYOZOAN EVOLUTION

Delayed clade replacement is not an evolutionary trend, but represents reoccupation of a vacant habitat. Laminar colonies of Palaeozoic trepostomes and cystoporates living on unconsolidated sediments became extinct, and were eventually replaced in that habitat by specialised groups of cheilostomes in the Late Cretaceous and Eocene (McKinney and Jackson, 1989). Trends may exist within the cheilostomes to produce such taxa; in fact, substantial reduction of colony size, probably paedomorphic, has occurred in interstitial bryozoans (Håkansson and Winston, 1985). The delayed replacement pattern itself, however, is not a trend.

Of the eight temporal patterns in bryozoan evolution listed by McKinney and Jackson (1989, Table 10.1), only three fit the definition of evolutionary trend. These include changes in colony form among species of adeoniform cheilostomes, increasing integration of zooids in cheilostome colonies, and phylogenetic patterns of increasing fit of zooidal characters to the growth-form model of Coates and Jackson (1985). The two latter trends were regarded by McKinney and Jackson 1989, p. 212 as 'only suggestive, because sampling was arbitrary'. Several of these patterns are 'noisy', however, that is they are embedded within high levels of variance. The remaining temporal patterns include increases and decreases in diversity, abundance, and proportions of species, shifts of species habitat, and the appearance of innovative character states. Shifts in diversity, abundance and proportion do not necessarily produce any morphological changes within a lineage; instead, they just move around what is already there. If new types of horse had not appeared sequentially during the Tertiary, there would have been no trend in horse evolution. Geographic shifting in the proportions of deciduous and coniferous trees is likewise not an evolutionary trend, even though it may be an important ecological pattern. Changing proportions or diversity of species sharing a morphological state could represent a trend if the changes were caused by convergent evolution within lineages, leading to increases in a particular evolutionary product. Shifts in habitat are not trends, unless accompanied by progressive morphological changes. Lastly, the appearance of a novel trait, such as frontal budding in cheilostomes, even if it is useful and proliferates, is not a trend. A trend must not only incorporate direction, it must be temporally sustained over intermediate character states. Otherwise, every character-state change in the fossil record would be a trend, and the cladogram in Figure 10.1 would incorporate at least 54 forward 'trends', as well as a large number of reversals.

WHICH MORPHOCLINES ARE TRENDS?

Of the 54 characters used to construct Figure 10.1, only nine represent polarised multistate morphoclines in skeletal morphology (starred in Appendix 10.1). These characters, therefore, are potential predictors of significant evolutionary trends that might be observed in the fossil record. Seven of the nine are gradients of increasing complexity of colony form:

simple uniserial colonies must precede multiserial colonies, which, in turn, lead to larger and more complex multiserial colony forms. Accompanying this gradient are increasing co-ordination of zooidal feeding, increase in number of zooids in a colony, elevation of colonies above the substrate, partitioning of larger colonies into subcolonies, integration of the colonial skeleton, and addition of regular growth cycles in the accretion of large zoaria. The remaining two characters are gradients in both lunaria and acanthostyles, which run parallel, in part, to the main gradient. Despite the formal separation of characters and character states, what is really present is a gradient in colony size and complexity. In general (i.e. excluding the larger cheilostomes), the left-hand branches of Figure 10.1 represent smaller and simpler colony forms, and the right-hand branches represent larger and more complex ones. Unfortunately, the suborders do not appear in cladogram order in the fossil record (Figure 10.2). The Ordovician suborders do not appear in any known order (Taylor and Curry, 1985; Taylor and Cope, 1987), but seem to appear all at once. Therefore some potential evolutionary trends may simply be stratigraphically unresolved, or evolutionary rates were sufficiently high to preclude the observation of progressive evolution in the fossil record. The fossil record of Triassic and Jurassic bryozoans is very poorly known (Taylor and Larwood, 1988), and its resolution may likewise produce some graded progression between suborders. Although simple uniserial anascans precede more complex cheilostomes in the fossil record (Dzik, 1975: Cheetham and Cook, 1983), better phylogenies are needed to sort out evolutionary trends, which are confounded by the manifold convergences among the 'higher' cheilostomes.

In general, trends are not present across higher taxa of bryozoans, either because of poor stratigraphic resolution, polyphyly, paraphyly, or because the evolutionary processes producing higher taxa were not gradualistic and progressive. The major vector of bryozoan evolution, however, is an increase in the size and complexity of colonies. Gradients in colony size and complexity nevertheless appear repeatedly in lower-level bryozoan taxa, and constitute some of the best examples of trends in bryozoan evolution. One should consider whether such trends are caused by long-term selection, and hence constitute classic anagenesis, or are simply an artefact of diversification itself (Gould, 1988).

TRENDS CONSTRAINED BY HETEROCHRONY

Heterochrony provides one of the simplest ways of producing an evolutionary trend, because ontogeny (and astogeny) represent pre-existing multistate gradients in homologous morphology within a developmental sequence. Heterochrony does not automatically produce trends, however, because sustained directionality is also required, not just the pre-pattern of a morphocline. Heterochrony in bryozoan evolution has been reviewed by Anstey (1987a), where a number of evolutionary trends are shown to be a consequence of sustained modifications of ancestral astogeny. The reader is advised that confusion exists in the bryozoan literature on the definitions of astogeny and 'colonial ontogeny', both of which have been used interchangeably.

Heterochronic trends long recognised in the bryozoan literature include three examples of astogenetic peramorphosis: a decrease in budding angle and zooid contiguity in colonies of Mesozoic uniserial tubuliporines (Cumings, 1910), and multistate transformation series from the ancestrula to the mature zooids in several cheilostome lineages, with reductions in the zone of early astogeny (Levinsen, 1909; Harmer, 1923; Dzik, 1975; Zimmer and Woollacott, 1977). Some ascophorans even retain an anascan-like ancestrula, or *tata*, but in others it has been lost and early astogeny has accelerated to attain mature states precociously. Lastly, the evolution of free-walled stenolaemates involves a progressive restriction of fixed-walled growth from the entire colony down to just the ancestrula (Larwood and Taylor, 1979), and may have occurred twice, once during the Ordovician and again perhaps in the Triassic. Although some extant species are known that have a combination of free and fixed walls, such forms are presently unknown in the Ordovician and Triassic, and this probable trend remains cryptic as far as the fossil record is concerned.

Anstey (1987a, Tables 3–4) lists 40 examples of heterochrony in Palaeozoic stenolaemates. Of these, 17 represent evolutionary trends in multistate characters among species of particular genera, including transformations in biometric exozonal characters, monticule characters, axial ratios, and cystiphragm characters. Fourteen of the trends are paedomorphoclines, and only three peramorphoclines, suggesting that trends may arise more easily as reductions of ancestral astogeny. Paedomorphic products generally include smaller, less densely packed colonies, less heavily calcified, and sometimes with larger zooecia. The remaining 23 examples of heterochrony do not constitute trends, because they represent a transformation between only two taxa, or in only a two-state transformation series. Biometric characters measured on a continuous scale (Anstey and Bartley, 1984) provide the richest source of bryozoan heterochronic trends, because they tend to display directional variation across both astogeny and phylogeny, particularly along paedomorphoclines. Reductions in colony size and complexity are usually reflected in stereological measurements.

A variety of examples suggest that both onshore and low-diversity habitats favoured or induced paedomorphosis, even within species (Anstey, 1987a; Pachut, 1989). A larger number of generic phylogenies are needed to obtain a more realistic estimate of the frequency of paedomorphoclines. The modest number of available examples, however, suggests that generic paedomorphoclines are the most common type of evolutionary trend in the bryozoan fossil record.

INCREASING DEPTH OF TREPOSTOME ZOOIDS

The maximum depth of the living chamber in trepostome bryozoans can be measured as the distance from the colony surface down to the last-formed solid diaphragm (Figure 10.4). Basal diaphragms are usually absent in living

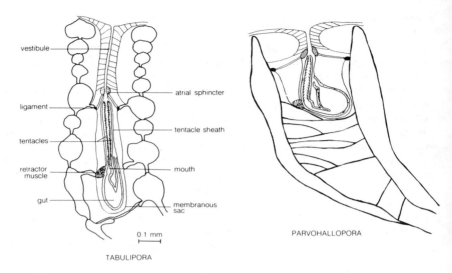

Figure 10.4. Reconstructed zooids of *Tabulipora* (left) and *Parvohallopora* (right), illustrating body depth and depth of vestibule. Reconstruction of *Tabulipora* was prepared by J.W. Bartley.

stenolaemates, which are very deep-bodied: an average zooid is a cylinder approximately eight times as deep as it is wide, and the vestibule, the space above the retracted lophophore, makes up about one-third of the total length of the zooid (McKinney and Boardman, 1985). The Late Palaeozoic genus *Tabulipora* (Figure 10.4) either lacks solid diaphragms, or has them so deep that a zooid of extant proportions can fit into the zooecium. Trepostomes which have long, cylindrical zooecia and lack solid diaphragms, or which have them at more than eight diameters, are here categorised as 'deep-bodied', indicating that their morphology is like that of extant, free-walled tubuliporates.

Surprisingly, Ordovician trepostomes generally have very shallow living chambers, only two to three diameters deep (usually less than 0.6 mm), and must have had much shorter to nearly non-existent vestibules. The chamber in *Parvohallopora* (Figure 10.4) is so cramped that the reconstruction has to shift the polypide sideways in the zooecium to keep a minimum vestibule and sufficient tentacle length to cover the zooecial diameter when extruded, analogous to the orientation of the polypides in box-like zooecia. Therefore there are two major grades of body depth in trepostomes: a deep-bodied type, like living stenolaemates; and a shallow-bodied type, unlike living stenolaemates. Tallying up the number of deep- and shallow-bodied genera (Table 10.1) shows that both grades are present within all three suborders and within 13 of the 18 families (all orders and suborders are shown in Figures 10.1 and 10.2). The halloporines and older amplexoporines are predominantly

Table 10.1 *Distribution of body depths in genera within trepostome families*

Suborder/family	Range	Shallow/ intermed.	Deep	% deep
I. Predominantly shallow-bodied families				
Esthonioporina:				
Orbiporidae	Ord.–Dev.	2	1	33
Halloporina:				
Dittoporidae	Ord.	2	1	33
Heterotrypidae	Ord.–Carb.	5	1	17
Halloporidae	Ord.–Sil.	4	0	0
Trematoporidae	Ord.–Carb.	8	1	11
Monticuliporidae	Ord.	5	0	0
Mesotrypidae	Ord.	2	0	0
Amplexoporina:				
Amplexoporidae	Ord.–Dev.	7	2	22
Atactotoechidae	Ord.–Carb.	10	0	0
Eridotrypellidae	Sil.–Carb.	7	0	0
Anisotrypidae	Sil.–Perm.	2	1	33
Aisenvergiidae	Dev.–Carb.	2	1	33
II. Predominantly deep-bodied families				
Esthonioporina:				
Esthonioporidae	Ord.	0	2	100
Amplexoporina:				
Crustoporidae	Sil.–Perm.	1	3	75
Dyscritellidae	Dev.–Trias.	2	3	60
Stenoporidae	Dev.–Perm.	2	8	80
Ulrichotrypellidae	Dev.–Perm.	2	3	60
Araxoporidae	Perm.	0	1	100

Table 10.2 *Measured depths of shallow- and intermediate-bodied trepostomes having living chambers floored by solid diaphragms, and ratios of depth to diameter*

	Ordovician	Silurian– Devonian	Carboniferous- Permian
Sample size	37	20	7
Mean depth (mm)	0.50	0.52	0.66
Standard error (mm)	0.04	0.04	0.11
Mean ratio	2.36	2.54	3.12
Standard error	0.16	0.24	0.50

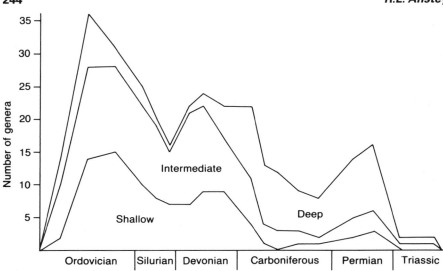

Figure 10.5. Diversity patterns of deep-, intermediate- and shallow-bodied trepostome bryozoan genera. Shallow is defined as less than 0.50 mm, and intermediate as 0.50–0.75 mm. Deep is defined as greater than 0.75 mm, or as having no solid diaphragms within eight zooecial diameters of the zoarial surface.

shallow-bodied. Furthermore, the deep-bodied forms increased in number throughout the Palaeozoic (Figure 10.5), and even the depth of shallow and intermediate forms increased throughout the Palaeozoic (Table 10.2).

Deep-bodied zooids have evolved, usually at the generic level, within almost three-quarters of the families. Therefore, like the frontal shields in cheilostomes, these grades are highly polyphyletic. Temporal polarities and frequency suggest that, in general, the shallow grade is primitive. Intermediate depths also exist (Figure 10.5), indicating the existence of multistate morphoclines. Although six Ordovician families each spawned at least one deep-bodied genus, the frequency of these transitions did not step up until the Carboniferous (Figure 10.5), when most of the new genera appearing were deep-bodied. No comparable replacement pattern is known between anascan and ascophoran cheilostomes. Increasing body depths may have been a response to predation, and bryozoan predators may have increased in the Middle Palaeozoic along with durophagous predators (Signor and Brett, 1984), which generated compensatory changes among molluscan, brachiopod and crinoid prey.

This evolutionary trend is not a simple case of clade replacement, because it occurs within suborders and within families. Similar patterns exist within lineages for cheilostome frontal shields (Banta and Wass, 1979). It is not a trend in the Halloporina or the Esthonioporina, each of which has only three deep-bodied genera. It is a trend within the Amplexoporina, which has 35 shallow and intermediate genera, and 22 deep-bodied genera (Table 10.1). Although shallow and intermediate-depth zooids extend through the Permian (Figure 10.5), there is clearly a change in the frequency of generic origins, and not just a change in variance (Gould, 1988). Furthermore, this is not a passive trend, but one requiring more calcification of the zooecium, and increasing difficulty of lophophore extrusion, analogous again to the transition from anascans to ascophorans. Both represent energetic costs to the animal, and suggest that some compensatory value was received in exchange. Finally, it parallels changes in predatory defences in Middle Palaeozoic molluscs, brachiopods, and crinoids. When free-walled stenolaemates reappeared in the Mesozoic Tubuliporata, they were apparently deep-bodied at the outset (although the Triassic record is in doubt), and have remained so to the present.

Increasing body depth in trepostomes may also represent a paedomorpho-cline. In shallow-bodied trepostomes with closely spaced solid diaphragms, diaphragms frequently become more closely spaced nearer the zoarial surface, suggesting that exozonal astogeny (ontogeny of some authors) includes a progressive shortening of the living chamber, or that growth cycles exist in diaphragm spacing and body depth. Therefore, increasing body depth may be a regression to earlier stages of colony growth, and could represent paedomorphosis in derived species.

TRENDS IN COLONY FORM

Many examples of bryozoan trends involve changes in growth form, usually from simple to complex, and from low colonies with few zooids to tall colonies with many zooids. The changes in architecture are like the evolution of buildings in human cities. Frontier settlements have simple, one-storey buildings. Towns and villages appear later with two and three-storey structures. When lateral space is fixed, the buildings grow upwards, and become architecturally complex, with new kinds of internal support, strengthening features, and other ways of stacking up usable space in three dimensions. Competition for space may exist between different kinds of building, and usually the larger ones win and displace the smaller ones. When cities are destroyed by disasters, they are often built up again along a parallel course of development, and recovery time depends on the depth of the devastation. After attacks, they may build up defences or defensive strategies. One may ask if this trend in human cities is driven by external selection, or is simply a passive consequence of allowing populations to grow and crowd modular units into a fixed amount of space.

Arenigian trepostomes (Taylor and Curry, 1985; Taylor and Cope, 1987) were predominantly encrusting and hemispherical species. Erect branching forms became more common in the Llanvirnian, and were joined by frondescent, bilaminate, and cribrate zoaria by the Caradocian. Similarly, the oldest ptilodictyines (bifoliate cryptostomes) were simple, explanate fronds, and more complex forms appeared in due course (Karklins, 1983); the same pattern occurred in the evolution of bifoliate fistuliporines (Utgaard, 1983).

The oldest cheilostomes, ctenostomes and tubuliporates are all simple uni-serial colonies, and in all three orders colony form diversified and became more complex as the lineages expanded. Within the fistuliporines, a greater diversity of colony forms was present after the Silurian (Anstey, 1987b). Lineages in which water-flow patterns can be mapped (Anstey 1981; 1987a; 1987b) also increased their variety of functional patterns over time. The rhabdomesines became larger in the Late Palaeozoic, and in so doing, developed a variety of convergences with trepostomes (Blake, 1980; 1983). Therefore most lineages began with simple colony forms, and increased the variety of growth forms as they diversified.

Do these trends in colony form represent the product of long-term selection? In terms of competition for space, Taylor (1984) found no differences between Silurian and Recent bryozoans in the ways in which species utilized and competed for substrate space. Both Taylor (1984) and Liddell and Brett (1982) found that simple, fixed-walled palaeotubuliporines were less successful in spatial competition than more complex, free-walled encrusting species. In successions on Ordovician cobbles, Wilson (1985) found that uniserial palaeotubuliporines were always overgrown by more complex cystoporates and trepostomes. Ordovician palaeotubuliporines, however, were more common in onshore, shallow water habitats from which more complex bryozoans may have been excluded (Anstey, 1986). Therefore, selection for spatial competition mechanisms may have been a factor early in the Ordovician, but once the standard ways of competing for and utilising space were achieved, very little subsequent long-term change can be attributed to such selection. Spatial competition may have been relaxed in low-diversity settings or stressful habitats, and gradients in spatial competi-tion may simply be diversity gradients. Trends away from simple uniserial colonies may have followed gradients for competitive interaction among encrusting species, and have been tracked in virtually all major bryozoan groups. Unfortunately, these transitions are embedded in the major radia-tions, and may not be temporally resolvable in the stratigraphic record.

Other than spatial competition, the only kind of selection that can really be examined in the fossil record is response to predation. The trend in body depth in trepostomes fits a predation model, and was probably a consequence of long-term selection, punctuated somewhat by the Givetian and Frasnian (Middle to Late Devonian) bryozoan extinctions. Trends in cheilostome frontal shields might also be predation responses, but are confused by homoplasy. The only known or suspected Recent bryozoan predators are pycnogonids, nudibranchs, and syllid polychaetes (Winston, 1986), none of which are skeletonised. Additional defensive trends might include increases in avicularia, which are known to deter syllids. Chemical defences, present in some living bryozoans, are obviously unpreservable. Trends in probable predation responses in the skeleton are the only patterns in the bryozoan fossil record that provide potential evidence of sustained selection over very long periods of time.

In some lineages, colony growth form is facultative, and may be a direct growth response to signals present in the physical environment (see, for example, Stach, 1936; Harmelin, 1975; Winston, 1976; McKinney and

Jackson 1989). Astogeny of a complex growth form, in fact, includes many intermediate stages, any of which can be fixed as the terminal stage of growth in a particular habitat (Anstey, 1987a). In many taxa, small pieces of colonies broken off or toppled by waves or storms may regenerate, and even grow into shapes that differ from the parent colony (see, for example, Blake, 1976; Hickey, 1988). Trends of increasing regeneration ability may be present in the fossil record, but are presently known only as being present or absent in isolated taxa, without intermediate states or a graded morphocline. Also, simple repair of injuries has to be sorted out from the growth-form plasticity frequently accompanying it. High levels of plasticity may be part of the life-history parameters of an opportunistic species, which can 'adapt' almost instantaneously to any kind of substrate, water depth or water movement. Plasticity seems to be a primitive state within many lineages, and is damped out as species acquire fixed growth forms and narrower ecological require-ments. Therefore, either the 'forward' or 'reverse' trend in colony size and complexity may involve some unknown degree of non-heritable plasticity, and represent only an ecological pattern and not evolution. Variance partitioning (Pachut, 1989) provides a technique of assessing heritability in fossil bryozoans, and separating evolution from non-heritable plasticity.

The role of simple geometric constraints has to be considered in trends in zooidal packing in bryozoans, leading to close-packed cerioid or 'honeycomb' colonies of closely appressed zooids, such as the trepostome *Amplexopora filiasa* (Anstey *et al.* 1976: Anstey, 1978; Anstey and Pachut, 1980; Anstey and Bartley, 1984). Hexagonal packing is a simple mechanical process that occurs in a wide variety of organic and inorganic systems, suggesting that it is just the normal way that colonies are formed. Is there a mechanism that 'pushes' zooids together until they end up in a hexagonal mosaic, or is there the opposite, some mechanism that forces them apart into non-cerioid colonies? Figure 10.1 suggests that non-cerioids are primitive, and that cerioids are derived (most of the free-walled suborders, except the fenes-trines, include close-packed colonies). Therefore the crowding of zooids into cerioid colonies is not the normal or primitive pattern, but indicates that evolution has acted to overcome the primitive constraint of zooids being widely spaced and loosely connected. Therefore forward gradients within lineages are overcoming this constraint, and the reverse gradients (paedo-morphoclines) may be passive, just 'letting it happen'. The high frequency of paedomorphoclines may reflect the passive character of this kind of trend. Some may even be just patterns of non-heritable plasticity, and not trends at all. Others may involve heterochronic changes, but 'easy' ones, minimal genetic changes producing global effects in the colonial phenotype. Among heterochronic phenomena, paedomorphosis alone simply requires a change in the rates or timing of development. Peramorphosis, requiring terminal additions to either ontogeny or astogeny, may not be as 'easy' to achieve, and consequently is a rarer pattern, or one primarily associated with the great radiations that generated *Baupläne*.

In bryozoans, major radiations have produced most of the morphological variance in bryozoan lineages, and not post-radiation 'stately progressions'. Gould's (1988) argument about increasing variance in lineages applies especially well to the Ordovician radiation of bryozoans, an event which seems to be over within just a stratigraphic moment. Likewise, the analogy with human architecture and Cope's Rule, suggesting a downhill flow towards increased size and complexity, and just filling unused ecological space, is really a characterisation of what happens in a radiation. The Ordovician radiation may never be stratigraphically resolved to everyone's satisfaction, and its evolutionary phenomena may remain cryptic. Radiations or origination events may involve a very different set of evolutionary phenomena from those that drive trends.

CONCLUSIONS

Evolutionary trends in bryozoans are difficult to decipher, first because 'trend' has been used in the non-technical sense of 'pattern' in important literature sources, and second, because bryozoologists have not favoured cladistic approaches to phylogeny, and assume that the failure of phenetic and conventional phylogenies is a failure of morphological phylogeny in general. Admittedly, homoplasy is a serious problem in bryozoan phylogeny, but many techniques exist that are capable of analysing it. Without good phylogenies, trends cannot really be recognised. 'Classic' trends in bryozoans involve morphoclines among polyphyletic grades. Real trends may be embedded in these, but they are not really known.

Clade replacements have been heavily confused with evolutionary trends. Simply put, trends are directional changes within clades. Clade replacements are long-term interactions between clades, and may even include sequential replacement of subclades within clades. Large-scale subclade replacement, if directional, may even be a higher-order analogue of evolutionary trends. Clade replacements constitute some of the most important patterns in the fossil record, and are thought of predominantly as a consequence of mass extinctions. The Ordovician extinctions initiated a long-term pattern of clade replacement between two major groups of stenolaemate suborders.

Trends are a higher-level analogue of phyletic gradualism, whereas radiations are an equivalent analogue of punctuated equilibrium. The most important evolutionary changes in bryozoan lineages take place during major radiations, and are not stratigraphically resolvable as evolutionary trends. Trends are long-term gradualistic phenomena that take place in post-radiation evolution, and, for bryozoans, predominantly include paedomorphoclines. These may represent facile evolutionary pathways, with minimal genetic or developmental obstacles to change, and are more frequent than peramorphoclines. The only palaeontological trends probably driven by long-term selection are those involving potential defences against predators. Adaptations for spatial competition were established very early and rapidly by bryozoans, and have remained stable to the present.

Increasing size and complexity of colony form is a complex gradient in bryozoan evolution (starred characters in Appendix 10.1), but belongs to the cryptic part of bryozoan evolution embedded in the Ordovician radiation. So much, in fact, is embedded there that one is tempted to dismiss trends as minor phenomena. The major features of bryozoan evolution arose in a punctuated fashion, twice in fact, during the Ordovician and Mesozoic radiations. A secondary radiation occurred during the Devonian, and replaced the previously predominant Ordovician lineages. Compared with radiations, trends have had little impact on bryozoan evolution. Their total effect on morphological evolution, except for predator resistance, has been minimal.

Analysis of trends in bryozoan evolution, has not, however, been disappointing. It has clarified the importance of the three great radiations in bryozoan evolution, and strongly indicates where future attention should be focused.

ACKNOWLEDGEMENTS

I thank Douglas H. Erwin for numerous discussions and his review of the manuscript; Donald L. Swiderski for discussions of bryozoan phylogeny and his work in preparing the cladogram in Figure 10.1; and John W. Bartley for his reconstruction of *Tabulipora* used in Figure 10.4.

REFERENCES

Anstey, R.L., 1978, Taxonomic survivorship and morphologic complexity in Paleozoic bryozoan genera, *Paleobiology*, **4**: 407–18.

Anstey, R.L., 1981, Zooid orientation structures and water flow patterns in Paleozoic bryozoan colonies, *Lethaia*, **14**: 287–302.

Anstey, R.L., 1986, Bryozoan provinces and patterns of generic evolution and extinction in the Late Ordovician of North America, *Lethaia*, **19**: 33–51.

Anstey, R.L., 1987a, Astogeny and phylogeny: evolutionary heterochrony in Paleozoic bryozoans, *Paleobiology*, **13**: 20–43.

Anstey, R.L., 1987b, Colony patterning and functional morphology of water flow in Paleozoic stenolaemate bryozoans. In J.P. Ross (ed.), *Bryozoa: present and past*, Western Washington University, Bellingham, WA: 1–8.

Anstey, R.L. and Bartley, J.W., 1984, Quantitative stereology: an improved thin section biometry for bryozoans and other colonial organisms, *J. Paleontol.*, **58**: 612–25.

Anstey, R.L. and Pachut, J.F., 1980., Fourier packing ordinate: a univariate size-independent measurement of polygonal packing variation in Paleozoic bryozoans, *J. Intl. Assoc. Math. Geology*, **12**: 139–56.

Anstey, R.L., Pachut, J.F. and Prezbindowski, D.R., 1976, Morphogenetic gradients in Paleozoic bryozoan colonies, *Paleobiology*, **2**: 131–46.

Banta, W.C. and Wass, R.E., 1979, Catenicellid cheilostome Bryozoa, I: Frontal walls, *Australian J. Zoology, Suppl. Ser.*, **68**: 1–70.

Blake, D.B., 1976, Functional morphology and taxonomy of branch dimorphism in the Paleozoic bryozoan genus *Rhabdomeson*, *Lethaia*, **9**: 169–78.

Blake, D.B., 1980, Homeomorphy in Paleozoic bryozoans: a search for explanations, *Paleobiology*, **6**: 451–65.

Blake, D.B., 1983, Introduction to the Suborder Rhabdomesina. In R.A. Robison (ed.), *Treatise on invertebrate paleontology, Part G (revised)*, Geological Society of America, Boulder, CO: 530–49.

Boardman, R.S., 1984, Origin of the post-Triassic Stenolaemata (Bryozoa): a taxonomic oversight, *J. Paleontol.*, **58**, 19–39.

Boardman, R.S., Cheetham, A.H. and Cook, P.L., 1983, Introduction to the Bryozoa. In R.A. Robison (ed.), *Treatise on invertebrate paleontology, Part G (revised)*, Geological Society of America, Boulder, CO: 3–48.

Cheetham, A.H. and Cook, P.L., 1983, General features of the class Gymnolaemata. In R.A. Robison (ed.), *Treatise on invertebrate paleontology, Part G (revised)*, Geological Society of America, Boulder, CO: 138–207.

Coates, A.G. and Jackson, J.B.C., 1985, Morphological themes in the evolution of clonal and aclonal marine invertebrates. In J.B.C. Jackson, L.W. Buss and R.E. Cook (eds.), *Population biology and evolution of clonal organisms*, Yale University Press, New Haven, CT: 67–106.

Cumings, E.R., 1910, Paleontology and the recapitulation theory, *Proc. Indiana Acad. Science*, **1909**: 305–40.

Dzik, J., 1975, The origin and early phylogeny of the cheilostomatous Bryozoa, *Acta Palaeontol. Polonica*, **20**: 395–423.

Erwin, D.H., 1989, Molecular clocks, molecular phylogenies, and the origin of phyla, *Lethaia*, **22**: 251–8.

Gould, S.J., 1988, Trends as changes in variance: a new slant on progress and directionality in evolution, *J. Paleontol.*, **62**: 319–29.

Håkansson, E. and Winston, J.E., 1985, Interstitial bryozoans: unexpected life forms in a high energy environment. In C. Nielsen and G.P. Larwood (eds.), *Bryozoa: Ordovician to Recent*, Olsen and Olsen, Fredensborg, Denmark: 125–34.

Harmelin, J.-G., 1975, Relations entre la forme zoariale et l'habitat chez les bryozoaires cyclostomes. Conséquences taxonomiques. *Doc. Lab. Geol. Fac. Sci. Lyon, hors-série*, **3**: 369–84.

Harmer, S.F., 1923, On cellularine and other Polyzoa, *J. Linn. Soc.*, **35**: 293–361.

Hickey, D.R., 1988, Bryozoan astogeny and evolutionary novelties: their role in the origin and systematics of the Ordovician moniculiporid trepostome genus *Peronopora*, *J. Paleont.*, **62**: 180–203.

Karklins, O.L., 1983, Introduction to the Suborder Ptilodictyina. In R.A. Robison (ed.), *Treatise on invertebrate paleontology, Part G, (revised)*, Geological Society of America, Boulder, CO: 453–88.

Larwood, G.P., 1985, Form and evolution of Cretaceous myagromorph Bryozoa. In C. Nielsen and G.P. Larwood (eds), *Bryozoa: Ordovician to Recent*, Olsen and Olsen, Fredensborg, Denmark: 169–74.

Larwood, G.P. and Taylor, P.D., 1979, Early structural and ecological diversification in the Bryozoa. In M.R. House (ed.), *The origin of major invertebrate groups*, Academic Press, London: 209–34.

Levinsen, G.M.R., 1909, *Morphological and systematic studies on the cheilostomatous Bryozoa*. Nationale Forfatteres Forlag, Copenhagen.

Liddell, W.D. and Brett, C.E., 1982, Skeletal overgrowths among epizoans from the Silurian (Wenlockian) Waldron Shale, *Paleobiology*, **8**, 67–78.

McKinney, F.K. and Boardman, R.S., 1985, Zooidal biometry of Stenolaemata. In C. Nielsen and G.P. Larwood (eds), *Bryozoa: Ordovician to Recent*, Olsen and Olsen, Fredensborg, Denmark: 193–203.

McKinney, F.K. and Jackson, J.B.C., 1989, *Bryozoan evolution*, Unwin Hyman, Boston.

Pachut, J.F., 1989, Heritability and intraspecific heterochrony in Ordovician bryozoans from environments differing in diversity, *J. Paleont.*, **63**: 182–94.

Ryland, J.S., 1970, *Bryozoans*, Hutchinson University Library, London.

Signor, P.W. and Brett, C.E., 1984, The mid-Paleozoic precursor to the Mesozoic marine revolution, *Paleobiology*, **10**: 229–45.

Stach, L.W., 1936, Correlation of zoarial form with habitat, *J. Geology*, **44**: 60–5.

Stanley, S.M., 1979, Evolution. In R.W. Fairbridge and D. Jablonski (eds), *The encyclopedia of paleontology*, Dowden, Hutchinson and Ross, Stroudsburgh, PA: 296–9.

Taylor, P.D., 1984, Adaptations for spatial competition and utilization in Silurian encrusting bryozoans, *Spec. Pap. in Palaeont.*, **32**: 197–210.

Taylor, P.D., 1985, Polymorphism in melicerititid cyclostomes. In C. Nielsen and G.P. Larwood (eds), *Bryozoa: Ordovician to Recent*, Olsen and Olsen, Frèdensborg, Denmark: 311–18.

Taylor, P.D., 1987, Fenestrate colony-form in a new melicerititid bryozoan from the U. Cretaceous of Germany, *Mesozoic Res.*, **1**: 71–7.

Taylor, P.D. and Cope, J.C.W., 1987, A trepostome bryozoan from the Lower Arenig of south Wales: implications of the oldest described bryozoan, *Geol. Mag.*, **124**: 367–71.

Taylor, P.D. and Curry, G.B., 1985, The earliest known fenestrate bryozoan, with a short review of Lower Ordovician Bryozoa, *Palaeontology*, **28**: 147–58.

Taylor, P.D. and Larwood, G.P., 1988, Mass extinction and the pattern of bryozoan evolution. In G.P. Larwood (ed.), *Extinction and survival in the fossil record*, Clarendon Press, Oxford: 99–119.

Utgaard, J., 1983, Paleobiology and taxonomy of the Order Cystoporata. In R.A. Robison (ed.), *Treatise on invertebrate paleontology, Part G, (revised)*, Geological Society of America, Boulder, CO: 327–57.

Wilson, M.A., 1985, Disturbance and ecologic succession in an Upper Ordovician cobble-dwelling hardground fauna, *Science*, **228**: 575–7.

Winston, J.E., 1976, Experimental culture of the estuarine ectoproct *Conopeum tenuissimum* from Chesapeake Bay, *Biol. Bull.*, **150**: 318–35.

Winston, J.E., 1986, Victims of avicularia, *Marine Ecology*, **7**: 193–9.

Zimmer, R.L. and Woollacott, R.M., 1977, Metamorphosis, ancestrulae, and coloniality in bryozoan life cycles. In R.M. Woollacott and R.L. Zimmer (eds), *Biology of bryozoans*, Academic Press, New York: 91–142.

APPENDIX 10.1
CHARACTERS USED IN CLADISTIC ANALYSIS OF BRYOZOAN SUBORDERS

In the analysis that produced Figure 10.1, the derived state of two-state characters is listed alone; polarised multistate character states are listed in order from primitive to derived. Some polarities reflect the use of entoprocts as an outgroup. Starred characters indicate potential trends.

1. Bivalved (cyphonautes) larva
2. Polyembryony
3. Intrazooidal development of ova
4. Bulbous protoecium
5. Diaphragmatic dilator muscles
6. Horse-shoe shaped lophophore and large body size
7. Tentacle number does not increase in ontogeny

  8. Co-ordinated extrazooidal feeding currents/ currents reflected in zoarial structure.
  9. Lophophore eversion by: (a) parietal muscle contraction; (b) annular contractions of membranous sac; (c) contraction of circular intrinsic wall muscles
10. Budding from stolozooid
11. Oral budding/ anal budding
12. Cystid precedes polypide/ polypide precedes cystid
*13. Budding pattern irregular/ intermediate/ highly geometric
14. Initial zooids from larva or directly from embryo
15. Ovicells
16. Ooecia
17. Kenozooids or mesozooecia
18. Avicularia
19. Vibracula
20. Rhizoids
21. Loss of stolons
22. Monticules ringed by large polymorphs
23. Pyriform zooids forming linear colonies
24. Articulated colonies
25. Zooid A. a stalked calyx: B, cylindrical: C, boxlike
26. Funicular integration
27. All species fully colonial
*28. Cerioid ('honeycomb') growth habit/ massive cerioids with overgrowths
29. Bifoliate cerioid growth habit
*30. Limited endozone/ well-developed endozone
31. One side of erect colony lacking autozooids
32. Anastomosing growth habit
33. Pinnately branched zoaria
*34. Monticular cormidia/ stellate cormidia
*35. Rugose monticules/ elongate ridge-like cormidia
*36. Fixed-walled skeleton/ free-walled skeleton
37. Boring into hard substrates/ skeleton
38. Skeleton present over membranous frontal wall
39. Laminated skeleton with laminae (a) parallel; (b) orally convergent; (c) orally divergent
40. Granular primary skeleton
41. Acicular primary skeleton
*42. Cyclic skeletal growth rates/ annular wall thickenings
43. Basal diaphragms
44. Cystiphragms
45. Hemisepta at zooecial bend
46. Longitudinal extrazooidal skeletal ridges
47. Polygonal extrazooidal skeletal ridges
*48. Lunaria/ lunaria with cores and ridges
49. Interzooecial vesicles or alveoli
50. Interzooecial pores
51. Pseudopores
52. Intrazooecial spines
53. Petaloid (septate) zooecia
*54. Acanthostyles/ bimodal acanthostyle size distribution

PART THREE

EVOLUTIONARY TRENDS
IN VERTEBRATES

Chapter 11

FISHES

John A. Long

INTRODUCTION

The history of fishes shows a great radiation and diversification during the Early Palaeozoic, with some groups undergoing bursts of rapid evolutionary change followed by total extinction at the end of the Devonian Period, while others survived on to modern times. In this chapter some of the important early groups of fishes will be treated, particularly with respect to factors, both extrinsic (environment, abundance of food, predator pressure) and intrinsic (genetic drift, heterochrony, biophysical constraints), which may have guided evolutionary trends. One of these groups, the placoderms, is of particular interest as trends in some orders are clear and unambiguous with respect to their guiding factors. Furthermore, they and the acanthodians are extinct groups, making us reliant *in toto* upon their fossil record for information on their biology and physiology. The chondrichthyans, a group virtually unchanged since the Devonian in their external appearance, have undergone subtle evolutionary change affecting the internal anatomy and histology of their tissues. The osteichthyes, or bony fishes, are the dominant group of fishes alive today, within which the lungfishes are perhaps one of the best-known groups represented throughout geological time and by three surviving genera. In a classic study, Westoll (1949) used lungfishes to demonstrate evolutionary trends in lower vertebrates. Since Westoll's work much new information has been gathered on both fossil and living lungfishes which allows us to assess the fundamental factors which directed their evolution. Similarly, the transition from fishes to tetrapods can be viewed in the light of fresh data on the ontogenetic stages in osteolepiform fishes, demonstrating the importance of heterochrony in this major step in vertebrate evolution. I will show that in each of these cases there is a variety of factors influencing evolutionary trends, although in each the role of extrinsic and intrinsic factors is different.

The terms 'paedomorphocline' and 'peramorphocline' (see Chapter 3 herein), as defined by McNamara (1986), cannot be applied here to lineages which do not comprise demonstrable successively-derived species. In the case of closely related taxa united by nested sets of synapomorphies, and in which heterochrony has played a pervasive role in the evolution of that set of closely related taxa, I here apply the terms 'paedomorphoclade' and 'peramorphoclade'. Examples of each are demonstrated below.

EVOLUTIONARY TRENDS IN ANTIARCH PLACODERMS

Romer (1945, p. 38) once described antiarchs as 'grotesque little creatures which looks like a cross between a turtle and a crustacean', and indeed anyone not familiar with fossil fishes could be taken aback by the image of a box-like armour-plated fish having two bone-covered arms extending from its shoulders. The acquisition of these arms, an evolutionary novelty in the metaphorical sense, is in itself an unusual story which demonstrates how extrinsic factors may direct evolutionary change in a totally unexpected direction.

The antiarchs (Figure 11.1) are characterised among the placoderms by having a long trunkshield with two median dorsal plates, a headshield with a single opening in its midline for the eyes and nares (called the orbital fenestra), and, in advanced forms, the presence of a brachial condyle and segmented bony pectoral appendage. Recent new discoveries from China of Silurian (Dr Zhang Gorui, pers. comm., 1989) and Early Devonian primitive antiarchs (the Yunnanolepiformes) have shown how the unusual pectoral appendage was acquired: very primitive types have unsegmented prop-like appendages which do not articulate with the trunk armour, but which inserted by a (presumably) cartilaginous attachment (the yunnanolepids). *Procondylepis* was an intermediate form with a primitive brachial process on the trunkshield (Zhang, 1984), and other forms with bisegmented appendages, such as *Liujiangolepis* (Wang, 1987) presumably had a brachial process which was enclosed by the anterior bones of the pectoral appendage, as in typical antiarchs. During the acquisition of these subsequent stages in pectoral appendage attachment, the head and trunkshield underwent little modification, indicating strong selection pressure on this particular structure. Shortly after its development, there was a great radiation of antiarchs all over the world (previously they were restricted only to one region, the South China Terrane—see Zhang, 1978; Young, 1981).

What factors, then, could have influenced the development of this novel external bony arm in these armoured fishes? Heterochrony does not come into play, as there are no other known placoderms or any other fishes which ever evolved a similar type of pectoral appendage. Extrinsic factors such as improved feeding ability, defence from predators or use in reproduction are most likely to have guided the development of the pectoral appendage. The

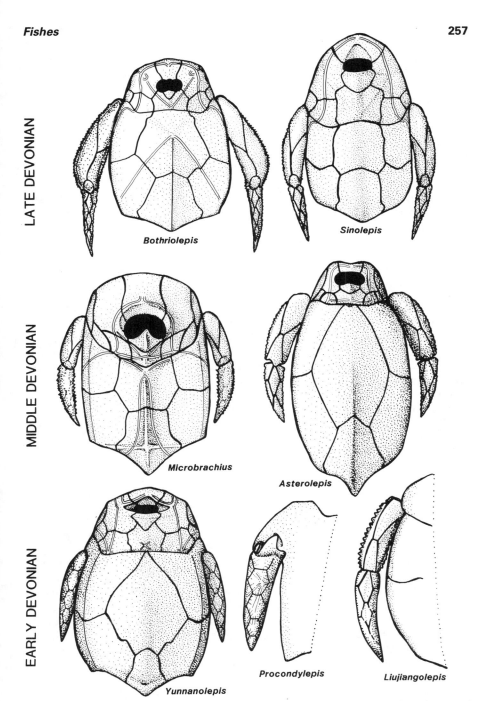

Figure 11.1. Antiarch dermal armour morphology. Note the relative sizes and shapes of pectoral fin appendages.

Late Devonian antiarch *Bothriolepis* is known from serially sectioned specimens to have ingested organic-rich mud as a source of food (Denison, 1941), and the overall body shape of antiarchs is suggestive of a benthic feeding detrital ingestor. The arms were most probably used in pushing the head of the fish down into the soft muddy sediment upon which it fed. Long segmented pectoral fins would be most effective at thrusting the fish forwards and downwards into the substrate, a reason perhaps for the ubiquitous success of *Bothriolepis*. There is even the possibility that the pectoral appendages in antiarchs were used to assist the fish in walking out of the water to find new feeding pools (this suggestion is supported by the possible presence of lungs in *Bothriolepis*—see Denison, 1941). Whatever the reason for these arms, it is clear that their origins are intimately related to external or environmental factors as there are no homologues in any other fish group.

The role of heterochrony in antiarch evolution is suggested by the juvenile phases of *Bothriolepis*. Although Stensiö (1948) documented ontogenetic changes in small to adult *Bothriolepis canadensis*, his specimens did not show many major morphological changes with growth. Werdelin and Long (1986) quantified Stensiö's observations to highlight anatomical regions showing the highest degree of allometric change for *Bothriolepis*. Young (1988) has provided new data on very small larval *Bothriolepis*. It seems that very small *Bothriolepis* specimens (Figure 11.2) have unusually large median ventral plates, which decrease in size relative to the size of the armour as the fish grows. This feature has only recently been discovered in sinolepid antiarchs, and suggests (assuming that all antiarchs grew in a similar fashion to *Bothriolepis*) the retention by adult sinolepids of the large median ventral plate (or its loss as a large median ventral fenestra) by paedomorphosis. The smallest juvenile *Bothriolepis* have pectoral appendages which do not extend beyond the trunk armour, only doing so in maturity (Figure 11.2). This could indicate that *Microbrachius* and *Wudinolepis* are either juvenile bothriolepids, or paedomorphically derived taxa (on the paedomorphoclade). In examining the armour of small *Bothriolepis* other paedomorphic features are seen in more primitive members of the bothriolepid lineage, such as the strongly convex anterior margin of the postpineal plate in *Microbrachius* (Figure 11.2), the ridges on the trunkshield of *Microbrachius*, (Figure 11.1), and the short pectoral fin length of *Dianolepis* (Zhang, 1984). By extrapolation of these trends to extremes seen in the features on *Bothriolepis* (Figure 11.2) it can be argued that several of the characters in *Bothriolepis* evolved by peramorphosis. Such features are the long pectoral appendages (possibly paralleled in *Sinolepis*) and the long premedian plate, the growth of which exceeds that of other headshield bones (Werdelin and Long, 1986).

In summary, it would appear that the factors influencing evolutionary trends in antiarchs are primarily intrinsic (heterochrony) in the derivation of new genera or new lineages (e.g. sinolepids), but extrinsic (environmental change and migrationary ability) for the radiation of large species groups (e.g. *Bothriolepis*, *Remigolepis* and *Asterolepis*). Peramorphosis has played a part in the origins of the advanced bothriolepid lineage, which could be termed a 'peramorphoclade' (from microbrachiids through *Dianolepis* through *Monarolepis* to *Bothriolepis*).

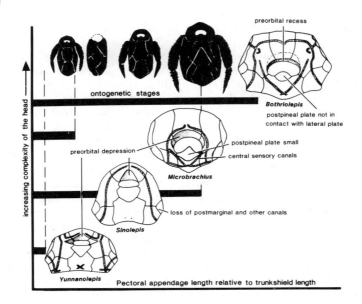

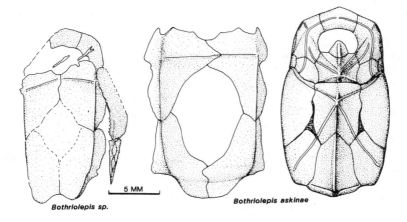

Figure 11.2.   Relative stages of development of the cranial features in antiarch placoderms plotted against relative proportions of pectoral fin to trunk-armour size. The black figures on the top line are ontogenetic stages for *Bothriolepis*, based on juvenile Mt Howitt material (bottom left) and juvenile *Bothriolepis askinae* (bottom right, after Young 1988). *Yunnanolepis* has a primitive skull and very short pectoral appendages. *Sinolepis* has a primitive skull and long pectoral appendages. *Microbrachius* has an advanced skull and very short pectoral appendages. *Bothriolepis* represents the most derived stage with advanced skull and long pectoral appendages. These data suggest that advanced antiarchs like *Bothriolepis* evolved through peramorphic processes enabling longer pectoral fin development.

EVOLUTIONARY TRENDS IN CHONDRICHTHYANS

The Chondrichthyes comprises the cartilaginous fishes: sharks, rays and holocephalans (chimaerids and ratfishes). Although most fossil forms are represented by teeth, scales and spines, there is a relatively good record of complete body fossils that allows their major evolutionary trends to be reconstructed. Figure 11.3 shows the diversity of forms seen in the group throughout the Phanerozoic. At first glance it appears that some sharks have undergone very little change since their first appearance. The Devonian *Cladoselache* has a similar shark-like body to modern sharks, the teeth are typically chondrichthyan in their appearance and histology, and the scales which cover the body are not too unlike modern placoid scales.

Despite these superficial similarities there has been a large degree of change in chondrichthyan lineages, in soft tissue anatomy, histology and biochemistry (Maisey, 1984; 1986). The interrelationships between the living lamnid sharks depend mostly upon such characters, most of which would be undetectable in fossils. In the Palaeozoic, changes most noticeable in chondrichthyans concern the evolution of the holocephalans and their subsequent radiation (cochliodontids, petalodonts, etc.—Zangerl, 1981), and the radiation of many now extinct lineages of selachians (e.g., edestids, caseodontids). The internal skeletal structure of the paired fins and caudal fin is a major character used in distinguishing groups, as is the morphology and histology of the teeth. This would indicate that the driving force in chondrichthyan evolution was the ability to feed effectively: body shape gives speed or benthic preference, tooth morphology and composition are guided by the types of food to be eaten. We have virtually no data on the ontogenetic changes in fossil sharks, and living sharks give no indications of the role of heterochrony in evolutionary lineages of fossil sharks. The evolution of sharks would therefore seem to be principally directed by extrinsic factors affecting subtle internal morphological changes (e.g., harder teeth by the evolution of multilayered enameloid—see Rief, 1977; 1978).

Major morphological changes occurred close to the Late Devonian–Carboniferous boundary when the holocephalans emerged. The basic selachian body plan changed into one adapted for benthic feeding, accompanied by radical changes in tooth morphology and structure. Teeth acquired hypermineralised pleromin (Ørvig, 1985), which provided increased hardness, enabling hard-shelled foods to be crushed. The earliest holocephalans are from the Namurian Bear Gulch Formation, Montana (Lund, 1986). Already the chimaerids had a modern chimaerid-like body plan (*Echinochimaera*, Figure 11.3), but the cochliodonts show the extremes reached in this trend for adaptation to benthic feeding. The Iniopterygians are another group thought to be aberrant holocephalans (Maisey, 1986), but Zangerl (1981) holds an alternative view that they are stem-group chondrichthyans.

The radiation of chondrichthyans reached its peak in the Carboniferous, following the decline of placoderms at the end of the Devonian (Figure 11.4). Latest Devonian marine placoderms included benthic feeders with durophagous dentition (e.g., ptyctodontids, mylostomatids). Also at this time the

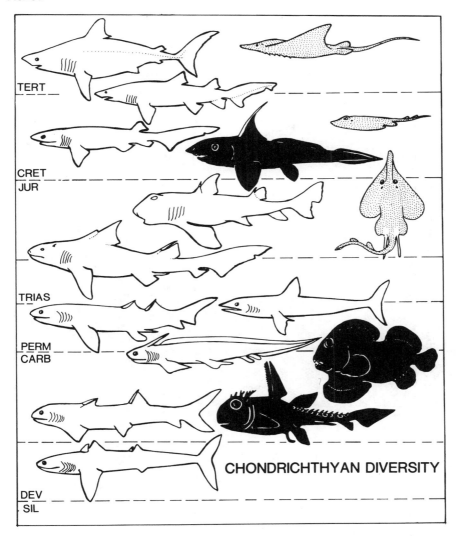

Figure 11.3. Chondrichthyan diversity through time. Major types of chondrichthyans are shown: selachians (uncoloured), holocephalans (black) and batoids (stippled).

lungfishes departed from the marine environment. It seems likely that certain chondrichthyans radiated into the niches left by the absence of these durophagous groups, or alternatively that they displaced durophagous placoderms and lungfishes by being more efficient benthic feeders.

Another major morphological change occurring in chondrichthyan evolution is seen in the origin of batoids. Early depressiform sharks with broad

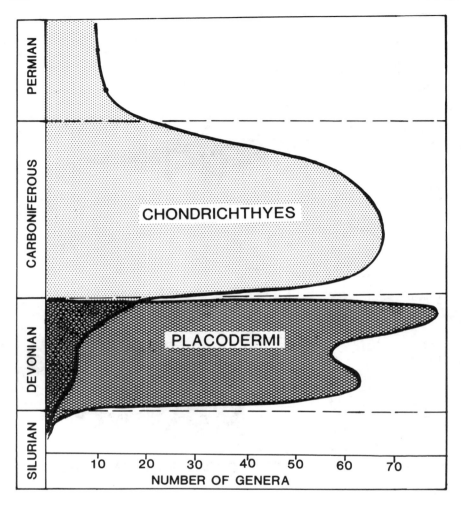

Figure 11.4 Graph of total number of genera of placoderms and chondrich-thyans in the Middle and Late Palaeozoic. Data from Denison (1978) and Zangerl (1981).

pectoral fins are known by the Late Jurassic (*Aellopos*) and true batoids had appeared by the Cretaceous (*Cyclobatis*). Since then they have changed very little. What factors may have influenced the origin of this group? The shift to a benthic lifestyle from a pelagic one had already been achieved within the sharks. The selection for more depressiform body shapes could have come about in several ways. The ability to be more camouflaged by burrowing below the sediment would result in selection for flatter individuals—an improved anti-predation strategy. The change in attitude of the mouth, from

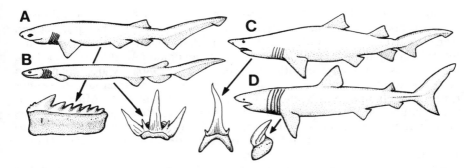

Figure 11.5. Selachians showing similar body morphology but widely differing dentitions. A, B, Hexanchiforms: A, *Hexanchus*; B, *Chlamydoselachus*; C, D, Lamniforms; C, *Eugomphodus*; D, *Cetiorhinus*.

forward- to directly downward-facing would have brought about increased efficiency at bottom feeding, further enhancing selection for flatter individuals. The evolution of more effective crushing dentitions could have brought about subsequent changes in body form. Such trends have occurred repeatedly in nearly all lineages of fishes, even if the same end result has been reached by morphologically different methods (e.g., drepanaspid and psammosteid agnathans; phyllolepid and gemuendinid placoderms, pleuronectiform teleosts).

The major trends in shark evolution have not been towards radical changes in body plan, but increased efficiency at feeding. This is reflected by the great diversity of shark tooth types, especially since sharks with identical body shapes can have widely differing dentitions, such as the living lamniforms *Cetorhinus* and *Odontaspis*, and the primitive living hexanchiforms *Hexanchus* and *Chlamydoselachus* (Figure 11.5). Biochemical changes reflect the ability of some sharks (e.g., carcharhinids) to invade environments of varying salinity, or the ability of some large predators (e.g., *Carcharodon*) to maintain higher body temperatures than surrounding water. Cladistic analyses of the interrelationships of living shark groups rely heavily on soft-tissue characters, including biochemical and histological features, because body and skeletal characters are not as variable between generic or familial level taxa (see, for example, Maisey, 1984: Compagno, 1977).

EVOLUTIONARY TRENDS IN SOME ACANTHODIANS

The acanthodian fishes have fin-spines preceding all the fins. They lived from the Silurian to Late Permian, but their greatest radiation took place from the Upper Silurian to the Early Devonian when three major groups emerged: the Climatiida, characterised by ventral armour on the shoulder girdle; the Ischnacanthida, characterised by teeth fused to gnathal bones, and the Acanthodida, characterised by one dorsal fin and the loss of teeth

(Denison, 1979). Evolutionary trends are not known for ischnacanthids as they are known from very few complete fishes, most taxa being described from gnathal bones. Acanthodids are a remarkably homogeneous group, and the only group definitely to survive past the Devonian (*Acanthodopsis* is here regarded as a possible acanthodid—see, Long, 1986a. Major changes in acanthodid morphology are seen in the loss of intermediate fin-spines (present only in the earliest form, *Mesacanthus*—see Miles, 1973), the enlargement of the pelvic fin, simplification of scale morphology (loss of ornament on crowns), and subdivision of the palatoquadrate into three parts with twin articulations to each side of the braincase (Long, 1986b).

Within the climatiids evolutionary changes can be demonstrated in the increasing complexity of the ventral shoulder girdle bones, and the increasing complexity of the fin-spine ornamentation and number of intermediate fin-spines (Long, 1986a). Figure 11.6 shows this trend with important taxa illustrated in stratigraphic order. The most primitive members of the climatiids, such as *Euthacanthus* have only one pair of pinnal plates that are not firmly connected to the pectoral fin-spines (Miles, 1973). The fin-spines are ornamented with simple linear ridges. The histology of *Euthacanthus* scales is also regarded as primitive for the group as it contains no Stranggewebe in the crown, has only thin growth zones on top, large pulp cavities and a flat base of cellular bone (Denison, 1979). The next morphological stage in climatiid evolution is represented by *Lupopsyrus* (Bernacsek and Dineley, 1977) which has a similar pectoral girdle to *Euthacanthus* except that there is a small median lorical plate joining the pinnals together. This condition is also found in *Parexus* but this genus is considered more apomorphic in having more complex ornamentation of the fin-spines and intermediate fin-spines between the pectoral spines and the pinnal plates. *Brachyacanthus* further develops this pattern by incorporating the intermediate spines onto an additional pair of pinnal plates, and the pectoral fin-spines have been firmly fixed to the pinnals. The most derived condition of the ventral pectoral armour in climatiids is typified by *Climatius* itself, which has several median lorical plates and a closely packed arrangement of the two pairs of pinnal plates, each of which bears accessory or intermediate fin-spines (Figure 11.6).

The heavily armoured climatiids were thought to represent the primitive condition for acanthodians, with the non-armoured ischnacanthids and acanthodidids derived from them by reduction of the dermal skeleton (Watson, 1937; Romer, 1945; Moy-Thomas and Miles, 1971). However, as the ischnacanthids appear to represent the primitive gnathostome condition better than climatiids, in having fewer specialisations (Long, 1986a), it

---

Figure 11.6. Evolutionary trends in acanthodians. In the climatiid lineage the acquisition of increasingly complex dermal ventral shoulder armour is here shown by a morphological transformation series (from *Euthacanthus* to *Climatius*). This trend is also supported by evidence from scale histology (simple in *Euthacanthus*, more complex in other forms shown).

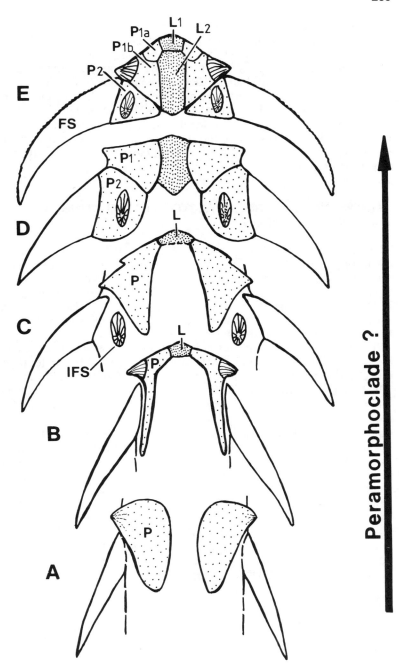

became apparent that the climatiids must be derived in the development of ventral dermal armour in the shoulder girdle. The fact that placoderm shoulder girdle armour does not resemble climatiid armour in plate shape or arrangement further supports this view. The functional advantages of developing a rigid shoulder girdle with fixed pectoral spines was a shared adaptation in some placoderms and climatiids. Unfortunately there is no ontogenetic data for these early acanthodians, although the role of heterochrony in the development of ventral shoulder girdle armour is indicated. In a transformational series where the character states become progressively more complex, the most likely influence is peramorphosis, by the addition of extra developmental stages.

INTRINSIC AND EXTRINSIC FACTORS IN DIRECTING LUNGFISH EVOLUTION

The lungfishes have been used by Westoll (1949) as a classic example of evolutionary trends, demonstrating their rapid rate of evolution during the Devonian with very little change after that period. Although this is still a valid observation much progress has since been made in determining evolutionary lineages, or monophyletic groups, within the Dipnoi. The fossil record of the group is excellent, and we have good environmental data on their habitats throughout geological time.

The earliest fossil lungfishes indicate that two discrete lineages had already been established, although both are characterised by at least 20 synapomorphies defining the monophyly of the group (Schultze and Campbell, 1986). These characters all more or less relate to one functional system, including the acquisition of a massive bite: long lower jaw symphysis, skull-roof table of small tightly packed bones for temporalis muscle attachment (Campbell and Barwick, 1982); loss of marginal tooth-bearing bones and modified cheek bone pattern, modified palate and histology of the dentition.

One lineage of lungfishes, represented by the Siegennian *Uranolophus* from North America, is characterised by a denticle-shedding dentition, whereas the second group had a palatal bite with a tooth-plated dentition. Primitive tooth-plated lungfishes are known from the Early Devonian of Vietnam (Thanh and Janvier, 1987) and North America (Denison, 1968). The dipnorhynchids from the Early Devonian of Australia include forms with massive dentine-covered palates (*Dipnorhynchus*) and possible intermediate types with rudimentary tooth rows (*Speonesydrion*) (Thomas and Campbell, 1971; Campbell and Barwick 1982; 1983; 1984a; 1985; 1986).

Figure 11.7. Changes in body and fin shape in two different lineages of lungfishes. Environmental changes are shown in the background. Numbers refer to convergent characters seen in both lineages: 1, smaller anterior dorsal fin; 2, longer second dorsal fin; 3, anterior dorsal fin extensive rostrocaudally; 4, anterior and posterior dorsal fins merge into one continuous fin; 5, anal fin continuous with caudal fin.

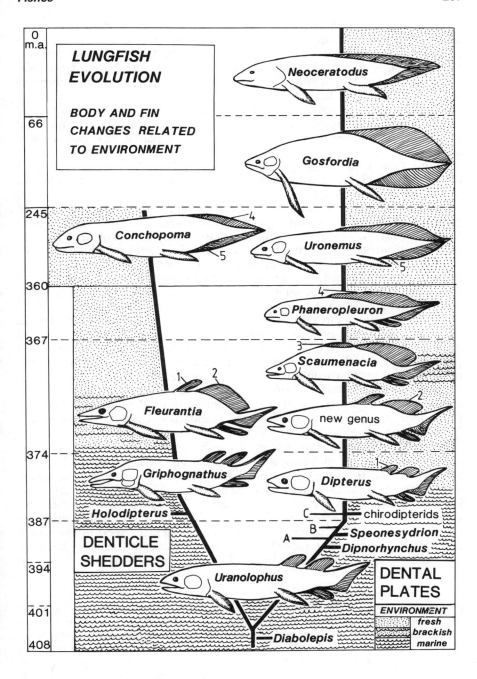

LUNGFISH EVOLUTION

BODY AND FIN CHANGES RELATED TO ENVIRONMENT

The major evolutionary trends exhibited in the bodies of lungfishes of both these lineages are seen in Figure 11.7. Both groups underwent similar changes to skull-roof and cheekbone patterns, and although the denticle shedders retained a similar body pattern throughout the Devonian, the latest member, *Gnathorhiza*, acquired a body plan convergent with the tooth-plated forms of that time. Since the decline of the denticle shedders in the Permian the surviving lineage of tooth-plated forms has undergone little evolutionary change apart from simplification of the skull-roof bone patterns, variation in dental morphology, and degeneration of the paired fins into sensory apparatuses.

Bemis (1984) has suggested that paedomorphosis has played a role in determining the direction of morphological change within lungfish evolution. The evidence for this is seen in the degeneration of the endoskeletal ossification (strong in fossil forms, absent in living forms), and the acquired body shape in post-Devonian lungfishes (which resembles the larval form in living lungfishes). Further evidence for heterochrony in Devonian lungfishes is seen by analogy with ontogentic changes in other Sarcopterygii. The osteolepiform *Eusthenopteron* shows greatest allometry in the region of the cheek and orbit. Juveniles have very short cheek regions relative to the length of the skull, and proportionately large orbits (Schultze, 1984). Comparison with four well-known Devonian tooth-plated forms, which are derived members of a monophyletic group (Figure 11.8), show a similar change in cheek and orbit proportions. As these genera are widely spaced in time and cannot be shown to be successive genera derived from each other, yet paedomorphosis appears to have played a major role in the evolution of the more derived taxa, it would be appropriate to term this trend in evolution as a 'paedomorphoclade'. If further material is found to demonstrate that successive species evolved in the same manner, thus filling the gaps between these four genera, the cluster of taxa could then correctly be termed a 'paedomorphocline'.

One of the major factors guiding lungfish evolution must have been the change from marine to freshwater habitats. Campbell and Barwick (1986) showed that all Early Devonian lungfishes inhabited marine environments, and by the Middle Devonian most were still marine, although some had invaded brackish to freshwater environments. More taxa became freshwater dwellers in the Late Devonian. None remained in marine environments after the Devonian. All of the major morphological changes seen in the tooth-plated lungfishes during the Middle to Late Devonian can be morpho-functionally related to the acquisition of air-breathing: changes in the size of

---

Figure 11.8. A possible paedomorphoclade in Devonian lungfishes, based on the assumption that juveniles had shorter cheeks and larger orbits than adults (as occurs in crossopterygians, *Eusthenopteron* – see page 272). Measurements of (A) the cheek length, (B) orbit, and (C) snout-opercular length are shown plotted for the four genera on the inserted chart. Chronological appearance of groups: Dipnorhynchids – Early Devonian; chirodipterids and dipterids – Middle Devonian, *Phaneropleuron* – Late Devonian.

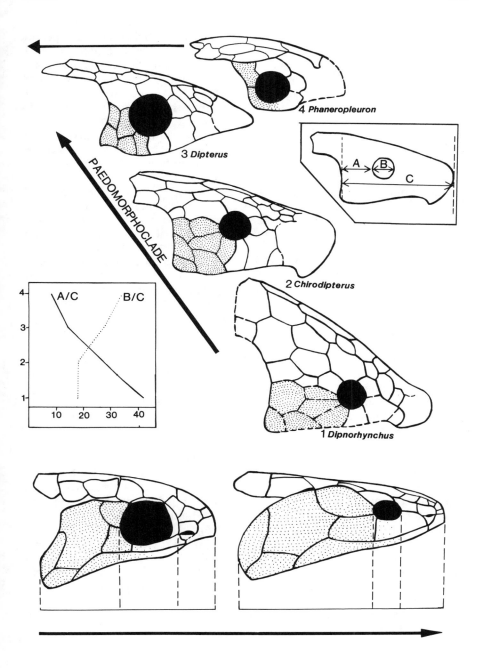

4 *Phaneropleuron*

3 *Dipterus*

PAEDOMORPHOCLADE

2 *Chirodipterus*

1 *Dipnorhynchus*

the parasphenoid length (to trap air in the mouth), modifications in the hyoid arch and pectoral girdle bones (to swallow the air bubble), and changes in body and fin shape for regular rises to the surface. Bishop and Foxon (1965) have shown the importance of each of these systems in the act of air-gulping in *Protopterus* and Campbell and Barwick (1986; 1988) have shown how these character sets relate to Devonian lungfish lineages and the independent acquisition of air-breathing in lungfishes (rather than as a shared character with tetrapods, — see, for example, Rosen *et al.*, 1981). An acquired feature of post-Carboniferous lungfishes is the ability to aestivate and survive long periods of drought until the next season's rainfalls. Permian lungfishes could do this (Carlson, 1968), and this is seen as an adaptation to climatic variations within the newly inhabited freshwater environment.

Most evolutionary trends in lungfish evolution during the Mesozoic and Tertiary centred on reduction of numbers of skull-roofing bones, by simplification through fusion and loss of bones (compared with reduction in numbers of skull elements in a number of reptile lineages described in Chapter 12), and the modification of dental morphology which allowed a wider variety of food types to be eaten. The main factors influencing the directionality of morphological change in lungfishes have been both intrinsic (heterochrony) and extrinsic (first, a dichotomy in feeding strategies; second, a major change in environments). During the major radiation of the Dipnoi during the Devonian, the rise of other fish groups, such as placoderms, other osteichthyans, and the chondrichthyans, would have provided continual selection pressure on the group to keep feeding ability in the marine environment at peak efficiency. This is demonstrated also by the acquisition of a highly complex electrosensory system in the snout of lungfishes for sensing prey in muddy environments (Campbell and Barwick, 1986; Cheng, 1989). The long-snouted rhynchodipterids, such as *Griphognathus*, were highly successful at nuzzling their long snouts along the sea-floor in search of food. They are one of the widespread Late Devonian lungfish groups which acquired cosmopolitan range.

Heterochronic change in lungfish body plan, skull proportions and internal skeletal ossification may have been catalysed by the initial shift from marine to brackish and freshwater environments. The genus *Dipterus* shows a high degree of change in orbit and cheek proportions relative to the marine chirodipterids and dipnorhynchids (Figure 11.8) and *Dipterus* was thought to be one of the first lungfishes to venture from the marine environment to brackish or fresh water habitats (Campbell and Barwick, 1986; 1988). The ability to breathe air through the development of lungs is an example of modification of an existing organ, the swim-bladder, found in all osteichthyans to regulate depth in the water column. The prime function of the organ was for gaseous exchanges so the ability to respire air to supplement oxygen from the gills was not a major step. Some modern teleosts can use the swim-bladder in this way.

THE FISH – TETRAPOD TRANSITION

Perhaps the greatest step in the evolution of the vertebrates was the transition of fishes into four-legged animals (tetrapods) which could leave the aquatic environment. Although some workers have recently resurrected an old idea that the lungfishes may have been the sister group to tetrapods (Rosen *et al.*, 1981; Gardiner, 1980; Forey, 1987), most scientists support the traditional viewpoint that the crossopterygians gave rise to the first amphibians. Recent work by Schultze (1986), Maisey (1986), Jarvik (1981), Long (1988; 1989), Campbell and Barwick (1984b) and others has reinforced this theory and pointed out the errors and inconsistencies in Rosen *et al.*'s work. In this review I shall accept the traditional view and show how morphologic trends and ontogenetic factors in one group of crossopterygians, the osteolepiforms, have influenced the rise of the first tetrapods.

Evolutionary trends in osteolepiform fishes include the loss of cosmine (e.g., in glyptopomids, eusthenopterids, platycephalichthyinids, panderichthyids —see Vorobyeva, 1977), the change from rhombic to round scales (e.g., in eusthenopterids), the changing proportions of parietal and frontoethmoid shields in the skull roof (all groups), and great variations in the arrangement of palatal and narial bones (e.g., in megalichthyinids—see Vorobyeva, 1975). From their earliest appearance in the Middle Devonian until their demise in the mid-Permian, the cosmine-covered osteolepidoids did not undergo much variation. The megalichthyinids are one of the few groups which can be diagnosed as monophyletic (Long, 1985a). They are presumed to have arisen from the osteolepidoids by specialisations of the palate (interpremaxillary bone, broad vomers in short medial contact) and nares (slit-like, enclosed posteriorly by a posterior tectal bone).

The eusthenopterids are believed to have arisen from osteolepidids through loss of cosmine and changing of scale shape from rhombic to round. The most primitive member, *Marsdenichthys*, retains the plesiomorphic extratemporal bone in the skull roof, but this is subsequently lost in all other genera (Long, 1985b). Trends in other osteolepiform groups are not known as only a few taxa comprise the platycephalichthyinids, panderichthyinids, lamprotolepids and rhizodopsids.

Only one osteolepiform, *Eusthenopteron*, has been studied to show ontogenetic variation in individuals (Schultze, 1984). Juveniles of the osteolepiform fish *Eusthenopteron* show more features in common with primitive tetrapods than any other crossopterygian group. These similarities indicate that paedomorphosis may have played an important role in the fish–tetrapod transition (Long, 1990).

The following characters are seen in juvenile *Eusthenopteron* and primitive tetrapods but are lost in mature *Eusthenopteron*. The cheek in juvenile *Eusthenopteron* has a deep postorbital bone which has a large orbital notch. In adults the postorbital is elongate with a much smaller portion of the anterior margin participating in the orbit (Figure 11.9). This compares well with the primitive amphibian *Crassigyrinus* (Panchen, 1985, Figure 11.9) and the East Greenland ichthyostegalids (Jarvik, 1952) in both the deep shape of the bone and the larger participation in the orbit.

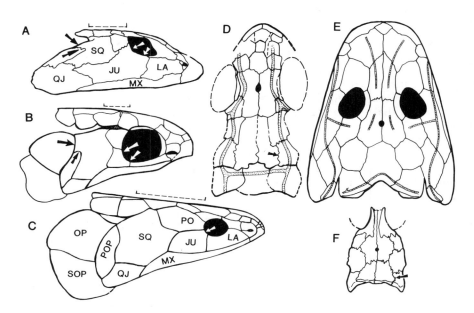

Figure 11.9.   Skulls in lateral view of (A) the amphibian *Crassigyrinus scoticus*; (B) juvenile and (C) adult of *Eusthenopteron foordi*. Skull roofs in dorsal view of (D) juvenile *Eusthenopteron foordi*; and the amphibians *Ichthyostega* (E) and *Crassigyrinus* (F). After Jarvik (1980), Schultze (1984) and Panchen (1985). Abbreviations: AT, anterior tectal; JU, jugal; LA, lachrymal; LR, lateral rostral; MX, maxilla; QJ, quadratojugal; OP, opercular; PO, postorbital; POP, pre-opercular; SOP, subopercular.

The squamosal bone in juvenile *Eusthenopteron* has a strongly indented notch for the preopercular bone, as in the cheeks of both *Crassigyrinus* and *Ichthyostega*. Mature *Eusthenopteron* have a tapering preopercular which does not protrude dorsally into the squamosal. In juvenile *Eusthenopteron* the squamosal has a free margin dorsal to the preopercular which contacts the opercular bone. In both *Ichthyostega* and *Crassigyrinus* there is a free posteriorly-directed margin above the preopercular (Figure 11.9). In mature *Eusthenopteron* the opercular meets the cheek along the entire margin of the preopercular, not contacting the squamosal. If the opercular were taken away in *Eusthenopteron*, as occurs in tetrapods, the margins of the cheek in juvenile *Eusthenopteron* would more closely match those of primitive tetrapods in having a posteriorly-directed margin on the squamosal. The cheek unit in juvenile *Eusthenopteron* has a shorter contact margin with the skull roof than in the adult, and in this way resembles primitive tetrapods.

In juvenile *Eusthenopteron* the jugal bone participates more in the orbit (approximately one-sixth of the circumference) than it does in the adult (approximately one-tenth to one-ninth). In three primitive tetrapods (ichthyostegalids, *Crassigyrinus*) the jugal has an orbital notch close to a quarter of the orbital circumference. Although the lachrymal bone does not participate in the orbit in the ichthyostegalids or *Crassigyrinus*, it does in the juvenile temnospondyl *Eugyrinus* (Milner, 1980) and certain other primitive amphibians (e.g., *Greererpeton*—see Smithson, 1982). It is likely that this condition was a juvenile feature of primitive ampihibians, matching the condition in osteolepiforms, but variable in adult forms.

The snout in juvenile *Eusthenopteron* appears to have fewer ossifications than the adult form. If this observation is correct that it would compare well with the condition in primitive tetrapods. The skull roof has more intricate suturing between many of the bones in juvenile *Eusthenopteron*. In the adult the sutures are straight or gently curved. In *Crassigyrinus* and parts of the ichthyostegalid skull (e.g., *Acanthostega*—see Clack, 1988) there are ir-regular and sometimes complex sutures between the skull-roof bones. The pineal foramen in juvenile *Eusthenopteron* appears to be a simple opening, although this may be an artefact of preservation. In mature *Eusthenopteron* and most other know osteolepiforms the pineal area has several plates around a small pineal foramen. In tetrapods the opening remains propor-tionately large and simple, lacking pineal bones. If the loss of these characters with increasing stages of ontogenetic development is assumed to have occurred throughout the osteolepiforms, then the juveniles of panderich-thyids, the group most closely allied to tetrapods (Schultze, 1970; 1986; Schultze and Arsenault, 1985), would predictably be even closer in cranial appearance to primitive tetrapods than *Eusthenopteron*.

The most distinctive tetrapod feature is the presence of digits on the paired limbs. If skull characteristics alone are compared it would be difficult, if not impossible, to distinguish an early amphibian from a panderichthyid osteo-lepiform (Schultze, 1986). The presence of a strong limb endoskeleton is a feature common to osteolepiforms, tetrapods and rhizodontiforms (Long, 1989). Strong selection pressure favoured those morphotypes that developed such a limb, because it allowed some individuals to cross a major adaptive threshold and push themselves out of the water to cross land and find new pools of water. The ability to lift the rib cage off the ground, through increased development of the girdles, limbs and digits, would also have greatly aided the development of aerial respiration. The presence of a true choana, or internal palatal nostril, is found only in osteolepiforms and tetrapods (Schultze, 1986; Long, 1988). Once the fish had acquired these abilities, the modification of the inner ear for receiving sounds through the medium of air would have soon developed, as shown by the presence of otic notches in the skulls of early amphibians.

The development of digits from a piscine fin may have resulted through peramorphosis (see Chapter 3). There are several examples of living sala-manders in which the number of digits and the number of phalanges in the digit may vary between the species (Alberch and Alberch, 1981; Hanken, 1982; Alberch and Gale, 1985; Oster *et al.*, 1988). Similarly, the fusion of wrist elements may occur as interspecific variations in the genus *Bolitoglossa*, as timing of developmental stages appears to be the strongest influence on this morphological feature. It is suggested from these studies that although the skull of early amphibians could have evolved from osteolepiform fishes through paedomorphosis, the development of limbs with digits probably arose through peramorphosis. Such dissociation of different growth rates between cranial and autopodial elements has been described in living salamanders (Alberch and Alberch, 1981), and would therefore be likely to have played an important role in the early evolution of the group as a whole.

CONCLUSIONS

The placoderms, an extinct shark-like group, show that evolutionary change in the antiarchs has also been influenced by heterochrony, with peramorphosis being a major factor in the origin of the most successful group, the bothriolepids. The origin and development of one of the vertebrates' most bizarre morphological traits, the segmented bony pectoral appendage, has no counterpart in any other vertebrate group, and its evolution would seem to be directed primarily by environmental factors; the ability to dig the fish into the substrate to take over a feeding niche not previously exploited, or alternatively to 'walk' out of ponds to find new feeding areas, thus going where no fish had gone before. In summary, the chondrichthyans show a different pattern of evolutionary trends to other fishes. Since the Devonian they have achieved an efficient body plan and biochemical balance in their degree of ossification of hard parts, and although they underwent some minor radiations in experimental body shapes (e.g., holocephalans, batoids), the main success of the group has not been due to major morphological changes in body shape or skeletal construction, but probably to more subtle changes in dentition and internal body chemistry. Whether heterochrony has directed these changes is hard to evaluate on the fossil evidence: there are no detailed ontogenetic studies of fossil sharks known to the author.

In climatiid acanthodians there was a trend to increase the complexity of the ventral shoulder girdle armour, while rigidly fixing the fin-spines to the armour. Again peramorphosis is suspected to have played a role in the development of these characters, although there is no direct evidence for this from fossil growth series.

Evolution in lungfishes has been influenced by heterochrony, particularly paedomorphosis, but major environmental changes, from marine to freshwater habitats, may have precipitated this. Since their initial burst of morphological diversity lasting for the duration of the Devonian Period, the lungfishes have changed very little; *Neoceratodus* may represent the longest living unchanged taxon of vertebrate alive today (?Triassic–Recent; Kemp and Molnar, 1981).

Dissociated heterochrony appears to have been a major influence on the origin of the first tetrapods from osteolepiform ancestry. Juvenile osteolepiforms share several cranial characters with primitive tetrapods which are lost in adult osteolepiforms, suggesting paedomorphosis. However, the continued development of the piscine fin into a functional tetrapod limb may have occurred by peramorphic processes. Evidence for heterochronic variations in amphibian limb structure is common in living salamanders.

It would be interesting to analyse the evolutionary trends in the many other groups of fishes, but lack of space precludes this. To attempt such a task requires detailed knowledge of the fossil record of the group, and some ontogenetic data on the growth of at least some primitive tax. Unfortunately there are few groups of fossil fishes for which we have ontogenetic data, but in recent years more palaeoichthyologists are recognising the value of such studies to fish phylogeny. Zidek (1976; 1985) has analysed growth in the Permian acanthodian *Acanthodes* and data from an Australian Devonian form *Howittacanthus* corroborates these results (Long 1986b). However, more new data on the growth and variations in early fossil fish groups are needed to evaluate the role of heterochrony in evolutionary trends of fishes. Perhaps this chapter may stimulate others to restudy primitive fish populations where growth sequences are preserved and add to the currently limited data base.

REFERENCES

Alberch, P. and Alberch, J., 1981, Heterochronic mechanisms of morphological diversification and evolutionary change in the neotropical salamander, *Bolitoglossa occidentalis* (Amphibia: Plethodontidae), *J. Morph.*, **167**: 249–64.

Alberch, P. and Gale, E.A., 1985, A developmental analysis of an evolutionary trend: digital reduction in amphibians, *Evolution*, **30**: 8–23.

Bemis, W.E., 1984, Paedomorphosis and the evolution of the Dipnoi, *Paleobiology*, **10**: 293–307.

Bernacsek, G.M. and Dineley, D.L., 1977, New acanthodians from the Delorme Formation (Lower Devonian) of N.W.T., Canada, *Palaeontographica*, **158 A**: 1–25.

Bishop, I.R. and Foxon, G.E.H., 1965, The mechanism of breathing in the South American lungfish, *Lepidosiren paradoxa*: a radiological study, *J. Zool.*, **154**: 263–71.

Campbell, K.S.W. and Barwick, R.E., 1982, The neurocranium of the primitive dipnoan *Dipnorhynchus sussmilchi* (Etheridge), *J. Vert. Paleont.*, **2**: 286–327.

Campbell, K.S.W. and Barwick, R.E., 1983, Early evolution of dipnoan dentitions and a new species, *Speonesydrion*. *Mem. Ass. Aust. Palaeont.*, **1**: 17–49.

Campbell, K.S.W. and Barwick, R.E., 1984a, *Speonesydrion* an Early Devonian dipnoan with primitive toothplates, *Palaeoichthyologica*, **2**: 1–48.

Campbell, K.S.W. and Barwick, R.E., 1984b, The choana, maxillae, premaxillae and anterior palatal bones of early dipnoans, *Proc. Linn. Soc. N.S.W.*, **107**: 147–70.

Campbell, K.S.W. and Barwick, R.E., 1985, An advanced massive dipnorhynchid lungfish from the Early Devonian of New South Wales, Australia, *Rec. Aust. Mus.*, **37**: 301–16.

Campbell, K.S.W. and Barwick, R.E., 1986, Palaeozoic lungfishes—a review, *J. Morph. Suppl.*, **1**: 93–131.

Campbell, K.S.W. and Barwick, R.E., 1988, Geological and palaeontological information and phylogenetic hypotheses, *Geol. Mag.*, **125**: 207–27.

Carlson, K.J., 1968, The skull morphology and estivation burrows of the Permian lungfish, *Gnathorhiza serrata*, *J. Geol.*, **76**: 641–63.

Cheng, H., 1989, On the tubuli in Devonian lungfishes, *Alcheringa*, **13**: 153–66.

Clack, J.A., 1988, New material on the early tetrapod *Acanthostega* from the Upper Devonian of East Greenland, *Palaeontology*, **31**: 699–724.

Compagno, L.J.V., 1977, Phyletic relationships of living sharks and rays, *Am. Zool*, **17**: 303–32.

Denison, R.H., 1941, The soft anatomy of *Bothriolepis*, *J. Palaeont*, **15**: 535–61.

Denison, R.H., 1968, Early Devonian lungfishes from Wyoming, Utah and Idaho, *Fieldiana (Geol.)*, **17**: 353–413.

Denison, R.H., 1978. *Placodermi. Vol. 2. Handbook of paleoichthyology*. Gustav Fischer Verlag, Stuttgart and New York.

Denison, R.H., 1979. *Acanthodii. Vol. 5. Handbook of paleoichthyology*. Gustav Fischer Verlag, Stuttgart and New York.

Forey, P.L., 1986, Relationships of lungfishes. *J. Morph., Suppl.*, **1**: 75–91.

Gardiner, B.G., 1980, Tetrapod ancestry: a reappraisal. In A.L. Panchen (ed.), *The terrestrial environment and the origin of land vertebrates*. Academic Press, London: 177–185.

Hanken, J., 1982: Appendicular skeletal morphology in minute salamanders, genus *Thorius* (Amphibia: Plethodontidae): growth regulation, adult size determination and natural variation, *J. Morph.*, **174**: 57–77.

Jarvik, E., 1952: On the fish-like tail in the ichthyostegalid stegocephalians with descriptions of a new stegocephalian and a new crossopterygian from the Upper Devonian of east Greenland. *Medd. Grøn.*, **114**: 1–90.

Jarvik, E. 1981, *Lungfishes, tetrapods, palaeontology and plesiomorphy* by D.E. Rosen, P.L. Forey, B.G. Gardiner and C. Patterson (review), *Syst. Zool.*, **30**: 378–84.

Kemp, A. and Molnar, R.E., 1981, *Neoceratodus forsteri* from the Lower Cretaceous of New South Wales, Australia, *J. Palaeont.*, **55**: 211–17.

Larsen, A., 1980, Paedomorphosis in relation to rates of morphological and molecular evolution in the salamander *Aneides flavipunctatus* (Amphibia: Plethodontidae), *Evolution*, **34**: 1–17.

Long, J.A., 1983, New bothriolepid fishes from the Late Devonian of Victoria, Australia, *Palaeontology*, **26**: 295–320.

Long, J.A., 1985a: A new osteolepidid fish from the Upper Devonian Gogo Formation, Western Australia, *Rec. West. Aust. Mus.*, **12**: 361–77.

Long, J.A., 1985b: The structure and relationships of a new osteolepiform fish from the Late Devonian of Victoria, Australia, *Alcheringa*, **9**: 1–22.

Long, J.A., 1986a, New ischnacanthid acanthodians from the Early Devonian of Australia, with comments on acanthodian interrelationships, *Zool. J. Linn. Soc.*, **87**: 321–39.

Long, J.A., 1986b, A new Late Devonian acanthodian fish from Mt. Howitt, Victoria, Australia, with remarks on acanthodian biogeography, *Proc. R. Soc. Vict.*, **98**: 1–17.

Long, J.A., 1988, Late Devonian fishes from the Gogo Formation, Western Australia, *Nat. Geog. Res.*, **4**: 438–452.

Long, J.A., 1989. A new rhizodontiform fish from the Early Carboniferous of Victoria, Australia, with remarks on the phylogenetic position of the group, *J. Vert. Paleont.*, **9**: 1–17.

Long, J.A., 1990, Heterochrony and the origin of tetrapods, *Lethaia*, **23**: 157–66.

Long, J.A., in prep. The long history of fishes on the Australian continent. In P. Rich *et al* (eds), *Fossil Vertebrates of Australasia*, Chapman and Hall, London.

Lund, R., 1986, The diversity and relationships of the Holocephali. In T. Uyeno, R. Arai, T. Taniuchi and K. Matsura (eds), *Indo-Pacific Fish Biology: Proceedings of the Second International Conference on Indo-Pacific Fishes*. Ichthyological Society of Japan, Tokyo: 97–106.

McNamara, K.J., 1986, A guide to the nomenclature of heterochrony. *J. Paleont.*, **60**: 4–13.

McNamara, K.J., 1988, The abundance of heterochrony in the fossil record in M.L. McKinney (ed.), *Heterochrony in evolution*. Plenum, New York: 287–325.

Maisey, J., 1984, Chrondrichthyan phylogeny: a look at the evidence. *J. Vert. Paleont.*, **4**: 359–71.

Maisey, J., 1986, Heads and tails: a chordate phylogeny, *Cladistics*, **2**: 201–56.

Miles, R.S., 1973, Articulated acanthodian fishes from the Old Red Sandstone of England, with a review of the structure and evolution of the acanthodian shoulder girdle. *Bull. Brit. Mus. Nat. Hist. (Geol.)*, **24**: 111–213.

Milner, A.R., 1980, The temnospondyl amphibian *Dendrerpeton* from the Upper Carboniferous of Ireland, *Palaeontology*, **23**: 125–141.

Moy-Thomas, J.A. and Miles, R.S., 1971, *Palaeozoic fishes*, 2nd edn, Chapman and Hall, London.

Ørvig, T., 1985, Histologic studies of ostracoderms, placoderms and fossil elasmobranchs. 5. Ptyctodontid tooth plates and their bearing on Holocephalan ancestry: the condition of chimaerids., *Zool. Scr.*, **14**: 55–79.

Oster, G.F., Shubin, N., Murray, J.D. and Alberch, P. 1988. Evolution and morphogenetic rules: the shape of the vertebrate limb in ontogeny and phylogeny, *Evolution*, **42**: 862–84.

Panchen, A.L., 1985, On the amphibian *Crassigyrinus scoticus* Watson from the Carboniferous of Scotland, *Phil. Trans. R. Soc.*, series B, **309**: 505–68.

Rief, W.E., 1977, Tooth enameloid as a taxonomic criterion: 1. A new euselachian shark from the Rhaetic–Liassic boundary. *Neues Jb. Geol. Paläont, Monats*, **1977**: 565–7.

Rief, W.E., 1978, Tooth enameloid as a taxonomic criterion: 2. Is *'Dalatias' barnstonensis* Sykes 1971 (Triassic, England) a squalomorphic shark?, *Neues Jb. Geol. Paläont, Monats*, **1978**: 42–58.

Romer, A.S., 1945, *Vertebrate paleontology* 2nd edn, University of Chicago Press, Chicago.

Rosen, D.E., Forey, P.L., Gardiner, B.G. and Patterson, C., 1981, Lungfishes, tetrapods, palaeontology and plesiomorphy, *Bull. Amer. Mus. Nat. Hist.*, **167**: 159–276.

Schultze, H.-P., 1970, Folded teeth and the monophyletic origin of the tetrapods, *Amer. Mus. Nov.*, **2408**: 1–10.

Schultze, H.-P., 1984, Juvenile specimens of *Eusthenopteron foordi* Whiteaves, 1881 (Osteolepiform Rhipidistian, Pisces) from the Late Devonian of Miguashua, Quebec, Canada. *J. Vert. Paleont.*, **4**: 1–16.

Schultze, H.-P., 1986, Dipnoans as sarcopterygians. *J. Morph. Suppl.*, **1**: 39–74.

Schultze, H.-P. and Arsenault. M., 1985, The panderichthyid fish *Elpistostege*: a close relative of tetrapods? *Palaeontology*, **28**: 292–309.

Schultze, H.-P. and Campbell, K.S.W., 1986, Characterization of the Dipnoi, a monophyletic group, *J. Morph. Suppl.*, **1**: 25–37.

Smithson, T.R., 1982, The cranial morphology of *Greererpeton burkemorani* Romer (Amphibia: Temnospondyli), *Zool. J. Linn. Soc.*, **76**: 29–90.

Stensio, E.A., 1948, On the Placodermi of the Upper Devonian of East Greenland II. *Palaeozoologica Groenlandica* **2**: 1–622.

Thanh, T.D. and Janvier, P., 1987, Les Vertébrés Dévoniens du Viêtnam. *Ann Paléont*, **73**: 165–94.

Thomson, K.S. and Campbell, K.S.W., 1971, The structure and relationships of the primitive Devonian lungfish — *Dipnorhynchus sussmilchi* (Etheridge), *Bull. Peabody Mus. Nat. Hist.*, **38**: 1–109.

Vorobyeva, E.I., 1975, Formenviefalt und Verwandtschaftsbeziehungen der Osteolepidida (Crossopterygii, Pisces), *Paläont. Zeits.* **49**: 44–54.

Vorobyeva, E.I., 1977, Morphology and nature of evolution of crossopterygian fish. *Tr. palaeont. Inst.*, **163**: 1–239.

Wang Shi-Tao, 1987, A new antiarch from the Early Devonian of Guanxi. *Vert. Palasiatica*, **25**: 81–90.

Watson, D.M.S., 1937, The acanthodian fishes. *Phil. Trans. R. Soc.*, series B, **228**: 49–146.

Werdelin, L. and Long, J.A., 1986, Allometry in the placoderm *Bothriolepis canadensis* and its significance to antiarch evolution, *Lethaia*, **19**: 161–9.

Westoll, T.S., 1949, On the evolution of the Dipnoi. In G.L. Jepson, G.G. Simpson and E. Mayr (eds), *Genetics, palaeontology and evolution*, Princeton University Press, Princeton, NJ: 121–84.

Young, G.C., 1981, Biogeography of Devonian vertebrates, *Alcheringa*, **5**: 225–45.

Young, G.C., 1984, Comments on the phylogeny and biogeography of antiarchs (Devonian placoderm fishes) and the use of fossils in biogeography, *Proc. Linn. Soc. N.S.W.*, **107**: 443–73.

Young, G.C., 1988, Antiarchs (placoderm fishes) from the Devonian Aztec Siltstone, Southern Victoria Land, Antarctica, *Palaeontographica*, **202 A**: 1–125.

Zangerl, R., 1981, *Chondrichthyes 1. Vol 3a. Handbook of paleoichthyology*. Gustav Fischer Verlag, Stuttgart and New York.

Zhang, G.-R., 1978, The antiarchs from the Early Devonian of Yunnan, *Vert. Palasiatica*, **16**: 147–186.

Zhang, G.-R., 1984, New form of Antiarchi with primitive brachial process from the Early Devonian of Yunnan. *Vert. Palasiatica*, **22**: 81–91.

Zidek, J., 1976, Kansas Hamilton Quarry (Upper Pennsylvanian) *Acanthodes* with remarks on the previously reported North American occurrences of the genus, *Univ. Kansas Paleont. Contr.*, **83**: 1–41.

Zidek, J., 1985, Growth in *Acanthodes* (Acanthodii, Pisces) data and implications, *Paläont. Zeits.* **59**: 147–66.

Chapter 12

REPTILES

Michael J. Benton

INTRODUCTION

The fossil record of reptiles provides a rich supply of morphological and phylogenetic data, and many of its components have been interpreted as trends. Certain lineages show trends of increasing body size, increasing brain size, digital reduction or crest development. Larger-scale trends are indicated in the step-by-step acquisition by therapsids of mammalian characters, and in the apparently sequential modification of theropod dinosaurs into birds. At a larger scale, the reptiles show supposed trends of increasing diversity and increasing breadth of adaptation.

The term 'evolutionary trend' has a broad range of meanings (see Chapter 1 herein), ranging from a rather non-committal sense almost synonymous with 'change through time', to a strong sense of progressive impelled modification in a single direction, a meaning not far from the much-reviled teleological (goal-directed) interpretations of certain evolutionists earlier this century. In the present chapter, various kinds of trend will be described, and the term is used only to indicate a pattern of evolutionary change which, in retrospect, heads in one direction. Examples will be given of large-scale trends that lasted for hundreds of millions of years, and of small-scale trends that occurred over a few million years. Attempts will be made to identify the causes, extrinsic (competition, predator pressure, progressive habitat modification), intrinsic (heterochrony, canalisation) or non-existent (statistical artefact, imagination).

LARGE-SCALE TRENDS

The phylogeny of tetrapods

Reptiles are traditionally one of the four classes of four-limbed vertebrates, the tetrapods. The class Reptilia is, however, a paraphyletic group: it can be defined in such a way that it arose from a single common ancestor, an amphibian, but it excludes two major groups of descendants, the classes Aves (birds) and Mammalia. The lower boundaries of these two classes define the upper limits of Reptilia, boundaries that are arbitrary and defined, for the class Reptilia, by the absence of characters, rather than by their possession. Hence, large-scale trends in reptiles must be considered in terms of the Tetrapoda, or the Amniota (i.e. Reptilia, Aves and Mammalia) as a whole.

The broad pattern of tetrapod phylogeny, radiations and extinctions have been outlined elsewhere (Benton, 1989) and will only be summarised here in order to illustrate some large-scale trends. The simplified phylogenetic tree (Figure 12.1 A) shows several apparent phases of radiation. The amphibians, mainly 'temnospondyls' and batrachosaurs, dominated Carboniferous coal forests, and continued as important aquatic animals during the Permian and Triassic. Several major reptilian lineages radiated in Late Carboniferous times, but seemingly at low diversity and small size. The mammal-like reptiles dominated Permian landscapes, 'pelycosaurs' in the Early Permian, and therapsids in the Late Permian.

Many new groups apparently arose and radiated in the Triassic: the archosaurs and cynodonts on land and the plesiosaurs, ichthyosaurs, and placodonts in the sea. Later in the Triassic, the turtles, crocodilians, pterosaurs, dinosaurs, and mammals began to radiate. The dinosaurs, pterosaurs, plesiosaurs and ichthyosaurs died out in the Late Cretaceous, and the birds and mammals became dominant.

When these phylogenetic data are plotted in the form of a graph of total familial diversity against time, some of the patterns become clearer (Figure 12.1 B). The most obvious feature is a marked rise in total diversity through time, which is interrupted by a number of declines. The key large-scale trend is diversity increase.

Figure 12.1 A. Phylogenetic tree of the Tetrapoda, showing the major groups of fossil and living amphibians and reptiles. Relationships (dashed lines), stratigraphic duration (vertical extent of balloon), and diversity (width of balloon) of each group are shown. Based on various sources given in Benton (1989). B. Standing diversity against time for tetrapods, with six postulated mass extinction events (nos. 1–6) shown, and their relative magnitudes (percentage falls in diversity) listed. Abbreviations: A, Ichthyostegidae; B, 'Pelycosauria'; Q, Quaternary; CARB, Carboniferous; CRET, Cretaceous; DEV, Devonian; PERM, Permian; TERT, Tertiary; TRIAS, Triassic.

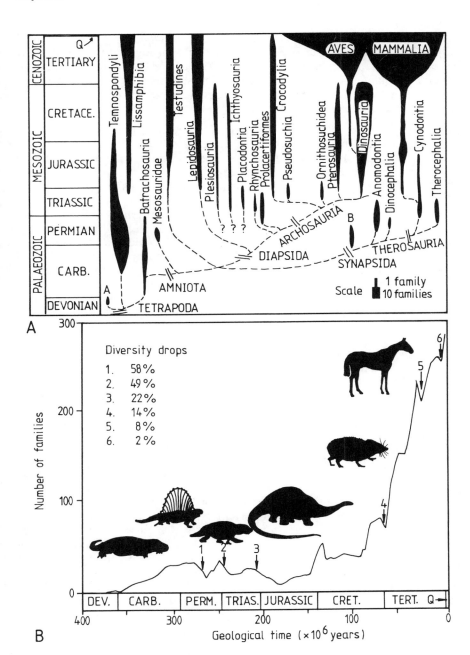

Diversity increase

All fossil records show increases in total diversity through time, whether they be records of marine invertebrates (Sepkoski *et al.*, 1981; Raup and Sepkoski, 1982), land plants (Niklas *et al.*, 1983), terrestrial vertebrates (Benton 1985a; see also Figure 12.1B). There are many possible explanations for this general large-scale trend (see also Benton, 1990):

(a) The trend is an artefact of the patchy quality of the fossil record and the way in which we study it (Raup, 1972):
  (i) the volume of sedimentary rock preserved unmetamorphosed increases towards the present;
  (ii) the area of exposure of such fossiliferous rock increases towards the present;
  (iii) palaeontologists devote much more attention to younger faunas and floras and hence name more species, genera, and families.
(b) The trend is real and the result of genuine biological factors, such as:
  (iv) reduction in the probability of extinction by the optimization of fitness (Raup and Sepkoski, 1982), or reductions in levels of diffuse competition within communities (van Valen, 1984), or increases in the species: family ratio through time (Flessa and Jablonski, 1985).
  (v) increase in the overall adaptive space occupied by a clade;
  (vi) subdivision of niches, so that later forms occupy narrower niches and have more specialised adaptations;
  (vii) increased endemicity as a result of abiotic changes;
  (viii) 'cladistic inevitability': if the groups under study are all clades (i.e. monophyletic groups, those that include all of the descendants of a single common ancestor), their diversity is likely to increase above one if they are to survive.

Most palaeontologists now seem to accept that the trend of increasing diversity is real, that explanations (i)–(iii) may affect the pattern, but they do not account for every aspect of it. Three main lines of argumentation have been employed. First, Valentine (1973) argued that, although factors (i)–(iii) would probably seriously affect graphs of total global species diversity through time, they become less important as sources of error at higher taxic levels. In other words, figures for total species diversity would be severely affected by the availability of suitable rocks, and the intensity and nature of palaeontological study. However, there are fewer genera and even fewer families at any time, and the discovery of one species establishes the presence of a genus or a family just as well as the discovery of a hundred closely related species in rocks of the same age. In other words, families approximate more closely in taxic scale the stratigraphic acuity and completeness of most of the fossil record than do species which equate to time-scales measured in thousands of years and with more complete representation.

The second line of argument, that rising diversity trends are real, was advanced by Sepkoski *et al.* (1981). They noted that five different data bases on marine invertebrates all yielded similar graphs of increasing diversity through time, even when the sources of data were semi-independent, such as marine trace fossil species, and families, genera, and species of body fossils compiled from different standard sources. The third argument was used by Signor (1982). He made estimates of the various postulated systematic sources of error in the known fossil record, such as (i)–(iii), and removed them as far as possible, by computer modelling, from the data. He found that the rising diversity trend was relatively little diminished by these manipulations, and that this was true for orders, families, genera and species.

Causes of diversity increase

If the rising diversity trend is real, is it possible to disentangle an explanation from hypotheses (iv)–(vii) noted above? Hypothesis (iv) may be true for marine invertebrates, since they appear to show a significant reduction in the probability of extinction through time (Raup and Sepkoski, 1982; van Valen, 1984; Flessa and Jablonski, 1985), whatever the reasons are for that. However, this is not the case for terrestrial tetrapods, in which Benton (1985a) found an overall *rising* trend in total extinction rates (0.008 more families dying out per million years; 5.0 per cent increase per stage; $p < 0.005$), and only a slightly declining per-taxon extinction rate (0.0001 fewer families dying out per family per million years; 0.08 per cent decline per stage; $p < 0.05$). The evidence does not suggest that tetrapods show a reduction in the probability of extinction.

Explanation (viii), the inevitability of an increase in diversity from a single ancestor, clearly plays a part, but this kind of stochastic, or random-walk type of model cannot account for the overall pattern. Hoffman (1986) developed a 'neutral' model for diversification, but had to maintain the rate of origination higher than the rate of extinction permanently in order to produce a realistic curve. A straightforward random-walk model with no such constraint would be as likely to decline as to increase.

This leaves three explanations, increase in adaptive space occupied (v), subdivision of niches (vi), and increasing endemicity (vii). These are all testable, and some preliminary analyses (Benton, 1990) suggest that all three factors have played a role in the diversification of the tetrapods.

The first known tetrapods, the families Ichthyostegidae and Acanthostegidae, were semi-aquatic piscivores that lived in and close to fresh waters. During the Carboniferous and Permian, many lineages of tetrapods became more fully terrestrial in habit, and various gliding and flying forms appeared in the Permian and Triassic. Fully marine forms arose in the Permian (mesosaurs), Triassic (ichthyosaurs, plesiosaurs, placodonts), Jurassic (crocodilians), Cretaceous (mosasaurs), Eocene (whales), and Oligocene (seals). Fliers became more diverse after the evolution of birds in the Jurassic and bats in the Eocene. Arboreal and burrowing habitats were occupied at low diversity from the Permian and Triassic onwards (Figure 12.2A). Furthermore, diets broadened to include insectivory and carnivory in the Carboniferous, broadly adapted browsing herbivory in the Permian, omnivory after the Late Permian, and ever-more specialised herbivorous and

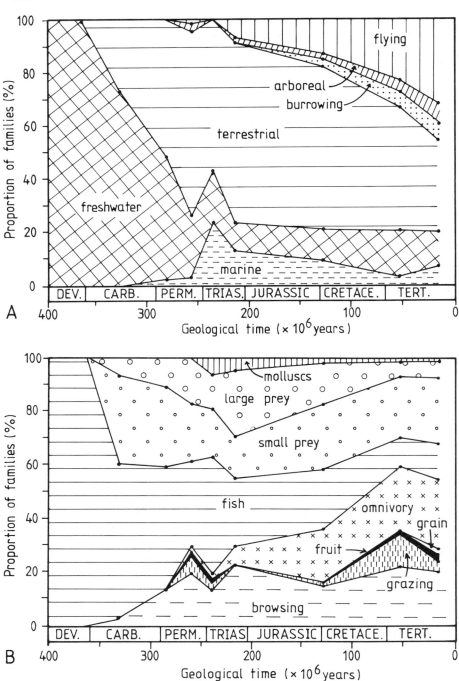

A

B

carnivorous modes (e.g., grazing, fruit, grain molluscs) after that (Figure 12.2B). These and other major adaptive expansions must have played a large part in increasing tetrapod diversity.

Subdivision of niches is suggested by an increase in the diversity of species within well-preserved tetrapod faunas. It is valid to compare an individual exceptionally preserved fauna (Lagerstätte) from the Carboniferous with one from the Eocene, since levels of preservation appear to be equivalent. In other words, Lagerstätten of all ages can show preservation of tiny animals, soft parts such as skin, scales and hair, and wholly soft-bodied organisms. A preliminary survey of 100 such faunas, spanning the past 350 million years in Europe and North America (Benton, 1990) shows a marked increase to mean faunal diversities of 18 in the Carboniferous and 35–51 in the Miocene and Pliocene (Figure 12.3A). In all cases, the species numbers are based on fossils found in a single locality or small fossil deposit, and the remarkable increase in standing diversity must in part be the result of specialisation and niche reduction.

Increasing endemicity, as the third possible factor, has also played a part (Benton, 1985b). Particularly since the Carboniferous, with increasing north–south climatic differentiation, tetrapod families have become more restricted in their geographical distribution. In Permian, Triassic and Jurassic times, many families of terrestrial tetrapods were essentially global in distribution, and it was expected that the break-up of Pangaea after the Triassic would have led to increased endemicity (Benton, 1985b). However, levels changed very little from Triassic times to the present (Figure 12.3B). It is likely that this large-scale increase in endemicity has allowed global familial diversity to increase. Levels of endemicity at lower levels (e.g. basin–basin) have yet to be investigated.

Global diversity increase is identified as a real trend in the evolution of tetrapods, and no doubt of most other major clades. This is a pattern that may be 'one-off' in all cases, dependent on specific historical circumstances, such as the break-up of Pangaea, latitudinal climatic diversification, the diversification of fishes (hence providing new adaptive space for predators), the diversification of angiosperms, and other major environmental changes. The specific effects of these events, and the relative roles of broad increases in adaptive space occupied, subdivision of niches, and increases in endemicity have yet to be assessed. Explanations for the global rising trend in the diversity of marine invertebrates and plants may be different. There is no evidence for an inevitable motor of change that drives diversity ever upwards.

Figure 12.2. Proportions of (A) broad habitat types and (B) diets of terrestrial and marine tetrapod families through time. The habitats and diets were determined from the primary literature, and they represent the activities of all, or the majority of, species within a family. Families are grouped into broader time units than stages in order to provide large enough samples throughout (N = 44–360, mean = 146 families). Abbreviations as for Figure 12.1.

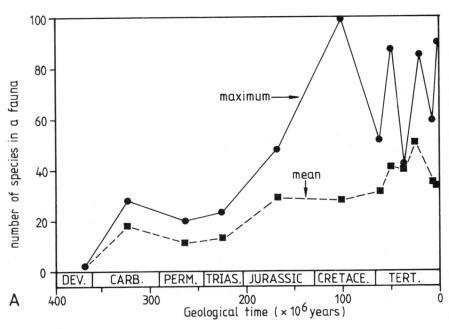

A

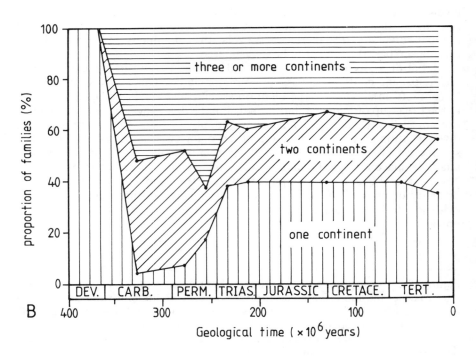

B

Size increase

A second major trend seen in reptilian evolution is body-size increase within all clades, at all levels of the taxonomic hierarchy. Among tetrapods as a whole, mean body size at any time has increased overall since the Devonian (Figure 12.4 A). The increase has, of course, been episodic, reset to lower levels by mass extinction events, which have generally affected large animals most. The overall increase in mean body size is not, however, as clear-cut as might have been expected, because the evolution of large animals has always been matched by diversification of smaller ones, too. Hence, the 'pulses' of evolution from modest to large size that follows major extinction events concern the largest taxa only and do not greatly affect the mean values. This is an example of a trend that is more to do with expansions of variance than with any genuine overall shift in mean body size (Gould, 1988).

Similar results are obtained from studies of particular major clades within the Tetrapoda, such as the Synapsida (Figure 12.4 B) and the Archosauria (Figure 12.4 C). The mean size of the synapsids declines with the advent of mammals, and that of the archosaurs declines with the advent of birds. The peak of maximum size represented by certain dinosaurs has not since been equalled.

The common finding of an evolutionary trend towards large body size in many lineages has become codified as Cope's Rule. Cope noted the tendency and he, and others, interpreted it as the expression of an inbuilt drive which could not be escaped. The trend has also been interpreted in terms of the selective advantages of large size, such as improved ability to capture prey or escape from predators, greater reproductive success, increased intelligence (large bodies have large brains), better stamina, expanded size range of possible food items, decreased annual mortality, extended individual longevity, and increased heat retention per unit volume (reviewed in Stanley, 1973). However, large animals suffer selective disadvantages, such as the need for large amounts of food, proneness to suffer when environments change, and small population sizes and restricted gene pools, all of which mean a great likelihood of extinction.

Stanley (1973) interpreted Cope's Rule in reverse: he argued that clades are always founded by small, often generalised, ancestors and hence the only way to evolve is up and towards specialization. Through time, the size ranges of clades extend to larger and larger values, an expansion of variance, even if not a real increase in mean size, as noted above. It is not easy to distinguish a

---

Figure 12.3. A: Maximum and mean number of tetrapod species in well-preserved terrestrial faunas. This graph is based on a sample of 100 faunas, with 1–20 faunas sampled per stratigraphic period (or epoch in the Caenozoic). Each 'fauna' is the species list of a single quarry or restricted sedimentary basin. Note the overall rise, with fluctuations in the maximum curve probably largely the result of variations in preservation quality. B: Variations in the ultimate geographic distribution of terrestrial tetrapod families arising during each time interval (as in Figure 12.2). Abbreviations as in Figure 12.1.

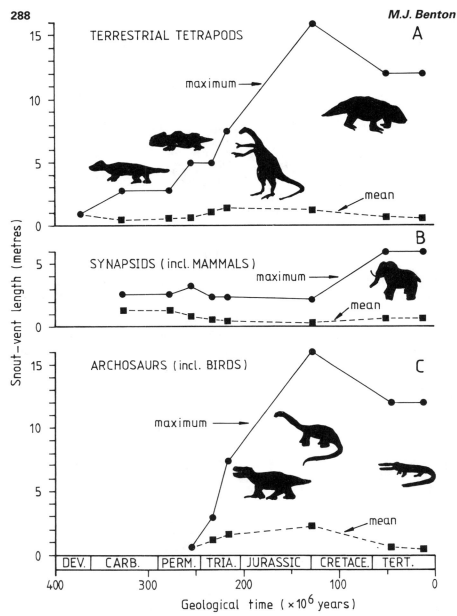

Figure 12.4. Size ranges of terrestrial tetrapods through time, plotted (A) for all taxa, (B) synapsids, including mammals, and (C) archosaurs, including birds. Maximum and mean measurements are indicated for grouped time intervals, as in Figure 12.2. Sizes are body lengths measured from the tip of the snout to the vent (i.e. excluding the tail), and they were recorded for each family according to a number of size classes. Mid-points of these size classes were used in the calculations (further details in Benton and Blacker, in prep.).

generally applicable explanation for increases in body size (reviewed by La Barbera, 1986).

SMALLER-SCALE TRENDS

Hundreds of examples of trends could be found from the fossil record of reptiles, some very small-scale, involving sequences of species within a genus, and others at a higher level, involving genera within a family, or families within an order. Four examples are given here, arranged in descending order of taxonomic and temporal scope.

Mammalian characters of synapsids

During the Late Carboniferous, Permian, and Triassic, the mammal-like reptiles were important terrestrial carnivores and herbivores; they included the ancestors of the mammals, which according to the traditional definition of Mammalia, arose in the latest Triassic (Kemp, 1982, 1988; Hopson and Barghusen, 1986). The mammal-like reptiles included a diversity of lineages during their history, all of which became extinct except for the one that eventually led to the mammals. In retrospect then, Kemp (1982; 1988) was able to show a progressive acquisition of mammal-like synapomorphies throughout the synapsid evolution, a 'trend' of increasing 'mammalness'. His data are converted here into a statistical expression of the cumulative addition of mammal-like synapomorphies through time, plotted for the postulated 'direct line' to mammals (Figure 12.5 A).

The anatomical changes involved affected all parts of the skull and skeleton: reduction and differentiation of the teeth, formation of a secondary palate, fusion of the orbit and lower temporal fenestra, reduction in complexity of the lower jaw, shift from the reptilian articular-quadrate jaw joint to the mammalian dentary-squamosal joint, modifications to neck and trunk vertebrae, loss of lumbar ribs, and modification of the limbs and limb girdles for erect gait. Each of these broad anatomical changes took place in several steps, and each could be interpreted as a trend, or the whole complex of changes from reptiles to mammals can be regarded as an integrated mosaic trend or what Kemp (1985) calls a 'correlated progression'.

If this sequence of changes can be regarded as a trend or trend complex, then the mechanistic interpretation by Kemp (1985) and others has been adaptational. Each facet of the change towards mammalness is seen as part of a progressive change in feeding, locomotory, thermoregulatory and sensory efficiency, all driven by the forces of natural selection. Suggested extrinsic factors to account for such a trend have included direct competition with other contemporary animals, such as the early archosaurs (see, for example, Charig, 1984); diffuse biotic interaction ('weak competition'), including predation pressure; differential environmental response, in which successful groups become modified as a result of climatic, vegetational, and other changes (Benton, 1987a); and chance, especially in terms of the apparently undirected selective effects of mass extinctions (relevant events indicated in Figure 12.5 A) and the opportunistic radiation of survivors (Benton, 1983; 1987a).

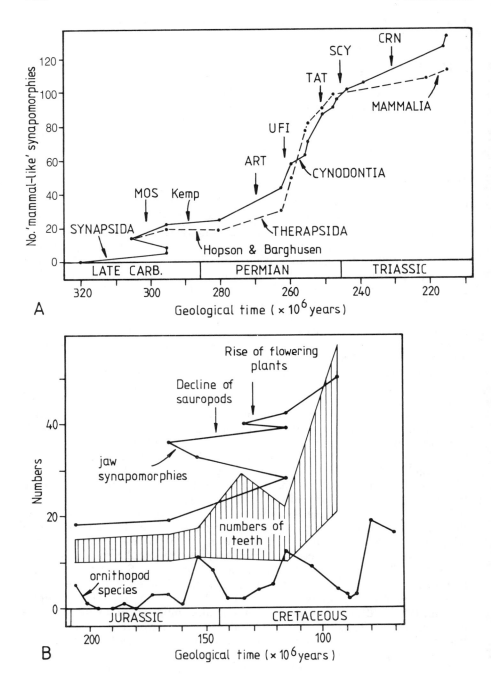

A   Geological time (× 10⁶ years)

B   Geological time (× 10⁶ years)

Herbivorous adaptation of ornithopod dinosaurs

The ornithopods, bipedal plant-eating dinosaurs, arose in the Late Triassic or Early Jurassic, and radiated at low levels during the Jurassic and Early Cretaceous, but were spectacularly successful in the Late Cretaceous. This success, as the duckbilled dinosaurs, or hadrosaurs, has generally been attributed to their increasingly efficient jaws and tooth-batteries (see, for example, Norman and Weishampel, 1985). Whereas early forms had low snouts, few spaced teeth, and shallow jaws, the hadrosaurs had high horse-like snouts, powerful batteries of up to 2000 teeth, deep well-muscled lower jaws, and specialised skull joints that allowed a form of chewing. These anatomical changes occurred in several stages during the 140 million years of the Jurassic and Cretaceous, and could be termed a trend.

This trend can be represented in several ways (Figure 12.5 B). First, a statistical analysis of the acquisition of synapomorphies of the ornithopods (cladistic analysis of Sereno, 1986) plotted against time, shows a pattern of jerky increase similar to the previous example. Unfortunately, the sequence of taxa as determined by the cladistic analysis does not match their temporal sequence, probably because of missing fossils, hence the zig-zag pattern of the graph. The second illustration of this trend is based on a single character: the range in numbers of tooth positions within each of the four jaw rami for key ornithopod taxa (Weishampel, 1984). There is an overall increase, from 10–15 in Early Jurassic taxa to 20–57 in the Late Cretaceous hadrosaurs, but much of the increase is produced by increase in variance. The lower limit rises from 10 to 20, but the upper limit extends dramatically from 15 to 57. The acquisition of dental and jaw characters and the rise in numbers of teeth is matched to some extent by an episodic rise in ornithopod species diversity (Weishampel and Norman, 1989), although much of this pattern merely reflects the distribution in time of dinosaur-bearing rock formations.

The trends in ornithopod teeth and jaw mechanics throughout the Jurassic and Cretaceous are generally interpreted as the result of responses to environmental stimuli, particularly changes in the available plant food, (Weishampel and Norman, 1989), and possibly to extinctions of other herbivores. Peaks in the evolutionary rates of ornithopod feeding

---

Figure 12.5. Two large-scale trends in reptilian evolution. A: The acquisition of mammal-like characters by synapsid reptiles of the Carboniferous, Permian and Triassic, based on the cumulative addition of synapomorphies at each node on the 'direct line' from the first synapsid to the first mammal. The taxa are those lettered A–H, J–R in Kemp (1982), with synapomorphies from Kemp (1982, 1988; solid line), and from Hopson and Barghusen (1986; dashed line). Dates are those of the oldest representative of the family, obtained from Benton (1987b). B: The trend to more specialised jaws and teeth (data from Sereno, 1986); ranges of the number of teeth in each jaw ramus of typical forms (data from Weishampel, 1984); and ornithopod species diversity, plotted stage by stage (data from Weishampel and Norman, 1989).

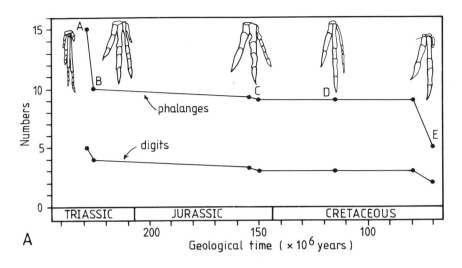

A

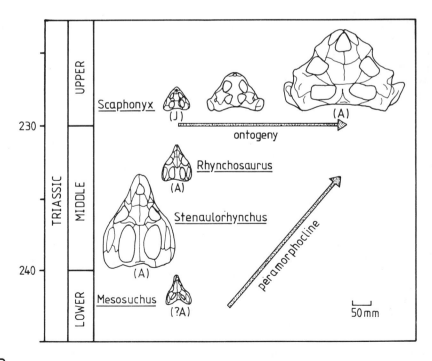

B

mechanisms and in global diversity follow the decline of sauropods, in the Northern Hemisphere at least, at the end of the Late Jurassic, and the rise of the angiosperms (flowering plants) during mid- to Late Cretaceous times.

Loss of digits in the theropods

A smaller-scale trend is seen in the theropod dinosaurs in which digits and phalanges were lost from the hands (Figure 12.6 A). The first theropods in the Late Triassic, such as *Coelophysis*, had four fingers and ten phalanges in all, having lost the fifth finger (equivalent to our little finger) seen in the first dinosaurs. Digit 4 was already reduced, having only one phalanx. Most Jurassic theropods retained four fingers until Late Jurassic times, when the number fell to three. This was typical of most Cretaceous forms, until the latest Cretaceous, when large carnosaurs such as *Tyrannosaurus* lost the third finger as well, leaving only two fingers and five phalanges in all.

It is hard to discern reasons, extrinsic or intrinsic, for such a trend. The theropods ranged greatly in size, from that of a turkey to the 14-metre length of *Tyrannosaurus*. Further, different taxa presumably used their hands for very different activities: grappling with prey, picking at carcasses, grasping eggs, capturing insects, picking their teeth, and so on. Hence, it would be hard to find a simple all-encompassing adaptive explanation for digital reduction.

Digital reduction events in Theropoda seem to have been unique. Hence, the reduction to four digits occurred at the Tetanurae node in the cladogram (Gauthier, 1986). The reduction to two digits occurs only in the Late Cretaceous tyrannosaurids, and some retain a remnant of digit 3.

One embryological explanation for digital reduction in modern tetrapods (Alberch and Gale, 1985) relates to the body size of embryos and adults. Small animals have fewer cells available for differentiation of the limb bud at early developmental stages, and hence can lose elements. It is hard to see how this kind of explanation could apply to *Tyrannosaurus*!

It might be possible to discern a heterochronic cause for this trend. Evidence from the embryology of modern birds (which are tetanuran theropod derivatives) show that digits 3 and 4 appear in early developmental

---

Figure 12.6. Two smaller-scale trends in reptilian evolution. A: Digital and phalangeal reduction in theropod dinosaurs. Numbers of fingers and phalanges in the hand of a selection of theropod dinosaurs, and the ancestral *Lagosuchus* (A). Theropods are *Syntarsus* (B), *Allosaurus* (C), *Deinonychus* (D), and *Tyrannosaurus* (E). B: The postulated peramorphocline (extended ontogeny) seen in the evolution of the rhynchosaur skull. Adult skulls of three Early and Middle Triassic rhynchosaurs and an ontogenetic series of three of the skulls of *Scaphonyx fischeri* are shown. The skulls are positioned vertically according to their occurrence in time (stratigraphic column on the left), and horizontally according to the ratio of posterior skull-roof width to mid-line skull-roof length. Abbreviations: A, adult; J, juvenile. Based on Benton and Kirkpatrick, 1989.

stages, but digit 4 disappears, and digit 3 is much reduced in adults. Further, a juvenile theropod, *Ornitholestes*, shows a tiny nubbin of digit 4, which is lost in the adult (Gauthier, 1986). It seems likely that the trend of digital reduction is peramorphic (ancestral adult morphology present in juvenile phase of descendant; see McNamara, 1986). It is not clear whether the 'add-on' developmental stages in which digits 4, or 3 and 4, of the juvenile are reduced and lost has resulted from earlier initiation of digital development (predisplacement), acceleration of the rate of development (acceleration), or delay in the onset of sexual maturity (hypermorphosis).

Hypermorphosis is often associated with an increase in adult size, which was generally the case at times of digital reduction in the course of theropod evolution. However, more study of the developmental sequences of theropods will be needed in order to decide which peramorphic mechanism applies.

Rhynchosaur skulls

The rhynchosaurs were important herbivores in the Triassic of much of the world. They were distantly related to the archosaurs (Figure 12.1 A), and they died out just before the radiation of the dinosaurs. During their relatively short span of 17 million years or so, rhynochosaurs evolved some remarkable specialisations in their skulls and teeth in particular. One striking trend was expansion of the posterior part of the skull.

The first rhynchosaurs had a typical reptilian skull shape, with a width to length ratio of 0.6 or 0.7. Later rhynchosaurs had much broader skulls, often broader than they were long, with width to length ratios of 1.0 to 1.2 (Figure 12.6 B). A developmental sequence of 13 skulls of the Late Triassic rhynchosaur *Scaphonyx fischeri* from Brazil demonstrated that this trend of skull expansion was peramorphic (Benton and Kirkpatrick, 1989). Although the youngest specimen in the sequence was six months to a year old, its skull width to length ratio of 0.8 was close to the ancestral adult condition. The peramorphic mechanism was identified tentatively as hypermorphosis since adult Late Triassic rhynchosaurs are generally larger than those of the Early or Middle Triassic.

PARALLEL EVOLUTION

Large-scale and medium-scale trends in tetrapod history have been hard to interpret in a clear-cut way, whether as real progressive changes induced by competition, predation, environmental change or chance. Smaller-scale trends often seem to result from heterochrony. Is there another way of testing among the various causal mechanisms?

Parallel evolution, the evolution of similar features in two or more lineages along approximately the same pathways, has occurred several times in the evolution of reptiles. The character patterns may provide a test between extrinsic and intrinsic causes since the underlying principles differ: natural selection/adaptation and heterochrony/canalisation, respectively. If parallel evolution is to be explained by extrinsic causes alone, the major morphological changes should occur roughly at the same time since they are presumably caused by particular new kinds of competitor or predator, or by particular changes in the physical environment. In detail, the morphological changes may seem very different, even though they have evolved towards the same function (i.e. analogies). Parallel evolution that is dominated by intrinsic constraints (heterochrony, canalisation) need not occur synchronously in all lineages, and the final results may be morphologically very similar (i.e. hard to distinguish from homologies) since they are channelled along a limited selection of developmental pathways. Two examples of large-scale parallel evolutionary trends among tetrapods will be considered: erect gait and skull-element reduction.

Erect gait

Erect (upright, parasagittal) gait is the derived posture seen in modern birds and mammals in which the limbs are held directly beneath the body and move backwards and forwards essentially in a vertical plane. It is not to be confused with bipedal (two-legged) gait. This posture is regarded as derived since living and fossil amphibians and most living and fossil reptiles have a sprawling gait, in which the limbs are held out to the sides and move in horizontal and vertical planes during walking.

The erect gait of mammals arose in ancestral mammal-like reptiles of the Triassic, while the erect gait of birds is traceable back through the dinosaurs to their Triassic thecodontian ancestors, the basal ornithosuchians. Erect gait arose at least twice in two other Triassic groups, the pseudosuchians (i.e. aetosaurs, rauisuchids, and poposaurids (?)) and the crocodylomorphs (the ancestors of crocodilians were small terrestrial bipedal insectivores with erect gait!) How synchronous were these four parallel changes from sprawling to erect gait in the Triassic, and how morphologically similar were they?

The changes are not obviously synchronous. Time-span ranges from 240–235 million years ago for the achievement of erect gait in the hind limb of cynodont mammal-like reptiles (Kemp, 1982) to about 220 million years ago for the achievement of erect gait in the early crocodylomorph *Saltoposuchus*. This part of the analysis reveals some problems of the kind palaeontologists always face. First, erect gait was achieved early on in the mammalian line, but only in the hindlimb—the first mammals of the latest Triassic still had a partially sprawling forelimb. Second, the first erect-limbed ornithosuchians and crocodylomorphs were essentially bipedal, during fast locomotion at least, and changes to the forelimbs may have been subject to different evolutionary forces. The first erect-limbed mammal-like reptiles and pseudosuchians were quadrupeds. Third, the origin of bipedality in crocodylomorphs may be much earlier than noted here if the poposaurids are the sister-group of crocodylomorphs (Benton and Clark, 1988). Conceivably, then, all dates would converge back on 235 million years ago. Hence, was the acquisition of erect gait synchronous or not?

Table 12.1 *Evolution of erect gait in four reptilian lineages (based on data from Kemp, 1982; Benton and Clark, 1988)*

Lineage	First appearance of erect gait	Type	Acetbulum	Primitive foot posture
Mammal-like reptiles	Middle Triassic	Buttress	Closed	Plantigrade
Ornithosuchians	Middle Triassic	Buttress	Open	Digitigrade
Pseudosuchians	Middle Triassic	Pillar	Closed	Plantigrade
Crocodylomorphs	Late Triassic	Buttress	Open	Digitigrade

In detail, the anatomical changes associated with the shift to erect gait are very different. Three groups share a 'buttress-erect' type of posture, in which the femur has an offset head that fits into the side of the near-vertical acetabulum (the hip socket) like the buttress of a church. The pseudo-suchians, on the other hand, have a 'pillar-erect' posture (Benton and Clark, 1988), in which the femur does not have an offset head, and it fits straight into a near-horizontal acetabulum, like a pillar supporting a heavy roof. Further, ornithosuchians and crocodylomorphs have an open acetabulum, and this is surrounded by all three hip bones in the ornithosuchians, but only the ilium and ischium in crocodylomorphs. The mammals and pseudosuchians retain a primitive closed acetabulum. The last two groups primitively had a planti-grade foot posture, in which the sole of the foot touches the ground, while the other two had a digitigrade posture, in which the foot rests only on the tips of the toes.

In general, then, the acquisition of erect gait may have been a response largely to extrinsic factors. It seems to have occurred broadly synchronously in a few unrelated groups, and it has caused different anatomical modifica-tions in general. An intrinsic element to the trend may be discerned, however, in some features of digital reduction, for example. The erect crocodylomorphs and ornithosuchians both show digital reduction, and the digit to become reduced in both cases is toe 5. This may suggest an element of canalisation.

Skull-element reduction

Most tetrapod lineages show some tendency to reduction in the number of skull bones. At the broad scale, the early amphibians had fewer skull elements than their fish ancestors, and the first reptiles had fewer than those amphibians. However, certain lineages of amphibians and reptiles lost further elements in comparison with their closest relatives. Examples include the extinct aistopods, living frogs, salamanders, and gymnophionans (caecilians), several lizard groups, amphisbaenians, and snakes (Table 12.2).

The reductions are obviously not synchronous, ranging in date from the Carboniferous (Visean, *c*.340 million years ago) to the Tertiary (dates

Table 12.2 *Loss of skull elements as a result of miniaturisation in diverse tetrapod groups. Data from Rieppel (1984), Benton (1987b), Carroll (1987), and Milner (1988)*

Lineage	Origin of group (age in millions of years)	Mean adult skull length (mm)	Elements lost*
AMPHIBIA			
Aistopoda	Visean (340)	15	pf,pp,sq,st,t
Anura (frogs)	Scythian (242)	15	ect,j,pf,prf,pp,t
Urodela (salamanders)	Bathonian (172)	10	ect,j,pf,pp,prf,t
Gymnophiona	Pliensbachian (195)	10	j,l,pf,po,prf,st
REPTILIA			
Acontinae	No fossils	15	l,po
Anguidae	Campanian (79)	10	sq
Pygopodidae	No fossils	8	j,l,po,sq,st
Dibamidae	No fossils	15	j,l,pf,po,sq,st
Amphisbaenia	Paleocene (60)	15	j,l,pf,po,sq,st
Serpentes (snakes)	?Early Cretaceous (120)	30	ep,j,l,pf?,sq

* Elements lost in extreme forms of the group in question.

Abbreviations: ect, ectopterygoid; ep, epipterygoid; j, jugal; l, lacrimal; pf, postfrontal; po, postorbital; pp, postparietal; prf, prefrontal; sq, squamosal; st, supratemporal; t, tabular.

unknown, lizard groups with no fossil record; less than 60 million years ago). Anatomically, many of the changes are remarkably similar: elements such as the jugal, postfrontal, postorbital, prefrontal, and supratemporal are lost even in widely different lineages. The similarities are stronger within major clades, such as the Lissamphibia (living amphibians) and the Squamata (lizards and snakes) (Table 12.2).

These observations suggest broadly intrinsic causation, especially canalisation. Rieppel (1984) has observed that all of these groups showing reduction in numbers of skull elements are miniaturised. The biological constraints of keeping the brain, eye and jaws large enough to be functional are associated with changes in relative skull proportions and losses of non-essential bony elements. In small frogs and salamanders,which may lack many skull elements, those that ossify late in development tend to be lost first (see, for example, Hanken, 1984; Trueb and Alberch, 1985), examples of the paedomorphic process of progenesis (see Chapter 3 herein). There is, however, variation among taxa in many cases that suggests the additional involvement of selective forces on the nature of skull-element reduction.

It should be noted that the losses of skull elements noted in Table 12.2 are often used as apomorphies in cladistic analyses of the groups, and yet they may not be homologous at all!

CONCLUSIONS

1. Trends occur in the fossil record of reptiles, and of tetrapods in general, at all scales, from their overall expansion in diversity, to changes in individual characters within lineages over a few million years.
2. Most large-scale trends seem to relate to major extrinsic causes, such as mass extinction events, changes in the physical environment, or the opening up of new adaptive space (e.g., new habitats, new sources of food).
3. Biotic factors such as competition and predation no doubt play a role in generating trends, but these are hard to imagine as remorseless driving forces on the geological time-scales involved (see also Benton, 1987a).
4. Most trends in tetrapods involve morphology and size at the lineage (species) level, but morphology, size and ecological strategies at higher taxic (major clade) levels.
5. Long-term trends, like the appearance of mammal-like characters in the ancestors of mammals, or modifications in herbivorous adaptations of ornithopod dinosaurs, seem to occur sporadically rather than in a gradual progression, with key changes happening in bursts, often associated with an opportunist radiation occurring after an extinction event of potential competitors.
6. Small-scale, species-level trends often involve clear evidence of heterochrony and canalisation constraining the pattern of change.
7. Parallel evolution may provide a test of the broad significance of extrinsic and intrinsic factors or trends. When extrinsic factors dominate, the parallel changes in more than one lineage may occur synchronously, and differ in morphological detail, whereas parallel trends dominated by heterochrony and canalization may show no synchrony at all, and changes should be morphologically similar. Examples of parallel evolution of erect gait (extrinsic) and of skull-element reduction (intrinsic) are given.

ACKNOWLEDGEMENTS

A large proportion of the original data in this paper was collected by Miss A.E. Blacker, funded by a Leverhulme Research Fellowship. I thank Heather Conn for typing the manuscript, and Libby Mulqueeny for the illustrations.

REFERENCES

Alberch, P. and Gale, E.A., 1985, A developmental analysis of an evolutionary trend: digital reduction in amphibians, *Evolution*, **39**: 8–23.

Benton, M.J., 1983, Dinosaur success in the Triassic: a noncompetitive ecological model, *Q. Rev. Biol.*, **58**: 29–55.

Benton, M.J., 1985a, Mass extinction among non-marine tetrapods, *Nature*, **316**: 811–14.

Benton, M.J., 1985b, Patterns in the diversification of Mesozoic non-marine tetrapods, and problems in historical diversity analysis, *Spec. Pap. Palaeont.* **33**: 185–202.

Benton, M.J., 1987a, Progress and competition in macroevolution, *Biol. Rev.*, **62**: 305–38.

Benton, M.J., 1987b, Mass extinctions among families of non-marine tetrapods; the data, *Mem. Soc. Géol. Fr.*, **150**: 21–32.

Benton, M.J., 1989, Patterns of evolution and extinction of vertebrates. In K.C. Allen and D.E.G. Briggs (eds), *Evolution and the fossil record*, Belhaven, London: 218–41.

Benton, M.J., 1990, The causes of the diversification of life. In G. Larwood and P. Taylor (eds), *Major evolutionary radiations*, Clarendon, Oxford, Syst. Ass. Spec. Vol. 39: in press.

Benton, M.J. and Clark, J., 1988, Archosaur phylogeny and the relationships of the Crocodylia. In M.J. Benton (ed.), *The phylogeny and classification of the tetrapods, Volume 1: amphibians, reptiles, birds*, Clarendon, Oxford, Syst. Ass. Spec. Vol. 35 A: 295–338.

Benton, M.J. and Kirkpatrick, R., 1989, Heterochrony in a fossil reptile: juveniles of the rhynchosaur *Scaphonyx fischeri* from the Late Triassic of Brazil, *Palaeontology*, **32**: 335–53.

Carroll, R.L., 1987, *Vertebrate paleontology and evolution*, Freeman, New York.

Charig, A.J., 1984, Competition between therapsids and archosaurs during the Triassic period: a review and synthesis of current theories, *Zool. Soc. Lond. Symp.*, **57**: 597–628.

Flessa, K.W. and Jablonski, D., 1985, Declining Phanerozoic background extinction rates: effect of taxonomic structure? *Nature*, **313**: 216–18.

Gauthier, J., 1986, Saurischian monophyly and the origin of birds, *Mem. Calif. Acad. Sci.*, **8**: 1–55.

Gould, S.J., 1988, Trends as change in variance: a new slant on progress and directionality in evolution, *J. Paleont.*, **62**: 319–29.

Hanken, J., 1984, Miniaturization and its effects on cranial morphology in plethodontid salamanders, genus *Thorius* (Amphibia: Plethodontidae): I. Osteological variation, *Biol. J. Linn. Soc.*, **23**: 55–75.

Hoffman, A., 1986, Neutral model of Phanerozoic diversification: implications for macroevolution, *N. Jb. Geol. Paläont. Abh.*, **177**: 219–44.

Hopson, J. and Barghusen, R., 1986, An analysis of therapsid relationships. In N. Hotton, III, P.D. MacLean, J.J. Roth and E.C. Roth (eds), *The ecology and biology of mammal-like reptiles*, Smithsonian Institution Press, Washington DC: 83–106.

Kemp, T.S., 1982, *Mammal-like reptiles and the origin of mammals*, Academic Press, London.

Kemp, T.S., 1985, Synapsid reptiles and the origin of higher taxa, *Spec. Pap. Palaeont.*, **33**: 175–84.

Kemp, T.S., 1988, Interrelationships of the Synapsida. In M.J. Benton (ed.), *The phylogeny and classification of the tetrapods, volume 2: mammals*, Clarendon Press, Oxford, Syst. Ass. Spec. Vol. 35 B: 1–22.

LaBarbera, M., 1986, The evolution and ecology and body size. In D.M. Raup and D. Jablonski (eds), *Patterns and processes in the history of life*, Springer-Verlag, Berlin: 69–98.

McNamara, K.J., 1986, A guide to the nomenclature of heterochrony, *J. Paleont.* **60**, 4–13.

Milner, A.R., 1988, The relationships and origin of living amphibians. In M.J. Benton (ed.), *The phylogeny and classification of the tetrapods, Volume 1: amphibians, reptiles, birds*, Clarendon Press, Oxford, Syst. Ass. Spec. Vol. 35 A: 59–102.

Niklas, K.J., Tiffney, B.H. and Knoll, A.H., 1983, Patterns in vascular land plant diversification, *Nature*, **303**: 614–16.

Norman, D.B. and Weishampel, D.B., 1985, Ornithopod feeding mechanisms; their bearing on the evolution of herbivory, *Am. Nat.*, **126**: 151–64.

Raup, D.M., 1972, Taxonomic diversity during the Phanerozoic, *Science*, **177**: 1065–71.

Raup, D.M. and Sepkoski, J.J. Jr, 1982, Mass extinctions in the marine fossil record, *Science*, **215**: 1501–3.

Rieppel, O., 1984, Miniaturization of the lizard skull: its functional and evolutionary implications, *Symp. Zool. Soc. Lond.*, **52**: 503–20.

Sepkoski, J.J., Jr, Bambach, R.K., Raup, D.M. and Valentine, J.W., 1981, Phanerozoic marine diversity and the fossil record, *Nature*, **293**: 435–7.

Sereno, P.C., 1986, Phylogeny of the bird-hipped dinosaurs (order Ornithischia), *Natn. geogr. Res.*, **2**: 234–56.

Signor, P.W., III, 1982,, Species richness in the Phanerozoic: compensating for sampling bias, *Geology*, **10**: 625–8.

Stanley, S.M., 1973, An explanation for Cope's Rule, *Evolution*, **27**: 1–26.

Trueb, L. and Alberch, P., 1985, Miniaturization and the anuran skull: a case study of heterochrony, *Fortschr. Zool.*, **30**: 113–21.

Valentine, J.W., 1973, Phanerozoic taxonomic diversity: a test of alternate models, *Science*, **108**: 1078–9.

Van Valen, L.M., 1984, A resetting of Phanerozoic community evolution, *Nature*, **307**: 50–2.

Weishampel, D.B., 1984, The evolution of jaw mechanics in ornithopod dinosaurs (Reptilia: Ornithischia), *Adv. Anat., Embryol., Cell Biol.*, **87**: 1–110.

Weishampel, D.B. and Norman, D.B., 1989, *Vertebrate herbivory in the Mesozoic; jaws, plants, and evolutionary metrics*, Spec. Pap. Geol. Soc. Am., 238: 87–100.

Chapter 13

MAMMALS

Christine M. Janis and John Damuth

INTRODUCTION

The fossil record of mammals and the study of trends

Mammals are an ideal group for tracking evolutionary trends at different hierarchical levels through time. They have an excellent fossil record, and the parts that preserve well (i.e. teeth and limb bones) are highly informative about probable mode of life. For many major groups there is a sufficient ecological diversity among extant species to permit the identification of detailed qualitative and quantitative functional relationships between ecology and morphology. Combinations of morphological features reflecting functional adaptation to particular kinds of ecological niche define ecomorphological types, or *ecomorphs*; distantly related species may represent the same ecomorph if they exhibit the same adaptive functional complex. Ecological inferences (e.g. diet, habitat) can thus be made on the basis of morphology and applied with confidence to extinct taxa (see, for example, Van Valkenburgh, 1987; 1988; Kappelman, 1988; in press; Solunias *et al.*, 1988; Solunias and Dawson-Saunders, 1988; Martin, 1989; Janis, 1990a; Scott, 1990). We do not need to rely upon direct (but questionable) analogy with the ecological characteristics of taxonomically related extant forms. (Solunias and Dawson-Saunders, 1988, provide an excellent example of the value of functional, ecomorphic analysis over simple taxonomic analogy. They show that African Miocene faunas, once assumed to contain savanna-dwelling animals because of the presence of bovids and hyaenids, actually comprised taxa with morphologies typical of a woodland or forest habitat. See also Kappelman, in press.) Understanding of the probable adaptive and palaeoecological significance of morphological features is crucial to the interpretation of many morphological trends.

The class Mammalia and its subclasses

Mammals are synapsid amniotes. Synapsids also include the mammal-like reptiles, the paraphyletic orders Pelycosauria and Therapsida, and apparently represent a very early offshoot of the amniote lineage in the Late Carboniferous. Mammals evolved from the cynodont therapsids in the latest Triassic (Rhaetian), a time period that also marked the virtual last of the therapsids (see Figure 12.1). While the mammal-like reptiles were the dominant large terrestrial vertebrates during the Permian and Early to Middle Triassic, they were eclipsed by the rise of the archosaurs in the Late Triassic (Benton, 1983a), and synapsid amniotes did not regain faunal dominance until the start of the Tertiary, following the extinction of the dinosaurs.

What constitutes a proper cladistic definition of a 'mammal' in the fossil record remains a problem: many of the characteristic mammalian features were either shared with·the latest cynodonts, or may have evolved in parallel within the Mammalia. Late Triassic mammals are generally recognised in the fossil record by their extremely small size and by the formation of wear facets on the teeth (that indicate precise occlusion with a rotary motion to the lower jaw) and a diphyodont pattern of tooth replacement.

Mammals are traditionally divided into three subclasses: Allotheria (extinct multituberculates), Prototheria (monotremes and numerous extinct Mesozoic groups), and Theria (including marsupials and placentals, as infraclasses, and various extinct precursors). The classic split into 'non-therian' and 'therian' mammals has recently been under attack; as originally defined, only therians were defined by unique features (suggesting that 'non-therians' were at best paraphyletic), and recent Mesozoic material has cast the traditional 'non-therian' status of monotremes under suspicion (Archer *et al.*, 1985). However, the 'trituberculate' therian mammals (those possessing the completely developed tribosphenic form of molar) appear to constitute a good monophyletic group that possesses all the 'typical' therian characteristics.

Early Mesozoic mammals are known world-wide, although in apparent low abundance and taxonomic diversity. The monotremes appear to have been limited to Australasia for their entire history. Multituberculates are known from Holarctic continents from the Jurassic to the Tertiary Oligocene, and from the Cretaceous of Argentina. Marsupials were apparently widely distributed in the Mesozoic and Early Tertiary (although from this time they are known primarily from North and South America; they appear in Europe in the Eocene, and their fossil record from Asia, Antarctica and Africa is extremely sparse). It is not known for certain when marsupials entered Australia. Since the Miocene they have been confined to South America and Australasia, although the opossum (*Didelphis*) immigrated to North America in the Early Pleistocene. Placentals are today found world-wide, although they were predominantly an Asian group in the Mesozoic, only first entered Australasia in the Miocene, and have no record from Antarctica.

MAJOR PATTERNS OF SYNAPSID EVOLUTION

Although the Tertiary is traditionally known as the 'Age of Mammals', and the patterns of evolution in later Tertiary mammals will be the focus of this chapter, the first two-thirds of the evolution of the class Mammalia occurred in the Mesozoic (Lillegraven *et al.*, 1979). Additionally, although mammal-like reptiles are officially classified in the (paraphyletic) class Reptilia, one cannot ignore the fact that the synapsids are a distinct lineage dating practically from the origin of the amniotes. Thus, we would like to present an overview of synapsid evolution, as portraying events leading up to the evolution of the present mammalian fauna. We have separated synapsid evolution into seven distinct pulses (cf. the 'dynasties' of Bakker, 1977, and the 'empires' of Anderson and Cruickshank, 1978) as shown in Figure 13.1.

Pulse 2 appears to be the first 'blossoming' of the synapsids, with a broader geographic distribution and a diversity of ecomorphs, including digging, semi-aquatic and insectivorous forms. Therapsids were almost certainly more active animals than pelycosaurs. The more erect posture of many early therapsids (Sues, 1986) may have functioned initially to facilitate breathing during locomotion, in response to selection pressure for greater locomotor stamina (Carrier, 1987). The postural change may also have been associated with the evolution of increased aerobic capacity, which would have been required to sustain high levels of locomotor activity. Continued selection for stamina may, in turn, have led in more derived therapsids to the evolution of endothermy, through progressively higher levels of active and resting aerobic metabolism (Bennett and Ruben, 1979; 1986). Pulse 2 could be subdivided into more primitive Early Triassic forms, and more advanced Middle Triassic forms (cf. Bakker, 1977), with the later forms comprising the more 'progressive' gorgonopsids, dicynodonts and theriodonts (exhibiting a larger temporal fenestra and an incipient bony secondary palate).

The Late Triassic cynodonts of Pulse 3 had many of the mammal-like features (see Figure 12.5A) that are suggestive of at least a degree of endothermy (Bennett and Ruben, 1986). However, despite the 'progressive' nature of this group, they were apparently ousted from the role of the dominant large tetrapods by the rise of the thecodont archosaurs (Benton, 1983a). The tiny earliest mammals of the latest Triassic (Pulse 4) appear following the replacement of both cynodonts and most other archosaurs by the dinosaurs. They were probably endothermic (Crompton, 1980) and were most likely scansorial, nocturnal, insectivorous forms. The only 'progressive' event during this pulse was the appearance of the specialised omnivorous/herbivorous multituberculates.

We see a distinct change in mammalian evolution with Pulse 5 in the early Late Cretaceous. This time period marks the first appearance of mammals above 1 kg in body weight and the first appearance of therians possessing a tribosphenic molar (and there was also an increase in the diversity of multituberculates and their dental complexity). Other apparently syn-chronous events include the initial radiation of the angiosperm plants and, presumably, a diversification of the insects that pollinate them. At this time, mammals apparently took over as the dominant small tetrapods, replacing the

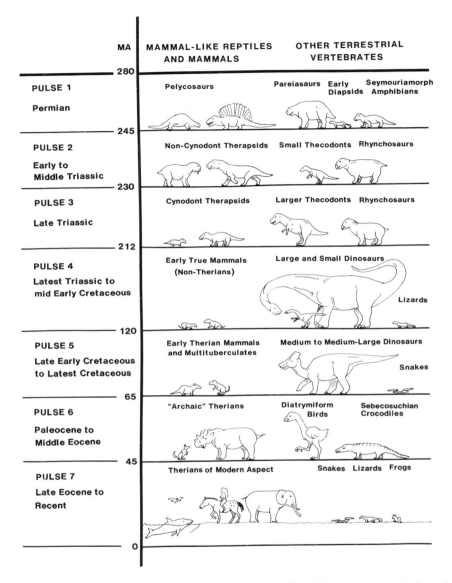

	MA	MAMMAL-LIKE REPTILES AND MAMMALS	OTHER TERRESTRIAL VERTEBRATES

Figure 13.1 'Pulses' of synapsid evolution. Note: Taxa illustrated are designed to give a flavour of the range of types of mammal and other vertebrates in each time period, and are not meant to be a comprehensive listing.
Body mass ranges are approximate estimates based on overall body size.

Pulse 1 (Permian): Pelycosaurs. Unspecialised medium- to large-sized amniotes (20–100 kg). Sprawling posture, ectothermic, tropical distribution.

Pulse 2 (Early to Middle Triassic): Therapsids. Probably higher metabolic rate than pelycosaurs: 'improved' posture (more of a para-sagittal stance) and earlier Mesozoic sphenodontids (lepidosaurs) and small dinosaurs. Thus, following the end-Cretaceous extinction event, mammals were in a prime position to form the 'replacement pool' for the next radiation of large tetrapods.

Following the extinction of the dinosaurs, Paleocene mammals of Pulse 6 radiated into larger body sizes and a greater diversity of dietary types. This pulse is marked by a number of 'archaic' mammalian orders that did not survive into the later Tertiary. Floral evidence from the Paleocene and Early Eocene indicates widespread closed-canopy tropical and subtropical forests throughout the Northern Hemisphere (Wolfe, 1978; 1985; Upchurch and Wolfe, 1987; Collinson and Hooker, 1987; Wing and Tiffney, 1987). Mammal faunas consistent with forest habitats—and containing similar taxa—are found from Baja California (30°N palaeolatitude) to Ellesmere Island (78°N palaeolatitude) in the Early Eocene of North America (McKenna, 1980; Flynn and Novacek, 1984). Thus, the adaptive diversity of Early Tertiary mammals was largely restricted to tropical and subtropical forest forms— species of small to medium body size, feeding on animal matter, fruit, and low-fibre leaves, with a paucity of true herbivores.

We consider the origins of the modern mammalian fauna of Pulse 7 to have been during the Late Eocene. The Middle Eocene represents the thermal maximum of the Tertiary, and also the end of the predominance of archaic types of mammal. From the Late Eocene to the present, the world has

---

Figure 13.1 – cont.
greater volume of jaw musculature. Wider geographical distribution (tropical and temperate zones). Size range 10–500 kg.

Pulse 3 (Late Triassic): Cynodont therapsids. Evidence for at least some degree of endothermy: diaphragm, secondary palate, differentiated dentition, masseter muscle in jaw. Size range 0.5–30 kg.

Pulse 4 (Latest Triassic to mid-Early Cretaceous): Early true mammals. Endothermic, widespread geographically but low diversity. Size range 30–500 g.

Pulse 5 (late Early Cretaceous to latest Cretaceous): First therian mammals. Concurrent with radiation of angiosperm plants, see appearance of therians with tribosphenic molars (cheek teeth that can crush as well as shear). Split of therians into placentals and marsupials happens at this time. Multituberculates (non-therians) also diversify and obtain more complex cheek teeth. Size range up to 5 kg.

Pulse 6 (Early Palaeocene to Middle Eocene): 'Archaic' therians plus didelphoid marsupials. Taxa seem mainly characteristic of tropical-type forest habitat world-wide. See first true carnivores and semi-aquatic herbivores. Size range up to 1000 kg.

Pulse 7 (Late Eocene to Recent): Modern therians, marsupials confined to Australasia and South America by Middle Miocene. Mammals inhabit diverse habitats, radiation of specialised groups such as bats, whales, cursorial mammals, hominoids. Size range 0.002–5000 kg (terrestrial) or to 100 000 kg (aquatic).

undergone a more or less one-way climatic change (see below), resulting in progressive net cooling, aridity and seasonality as demonstrated by patterns of vegetational zonation. These events produced an increasingly diverse, vegetationally heterogeneous and climatically complex world, and the subsequent 'progressive' radiation of mammals can be seen as a parallel, although more dramatic, version of Pulse 2, the initial radiation of the therapsids into a diversity of body sizes and morphological types.

Overall controls at the level of these major pulses must be seen as having been primarily extrinsic, related to broad-scale ecological events and the waxing and waning fortunes of broadly competing reptile groups (see Chapter 12 herein). Two important cross-species trends seen at this level stand out as possible examples of progression towards refinement of general adaptations. The first is a progressive increase in locomotor stamina, activity and metabolic levels, culminating in endothermy. This was presumably a response to long-term selection pressures for locomotor efficiency and success in predator–prey interactions. The second is the progressive changes in the skull, jaws and dentition of advanced cynodonts that led to a novel jaw articulation, incorporation of some elements of the former articulation into the middle ear, and the development of precise dental occlusion in mammals. Here, the long-term selection pressures were presumably for increased hearing acuity (Allin, 1986), which ultimately necessitated changes in musculature and jaw articulation to accommodate mammal-like chewing (Crompton and Hylander, 1986). Although in both cases general selective forces can be inferred, the evidence, particularly in the case of the cranial complex, indicates that the characters evolved in a mosaic fashion, and the detailed outcome could not have been predicted in advance (Kemp, 1982; Levinton, 1986). In addition, both trends cross a major shift in adaptive zone (*sensu* Van Valen, 1971). The particular set of features characteristic of the mammals may owe much to the fact that they went through a small-size bottleneck and a shift to nocturnal insectivory (McNab, 1983)—and should not be regarded as the endpoint of a unitary progressive series of adaptations to a single adaptive zone.

THE ROLE OF BIOGEOGRAPHY AND CLIMATE IN TERTIARY MAMMAL EVOLUTION

Pulse 7 of mammalian evolution took place in a world that was very different from the previous conditions of synapsid evolution. Although the break-up of the supercontinent Pangaea commenced in the Early Mesozoic, it was not until the Early Tertiary that the present-day condition of a number of fairly separate continental masses came about. For most of the Tertiary, the major terrestrial biogeographic regions were even more isolated than at the present. South America was separated from North America until the Pliocene; Europe and Asia had significant barriers to dispersal until the Early Oligocene; African faunas were apparently largely isolated from Holarctica until the Early Miocene; and, although Australia and Antarctica had separated from South America by the Early Tertiary, Australia was sufficiently distant from Asia to prevent immigration by island hopping at least until Miocene times.

Thus, in the later Tertiary all the continental masses contained their own 'seed fauna' of mammals, and much of the taxonomic diversification throughout mammalian history has been the result of convergent evolution of similar ecomorphs from different taxonomic sources on different continents (Kurtén, 1973). In fact, the present taxonomic diversity of mammals is considerably lower than that of the Middle Tertiary: although Australia is still fairly isolated and contains its own unique fauna, the endemic mammals of Africa and South America have suffered a considerable reduction in diversity since the immigration of Holarctic taxa.

The second difference in the world in Pulse 7 has been the growing impact of higher-latitude seasonality and vegetational zonation. This is also inter-related with biogeography, as changing continental positions and resulting changes in oceanic circulation (as evidenced, among other things, by the formation of the Antarctic ice-cap in the Oligocene–Miocene) were probably the major influences in the climatic changes of the later Tertiary (see Prothero, 1985; 1989). The vegetational zonation seen today, ranging from tundra and taiga at high latitudes to tropical forest at the Equator, and with a significant proportion of the tropical and subtropical land mass consisting of desert, probably represents a more heterogeneous global vegetation than at any other time in tetrapod history (Wolfe, 1978; 1985). Thus, the diversification of the modern mammals is associated with a greater number of available habitat types and potential ecological niches than seen in immediately previous times. A major point, which will be illustrated many times later in the text, is that it has been this more or less unidirectional climatic change in the Tertiary that has driven much of mammalian evolution and diversification (see also Gingerich, 1987).

GENERAL PATTERNS IN TERTIARY MAMMAL EVOLUTION

Diversification and clade shape

The apparently exponential (but fluctuating) net increase in total mammalian generic diversity throughout the Caenozoic (Gingerich, 1987) is probably largely an effect of a progressively more representative fossil record for younger time intervals (the magnitude of any real diversity increases, and the possible effects of changing levels of endemism, are hard to assess; we expect some increase in diversity from the increasingly heterogeneous global environment, as discussed above). Individual faunas of the Cretaceous and Paleocene show essentially modern levels of species richness (Gingerich, 1987). Moreover, different regions exhibit different histories of diversity: the climax of faunal diversity in Africa, and probably also in Australia (although the Tertiary record is poor), was in the Pleistocene; in North America, it was in the Middle Miocene.

At the level of genera within orders, most mammalian clades increase in diversity throughout their durations (Gingerich, 1987). At the level of genera within families, there is a trend from 'bottom-heavy' clades (those whose peak of diversity is early in their history) early in the Caenozoic radiation to 'top-heavy' clades later in the radiation (Gould *et al.*, 1987). Such a trend, from the early-originating to the late-originating clades within a taxonomic group, is thought to be typical of animal groups diversifying initially into 'empty' adaptive space, and its appearance among Tertiary mammals may reflect nothing more than the one-time situation of the Mammalia, radiating after the mass extinctions of the terminal Cretaceous. The causes of this general trend of change in clade shape are not clear for any group but presumably reflect an initially high opportunity for diversification, either from decreased evolutionary competition in an ecological vacuum or from new opportunities afforded by an altered physical environment (Gould *et al.* 1987; Kauffman, 1989; Jablonski and Bottjer, in press).

Extinction

Competitive displacement versus replacement
We sometimes see in the fossil record one taxonomic group achieve dominance in an adaptive zone that was formerly occupied by other, adaptively similar taxa. In cases where temporal overlap is involved, the issue arises as to whether the later group actually exterminated the earlier one by competition (competitive *displacement*) or whether the expansion of the later group is merely opportunistic, following the otherwise unrelated demise of the earlier group (*replacement*). If displacement is common, then trends towards progressive improvement of adaptation would be associated with successive radiations of new taxa. If replacement is the usual case, successive radiations imply nothing about trends in relative adaptedness. The difficulties of distinguishing competitive displacement from simple replacement in the fossil record are discussed at length by Benton (1983b) and Mass *et al.* (1988).

There are a number of classical examples in the mammalian fossil record that have been described as competitive displacement, with a 'superior' group ousting an 'inferior' one. One is the apparent success of placental mammals in replacing marsupials on most continents throughout the Tertiary, frequently ascribed to the differences in reproductive mode between the two taxa. Marsupial young are born at a relatively earlier stage than those of placentals and may be housed in a pouch to complete their development (although this is probably not the primitive marsupial condition). This fact, coupled with the fact that some placentals appear to have outcompeted some marsupials when the two have come into contact, has led to the assumption that the placental mode of reproduction is generally superior. Some have argued that the lower maternal investment in the offspring at birth constitutes an advantage for marsupials under some circumstances (Kirsch, 1977; Parker, 1977), but it is not clear if this is reflected in any macroevolutionary pattern. What seems to be true, however, is that placentals are at a competitive advantage in certain kinds of niche: where food resources permit increased metabolic rates (meat, seeds and grass, as opposed to invertebrates, fruit or leaves); at small body masses; and in very cold climates. This is because placentals can use their higher rates of metabolism in these circumstances to achieve higher rates of population growth (McNab, 1986; Lillegraven *et al.*, 1987). Marsupials have coexisted continuously with placentals in South America, but only in dietary niches whose occupants, placental and marsupial, are characterised by low metabolic rates. Throughout the course of the Tertiary, areas characterised by colder climates and the availability of grass as a resource both increased, expanding the scope of some niches in which placentals were favoured. Reproductive mode may also have limited the potential adaptive diversity of marsupials (e.g., marsupials have never produced fully aquatic forms, nor forms with large brains—see Lee and Cockburn, 1985; Lillegraven *et al.*, 1987). We thus see a pattern of *selective* displacement, combined with greater placental adaptive diversification, in the history of interaction of these two groups.

Another example is the supposed displacement in the Miocene of perissodactyls by artiodactyls as the dominant large herbivorous ungulates. This has traditionally been attributed to the superiority of the foregut system of fermentation in ruminant artiodactyls. However, the patterns of radiation and extinction in these two orders do not exhibit the inverse relationships that the hypothesis of displacement requires (Cifelli, 1981). Rather, replacement took place in the context of new resource types, to which artiodactyls could more easily adapt than perissodactyls. Patterns of climatic change and concomitant changes in vegetation structure and resource quality and abundance can explain the reduction of diversity of perissodactyls, which are better adapted than artiodactyls for feeding on high-fibre, tropical non-deciduous foliage; this interpretation is supported by the fact that the replacement took place at an earlier time in the higher latitudes than in the lower latitudes (Janis, 1990b). The perissodactyls that remained in abundance until the present day 'escaped' this constraint by either adopting a specialised high-fibre grazing diet (horses) or developing very large body size (rhinos).

In a third example, patterns of diversification and extinction in the fossil record support competitive displacement of both multituberculates and certain plesiadapiform primates by rodents in the Paleocene and Eocene of North America (Krause, 1986; Maas *et al.*, 1988). However, it is not known why rodents should have been generally adaptively superior.

Consideration of these examples suggests that displacement, and thus association of adaptive advance within an adaptive zone with successive radiation of new taxa, is not the rule in mammalian history, at least at the taxonomic levels considered here. Even where displacement is supported by patterns in the fossil record we may not be able to infer the reasons for the success of the winning taxon.

Extinction during intercontinental invasion and faunal interchange
If there are general tendencies or expected outcomes to interactions among faunas of different continents or regions, then large-scale changes in mammalian taxonomic diversity may exhibit directionality as a result of changing relations between biogeographic regions—whether or not these changing

relations themselves have an evident directional component (such as in the break-up of Pangaea). For example, it has been suggested that species from larger and more species-rich regions should have been selected for high competitive ability and thus have disproportionate success when invading smaller regions with 'naive' indigenous faunas (Webb, 1985); equilibrium island biogeographical theory (MacArthur and Wilson, 1967) extended to the scale of intercontinental faunal interchanges predicts that faunal impact and disturbance would be proportional to the numbers of taxa in the respective source regions (Marshall et al., 1982). Taken together, these theories would predict a gradual homogenisation of the world fauna, it becoming increasingly dominated by taxa from the larger land masses, with a net increase in the average level of adaptedness.

The infamous present-day example of the introduction of rabbits into Australia often serves as a model for support of the hypothesis that a new immigrant taxon will run hog-wild over the indigenous fauna. Although many examples of highly successful intercontinental invasion are found in the fossil record, Maas et al., (1988) rightfully point out that it is difficult to recognise cases in which successful outcomes of invasions *did not* occur in the fossil record. The fossil record will inevitably be biased towards preserving evidence of the luck of the winners.

Some well-known invader success stories include hypsodont horses from North America into the Old World in the Late Miocene; ruminants from Eurasia into North America, first in the Early Miocene and again in the Plio-Pleistocene; proboscideans from Africa to Holarctica in the Miocene; rodents from Eurasia into North America in the Late Palaeocene; dingos (probably by human agency) from Eurasia into Australia in the Holocene; felids from North America into South America in the Pliocene. To help set the record straight, we would like to note a few instances of 'invader failure', or at best only partial success, that are recorded in the fossil record. In these cases, invading taxa became established but neither radiated significantly nor, as far as can be determined, caused the extinction of major portions of the native fauna: the brachydont horse *Anchitherium* from North America into the Old World in the Early(?) Miocene; the supposed bovid *Neotragoceros* from Eurasia into North America in the Late Miocene; camelids from North America into the Old World in the Pliocene; entelodonts and anthracotheres from Eurasia into North America in the Oligocene; edentates from South America into North America in the Miocene; muroid rodents from Asia into Australia in the Miocene; hyraxes from Africa into Eurasia in the Pliocene.

It seems that there is no 'law' of immigrant winners or losers here, or any consistent relationship between size of source area and immigrant success. Each case must be interpreted on the basis of its own particular circumstances.

A related phenomenon is the merging of faunas due to the breakdown of major biogeographical barriers. Cases where this has been recorded in the fossil record are: the Grande Coupure of the earliest Oligocene, where the barrier of the Turgai Straits was removed between Europe and Asia, and the immigration of the Asian faunas apparently resulted in the demise of many European taxa; the merging of African and Eurasian faunas sometime in the

Late Oligocene or Early Miocene, with the extinction or reduction in diversity of many of the African taxa; and the formation of the Isthmus of Panama in the Pliocene, leading ultimately to the replacement of many South American taxa by descendants of immigrant North American forms (see below). In these cases, at least, it appears that the faunas of the larger source areas were the more successful; this is inconsistent with the variety of outcomes discussed above for less extensive immigration events. Are there different rules in effect for 'occasional' invasions as opposed to large-scale faunal interchanges?

The interchange between North and South America is the only one of these episodes that has been examined in detail. While the numbers of genera dispersing in each direction were proportional to the taxic richnesses of the respective source continents, the subsequent diversification of North American forms in South America was enormously greater than that of the South American immigrants in North America (Marshall *et al.*, 1982). The forms derived from North American taxa have often been suspected of being generally competitively superior to the South American species (Marshall *et al.*, 1982), but the reasons why this might have been so are unclear. Until adequate comparisons can be made and specific causal models evaluated, it seems premature to conclude that there are any simple, general rules applying to the outcome of faunal interchanges.

Mass extinctions of mammals

There appear to be three major periods of mammalian mass extinction during the later Tertiary: the Late Eocene, the Middle to Late Miocene, and the Late Pleistocene. The first of these two fit in with the 26-million-year cycles of extinction postulated by Raup and Sepkoski (1984). In both cases there is clear evidence of a non-random pattern of extinction that can be explained by climatic events, and both extinctions appear to have occurred gradually over time, rather than geologically instantaneously. The Late Eocene extinctions affected mainly the remaining archaic mammals and those typical of tropical environments at higher latitudes, and can be explained by higher-latitude cooling resulting, ultimately, from continental movements (see Prothero, 1985; 1989). The Miocene extinctions appear to be the result of an increased seasonality of rainfall, culminating with the Messinian crisis in the Late Miocene, and the mammals most severely affected were those dependent on moist woodland habitats, such as browsing ungulates. The Late Pleistocene extinctions, famous for the 'megafaunal' extinctions of large carnivores and herbivores, resemble the other global extinctions as to the apparently non-random pattern of taxonomic losses. The Pleistocene extinctions have attracted considerable attention, in part from the suggestion that human hunting caused the extermination of large herbivores (see, for example, McDonald, 1984). However, Pleistocene climatic changes were no less severe than those associated with earlier Tertiary extinction episodes, when humans could not have been involved. The Pleistocene megafauna largely survived in Africa, and it has been suggested that humans, having evolved in Africa, were coadapted with the large herbivores and thus had little effect upon them

(Martin, 1984). However, we note that Africa is the one place where extensive tropical savannas survived Pleistocene climatic changes.

Anagenesis—trends within lineages

The most detailed studies of long-term phyletic evolution in the mammalian fossil record involve Early Tertiary mammals of North America. Here, large sample sizes and tight stratigraphic control have permitted detailed examination of morphological evolution over long spans of time (typically, 0.5–8 million years), within species lineages and during evolutionary transitions between previously named species and even genera (Gingerich, 1976; 1985; 1987; Bookstein *et al.*, 1978; Rose and Bown, 1984; Bown and Rose, 1987). For a given character, such as molar area, trends of both increase and decrease, of various durations, and periods of evolutionary stasis are observed in a variety of species. There is no tendency for phyletic evolution to proceed in any preferred direction. Bown and Rose (1987) have shown that evolution of the kinds of character complex that usually are diagnostic of new taxa proceed gradually and in a mosaic fashion. Such sequences frequently show very slow net rates of morphological change, which would correspond to extremely weak selection pressures apparently operating more or less consistently over millions of years (Lande, 1976; Charlesworth, 1984). Given strong selection pressures, morphological evolution could have occurred at faster rates, and certainly has done so at other times (see, for example, Lister, 1989). Short-term fluctuating episodes of strong directional selection could effect slow changes, but it is difficult to see how this could result in the smooth long-term trends that are observed. In any case, if the patterns and net rates of evolution exhibited by these Early Tertiary mammals are typical, the majority of mammalian clades would not show sustained, concerted phyletic evolution in any direction throughout their duration. Most species, once established, would tend not to change drastically but rather would occupy essentially their original niches until replaced or the niches disappeared (but see the next section on convergences). The major adaptive changes or diversification/extinction patterns underlying many large-scale trends in the mammals thus must reflect circumstances of unusual opportunity for speciation and adaptive change (such as altered environments, or entry into empty—or vacated—adaptive zones) rather than mere extrapolations of ordinary phyletic change. However, these studies of Early Tertiary lineages also suggest that natural selection is the primary orienting force of phyletic change and that it is active in the morphological divergences associated with speciation events.

CONVERGENCES AND PARALLELISMS IN MAMMALIAN EVOLUTION

Intercontinental convergences

Several forms of Tertiary mammal specialisation appear to have evolved only once in the Mammalia: the evolution of obligate aquatic forms in the Cetacea; the evolution of volant forms in the Chiroptera; the evolution of large ricochetal forms in the Phalangeroidea (kangaroos); and the evolution of bipedal striding accompanied by an upright posture in the primates (hominids).

However, most major forms of mammalian specialisation can be multiply illustrated by taxa of diverse origins filling these same ecological roles on different continents, as summarised in Table 13.1. We regard the widespread occurrence of detailed ecomorphic convergences to be prima-facie evidence for sustained, adaptive, phyletic trends at lower hierarchical levels, though the actual evolutionary sequences from more generalised ancestral forms are not observed. We emphasise some particularly stunning examples of convergence here. Specialised insect-borers (Table 13.1, ecomorph 5), with chisel-like incisors and one or more digits modified into an extremely long, thin tool to dig insects out of tree trunks, comprise an ecomorph which is exemplified by the aye-aye (*Daubentonia*) of Madagascar, but is also seen to a lesser extent today in the striped possum (*Dactylopsila*) in Australia and in the 'insectivoran' apatemyids of the Eocene in the Northern Hemisphere (von Koenigswald and Schierning, 1987). It has been suggested that this particular mammalian ecomorph evolves only in the absence of woodpecking birds (von Koenigswald and Schierning, 1987). Next, consider the ecomorph consisting of large herbivores with prominent claws on the front limbs (Table 13.1, ecomorph 2f). No similar ecomorphs exist today, but analysis of the post-cranial skeleton suggests that many of these taxa browsed by standing on their hind legs and pulling down vegetation with their forelimbs (see Coombs, 1983, for review). Finally, consider gliding forms (Table 13.1, ecomorph 9). Note the strange fact that such ecomorphs are absent from the tropical forests of South America, which instead harbour a large variety of arboreal forms with prehensile tails (Emmons and Gentry, 1983). A further interesting morphological convergence (not listed in Table 13.1) is the independent evolution in multituberculates, marsupials and placentals of specialisations within the tarsus allowing for backwards rotation of the foot, enabling arboreal forms to descend tree trunks head first (Szalay and Decker, 1974; Krause and Jenkins, 1983; Jenkins and McLearn, 1984).

Intracontinental iterative evolution

A special case of convergent or parallel evolution is the repeated, sequential replacement of the occupants of a particular adaptive zone in a region with species belonging to the same ecomorphological type, although not direct lineal descendants. There is often a gap in time between the occurrence of these equivalent ecomorphs, suggesting parallel or convergent evolution

Table 1: Examples of convergent evolution in mammals.

HOLARCTICA	AFRICA	MADAGASCAR	S. AMERICA	AUSTRALIA
1. CARNIVOROUS MAMMALS				
CARNIVORA	CARNIVORA	CARNIVORA	DIDELPHOIDS (M)	DASYUROIDS (M)
				PHALANGEROIDS (M)
		viverrids	borhyaenids*	marsupial 'lion',
				(*Thylacoleo**)
CREODONTA*	CREODONTA*			
UNGULATES:				
CETACEA				
mesonychids*				
1a. Long-legged Cursorial Carnivores: († = highly cursorial pursuit predators)				
CARNIVORA	CARNIVORA		DIDELPHOIDS (M)	DASYUROIDS (M)
dogs†, bear dogs*	hyaenas†		some borhyaenids*	marsupial 'wolf'
dog bears*	cheetah†			(*Thylacinus**)
CREODONTA*	CREODONTA*			(?still extant)
some hyaenodontids*	some hyaenodontids*			
ARCHAIC UNGULATES*				
some mesonychids*				

Table 1 (continued)

1b. Sabre-toothed Carnivores: († = Recent forms that may be evolving in this direction)

HOLARCTICA	AFRICA	MADAGASCAR	S. AMERICA	AUSTRALIA
CARNIVORA	CARNIVORA	CARNIVORA	DIDELPHOIDS (M)	PHALANGEROIDS (M)
sabre-toothed felids*	sabre-toothed felids*	fossa† (viverrid)	borhyaenid*	*Thylacoleo**
nimravids*			(*Thylacosmilus**)	
clouded leopard†				
CREODONTA*				
some hyaenodontids*				

1c. Semiaquatic Carnivores/Omnivores:

HOLARCTICA	AFRICA	MADAGASCAR	S. AMERICA	AUSTRALIA
INSECTIVORA	INSECTIVORA	INSECTIVORA	DIDELPHOIDS (M)	MONOTREMES
desmans	otter shrew	otter tenrec	water opossum (Yapok)	platypus
water voles				
pantolestids*				
RODENTIA				RODENTIA
muskrats				Australian water rats
CARNIVORA				
otters				

Table 1 (continued)

HOLARCTICA	AFRICA	MADAGASCAR	S. AMERICA	AUSTRALIA

1d. Fully Aquatic Carnivores: († = obligate aquatics)

HOLARCTICA	AFRICA	MADAGASCAR	S. AMERICA	AUSTRALIA
mink				
polar bear				
CARNIVORA				
seals & sea lions				
CETACEANS				
whales† & dolphins†				

2. SPECIALISED HERBIVORES WITH CELLULOSE FERMENTATION

HOLARCTICA	AFRICA	MADAGASCAR	S. AMERICA	AUSTRALIA
	UNGULATES:		UNGULATES:	PHALANGEROIDS (M)
	PERISSODACTYLS	PROBOSCIDEANS	NOTOUNGULATES*	
	ARTIODACTYLS	HYRACOIDS	LITOPTERNS*	
		ARSINOTHERES*	PYROTHERES*	
	SOME RODENTS	SOME RODENTS	SOME RODENTS	

2a. Cursorial (running) Forms with Long Legs: († = richochetal)

HOLARCTICA	AFRICA	MADAGASCAR	S. AMERICA	AUSTRALIA
PERISSODACTYLS	HYRACOIDS		LITOPTERNS*	PHALANGEROIDS (M)
horses	pliohyracines*		proterotheres*	kangaroos†

Table 1 (continued)

HOLARCTICA	AFRICA	MADAGASCAR	S. AMERICA	AUSTRALIA
ARTIODACTYLS			NOTOUNGULATES*	PHALANGEROIDS (M)
camelids			notohippids*	rat kangaroos†
ruminants (antelope, deer, etc.)				

2b. Small-bodied Bounding Forms: († = richochetal)

HOLARCTICA	AFRICA	MADAGASCAR	S. AMERICA	AUSTRALIA
LAGOMORPHS	RODENTS		RODENTS	
rabbits & hares	spring hare†		mara & agoutis	
ARTIODACTYLS	ARTIODACTYLS		NOTOUNGULATES*	
tragulids	tragulids		typotheres*	

2c. Small-bodied, Rock Dwelling Forms:

HOLARCTICA	AFRICA	MADAGASCAR	S. AMERICA	AUSTRALIA
RODENTS	RODENTS		RODENTS	PHALANGEROIDS (M)
marmots	gundis & dassie rats		vizcachas,	ring-tailed possums
LAGOMORPHS	HYRACOIDS		chinchillas	rock wallabies
pikas	rock hyraxes		*Kerodon*	

2d. 'High-browsing' Forms, with Long Neck or Equivalent:

HOLARCTICA	AFRICA	MADAGASCAR	S. AMERICA	AUSTRALIA
ARTIODACTYLS	ARTIODACTYLS		LITOPTERNS*	PHALANGEROIDS (M)
aepycameline* camelids	giraffes		macraucheniids*	sthenurine* kangaroos

Table 1 (continued)

HOLARCTICA	AFRICA	MADAGASCAR	S. AMERICA	AUSTRALIA
	giraffe antelope (gerenuk)			(Could elevate arm above head)

2e. Large-bodied ('Graviportal') Terrestrial Forms:

HOLARCTICA	AFRICA	MADAGASCAR	S. AMERICA	AUSTRALIA
PERISSODACTYLS	PROBOSCIDEANS		NOTOUNGULATES*	PHALANGEROIDS (M)
rhinos	elephants		toxodontids*	diprotodontids*
brontotheres*	mastodons*		PYROTHERES*	(= giant 'wombats')
	gomphotheres*		EDENTATES	
	ARSINOTHERES*		glyptodonts*	

2f. Large-bodied Clawed Forms:

HOLARCTICA	AFRICA	MADAGASCAR	S. AMERICA	AUSTRALIA
PERISSODACTYLS			NOTOUNGULATES*	PHALANGEROIDS (M)
chalicotheres*			homalodotheres*	palorchestids*
			EDENTATES	
			ground sloths*	

2g. Medium to large-bodied Semiaquatic Forms: († = fully aquatic)

HOLARCTICA	AFRICA	MADAGASCAR	S. AMERICA	AUSTRALIA
PERISSODACTYLS	PROBOSCIDEANS	ARTIODACTYLS	RODENTS	PHALANGEROIDS (M)
tapirs	moeritheres*	hippo*	capybara	?diprotodontids*
brachypotherine* rhinos			coypu	

Table 1 (continued)

HOLARCTICA	AFRICA	MADAGASCAR	S. AMERICA	AUSTRALIA
amynodontid* rhinos				
	ARTIODACTYLS			
ARTIODACTYLS	hippos			
some oreodonts*	anthracotheres*			
HYRAXES	SIRENIANS†			
Pliohyrax*	sea cows & manatees			
ARCHAIC UNGULATES*	DESMOSTYLIANS*			
pantodonts*				
uintatheres*				
RODENTS				
beaver				

2h. **Forms Possessing Horns (or Equivalent Structures):** († = dermal horn made of keratin)

HOLARCTICA	AFRICA	MADAGASCAR	S. AMERICA	AUSTRALIA
PERISSODACTYLS	ARSINOTHERES*		NOTOUNGULATES*	PHALANGEROIDS (M)
rhinos†			toxodontids*	kangaroos
brontotheres*				(Males have large arms, for
ARTIODACTYLS				'boxing')

Table 1 (continued)

HOLARCTICA	AFRICA	MADAGASCAR	S. AMERICA	AUSTRALIA
protoceratids*				
some oreodonts*				
one pig*				
ruminants (6 times in				
parallel)				
ARCHAIC UNGULATES				
uintatheres*				
RODENTS				
mylagaulids*				

3. ARBOREAL FRUGIVORE/HERBIVORES

HOLARCTICA	AFRICA	MADAGASCAR	S. AMERICA	AUSTRALIA
PRIMATES	PRIMATES	PRIMATES	PRIMATES	PHALANGEROIDS (M)
Old World monkeys	Old World monkeys	lemurs	New World monkeys	possums
	HYRACOIDS		DIDELPHOIDS (M)	tree kangaroos
	tree hyrax		some opossums	koala
			EDENTATES	
			tree sloths	

Table 1 (continued)

HOLARCTICA	AFRICA	MADAGASCAR	S. AMERICA	AUSTRALIA
4. ARBOREAL BRACHIATING FORMS				
PRIMATES		PRIMATES	PRIMATES	
gibbons		*Palaeopropithecus*	spider monkeys	
5. ARBOREAL INSECT BORERS (WITH ELONGATED DIGIT ON HAND)				
INSECTIVORA		PRIMATES		PHALANGEROIDS (M)
apatemyids*		aye-aye		striped possum
6. GUM EATING ARBOREAL FORMS				
PRIMATES	PRIMATES	PRIMATES	PRIMATES	PHALANGEROIDS (M)
lorises	bushbabies	dwarf lemurs	marmosets	petaurid possums
				Leadbeater's possum

7. MYRMECOPHAGEOUS (ANT/TERMITE-EATING) FORMS: († = less specialised forms that may be evolving in this direction)

HOLARCTICA	AFRICA	MADAGASCAR	S. AMERICA	AUSTRALIA
PHOLIDOTA	PHOLIDOTA		EDENTATES	DASYUROIDS (M)
pangolins	pangolins		anteaters	numbat
	TUBILIDENTATA		armadillos†	MONOTREMES
	aardvark			echidna

Table 1 (continued)

HOLARCTICA	AFRICA	MADAGASCAR	S. AMERICA	AUSTRALIA
	CARNIVORA	CARNIVORA		
	aardwolf† (hyena)	falanouc† (viverrid)		
	bat-eared fox†			

8. NECTAR EATING FORMS

HOLARCTICA	AFRICA	MADAGASCAR	S. AMERICA	AUSTRALIA
CHIROPTERA	CHIROPTERA	CHIROPTERA	CHIROPTERA	CHIROPTERA
				PHALANGEROIDS (M)
				honey possum
				feather-tailed possums

9. GLIDING FORMS

HOLARCTICA	AFRICA	MADAGASCAR	S. AMERICA	AUSTRALIA
RODENTS	RODENTS			PHALANGEROIDS (M)
flying squirrels	scaly-tailed squirrels			flying possums
'DERMOPTERANS'				
flying 'lemur' (colugo)				
Paromomyidae* (*Ignacius**)				
(relationships to colugo				
uncertain)				

Table 1 (continued)

HOLARCTICA	AFRICA	MADAGASCAR	S. AMERICA	AUSTRALIA
10. SUBTERRANEAN BURROWING FORMS				
INSECTIVORA	INSECTIVORA	INSECTIVORA	DIDELPHOIDS (M)	DASYUROIDS (M)
true moles	golden mole	mole tenrecs	*Necrolestes* *	marsupial mole
RODENTS	RODENTS			
gophers	root rats			
mylagaulids*	African mole rats			
Asiatic mole rats				
PHOLIDOTA				
epoicotheres*				
11. FORMS POSSESSING SPINES [OR OTHER DERMAL ARMOUR(†)]				
RODENTS	RODENTS		RODENTS	MONOTREMES
New World porcupines	Old World porcupines		New World porcupines	echidna
INSECTIVORA		INSECTIVORA		
hedgehogs		hedgehog tenrecs		
PHOLIDOTA	PHOLIDOTA		EDENTATA	
pangolins†	pangolins†		armadillos†	

Table 1 (continued)

HOLARCTICA	AFRICA	MADAGASCAR	S. AMERICA	AUSTRALIA
			glyptodonts†*	

12. RICHOCHETAL SEED-EATING FORMS

HOLARCTICA	AFRICA	MADAGASCAR	S. AMERICA	AUSTRALIA
RODENTS			DIDELPHOIDS (M)	RODENTS
kangaroo rats			argyrolagids*	Australian hopping mice
jerboas				DASYUROIDS (M)
PRIMITIVE EUTHERIANS*				some marsupial "mice"
zalambdalestids*				

13. SMALL-BODIED, BOUNDING INSECTIVOROUS/OMNIVOROUS FORMS

HOLARCTICA	AFRICA	MADAGASCAR	S. AMERICA	AUSTRALIA
	MACROSCELIDEA			PERAMELINA (M)
	elephant shrews			bandicoots

* = extinct taxon; (M) = marsupial taxon. Geographic region refers primarily to area of origin, rather than present-day distribution, except for Madagascar. Ecomorphs are numbered and some are further subdivided. Taxa of superfamilial rank and higher are indicated in upper-case, contained taxa of lower rank or specific examples are indented and given in lower-case.

following local extinction rather than replacement by competition. Particularly striking examples of such *iterative evolution* (Simpson, 1953) have been described from the Tertiary of western North America by Martin (1985; see also Figure 13.2). Here, suites of ecomorphs (saber-toothed cats, fossorial rodents, and selenodont artiodactyls) become extinct, then re-evolve from more distantly related lineages (or immigrate and radiate) according to a repeated, regular pattern, which Martin interprets as wet–dry climatic cycles lasting approximately 2.3 million years each. However, no iterative pattern of replacement is evident in many of the larger North American ungulates, such as equids and camelids. The causes of these repeating patterns are not well understood, but it is likely that extrinsic factors are affecting an array of quite different mammalian lineages simultaneously, and the sequence of ecomorphs is consistent with some form of climatic or environmental change.

'PROGRESSION' IN MAMMALIAN EVOLUTION: FACT AND FANTASY

A number of trends in mammalian evolution have been perceived as 'progression', implying that a given trend represents increasing levels of adaptedness in successive species to the requirements of a particular adaptive zone. Many of these trends are standard text book examples, and yet the hypotheses for the mechanisms involved are often superficial at best. There is no reason to assume that, because some lineage has survived to the present day, it (or its extant members) are somehow superior to extinct forms or necessarily represent the culmination of a logical or adaptive progression (see, for example, Langer, 1988). The Recent is as strange and idiosyncratic a slice of time as any other. There is nothing more 'special' about surviving from the Late Miocene to the Recent than there is about surviving from the Eocene to the Miocene. Much of what has been seen as progress in mammalian evolution may merely be ascribed to habitat change, which cannot be foreseen by a taxon or by the evolutionary process. The classic example of the 'progression' of *Hyracotherium* to *Equus* focuses undue attention on the adaptive adventures of a seemingly goal-directed lineage rather than considering the motor of equid evolution to have been the change in dominant habitat type in North America from tropical-type forest to prairie (see Figure 13.3).

Several overall trends in mammalian evolution have been ascribed explicitly to later Tertiary climatic change. One is the correlation of primate diversity in higher-latitude faunas with changing palaeotemperature (see Gingerich, 1984). Another is the development of 'savanna-adapted' faunas in the Miocene of North America (Webb, 1977) and South America (Webb, 1978; Pascual et al., 1985), although the North American faunas were not identical in ecological diversification to those of present-day East Africa (Janis, 1982, 1984; see also Figure 13.4). We review below a number of classical cases (or supposed cases) of directionality in large-scale mammalian evolution that can more plausibly be ascribed to environmental forcing or to adaptive diversification than to progression (as we define it here).

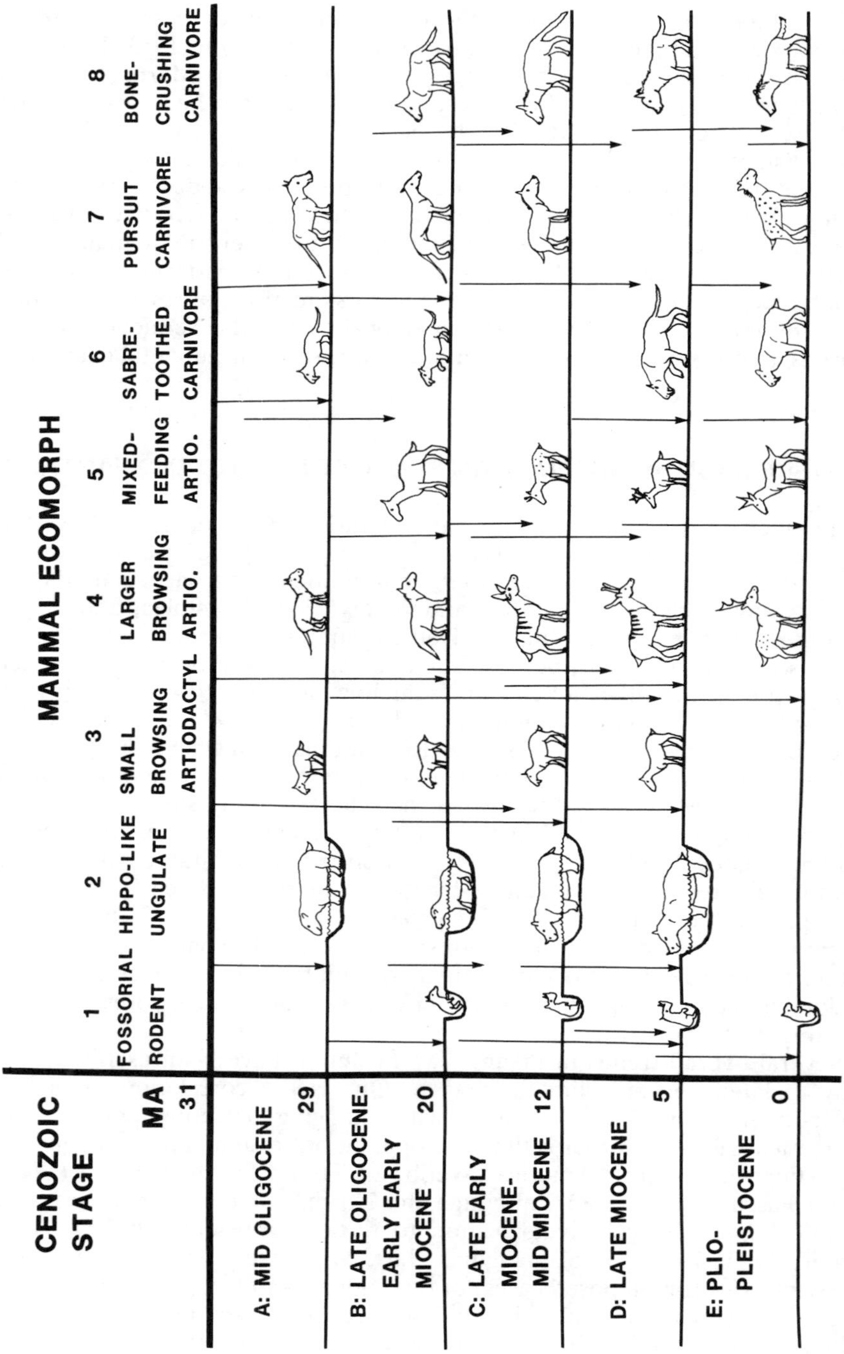

CENOZOIC STAGE
MA
MAMMAL ECOMORPH
1 FOSSORIAL RODENT
2 HIPPO-LIKE UNGULATE
3 SMALL BROWSING ARTIODACTYL
4 LARGER BROWSING ARTIO.
5 MIXED-FEEDING ARTIO.
6 SABRE-TOOTHED CARNIVORE
7 PURSUIT CARNIVORE
8 BONE-CRUSHING CARNIVORE
31
A: MID OLIGOCENE
29
B: LATE OLIGOCENE-EARLY EARLY MIOCENE
20
C: LATE EARLY MIOCENE-MID MIOCENE
12
D: LATE MIOCENE
5
E: PLIO-PLEISTOCENE
0

The evolution of large brain size

Mammals (along with birds) have large brains for their body size relative to most other vertebrates (Jerison, 1973). Humans are the most highly encephalised vertebrates (approached only by some cetaceans), and it is common for researchers to regard larger relative brain size as signifying greater 'intelligence'. The evolution of mammalian brain size is thus often expressed as a story of progressive increase in intelligence across the whole class (Jerison, 1973). One would expect that increased intelligence would be a general adaptation conferring greater fitness in all ecological contexts, so the expected macroevolutionary pattern would be more or less steady increase in relative brain size in all clades. However, it is certainly not the case that all mammal lineages have shown an increase in relative brain size over time. Most extant marsupials, and many placentals (for example, edentates, most insectivores and some rodents), have retained a similar size of brain to the mammals of the Mesozoic and Early Tertiary. An increase in relative brain

---

Figure 13.2 Patterns of iterative evolution on the Great Plains of North America.

Taken in part from text in Martin (1985).

Arrows show time duration of taxon or related taxa to the right of the arrow, and indicate where temporal overlap of similar ecomorphs occurred.

1. Fossorial rodent: 1B, *Palaeocastor* (castorid). 1C, *Mylagaulus* (mylagaulid). 1D, *Epigaulus* (mylagaulid). 1E, *Geomys* (geomyid). [=Recent ecomorph.]

2. Hippo-like semi-aquatic ungulate: 2A, *Metamynodon* (amynodont rhino). 2B, *Promerycochoerus* (oreodont). 2C, *Teleoceras* (brachypotherine rhino). 2D, *Teleoceras*. [No Recent ecomorph.]

3. Small browsing selenodont artiodactyl: 3A, *Leptomeryx* (traguloid ruminant). 3B, *Leptomeryx*. 3C, *Blastomeryx* (moschid ruminant). 3D, *Pseudoceras* (gelocid ruminant). [No Recent ecomorph.]

4. Larger browsing selenodont artiodactyl: 4A, *Merycoidodon* (oreodont, group 1). 4B, *Mesoreodon* (oreodont, group 2). 4C, *Dromomeryx* (dromomerycine dromomerycid ruminant). 4D, *Cranioceras* (cranioceratine dromomerycid ruminant). 4E, *Odocoileus* (cervid ruminant). [=Recent ecomorph.]

5. Small- to medium-sized, mixed-feeding selenodont artiodactyl: 5B, *Stenomylus* (camelid). 5C, *Aletomeryx* (aletomerycine dromomerycid ruminant). 5D, *Merycodus* (merycodontine antilocaprid ruminant). 5E, *Stockoceras* (antilocaprine antilocaprid ruminant). [Recent ecomorph is the related *Antilocapra* (pronghorn).]

6. Dirk-toothed sabre-toothed carnivore: 6A, *Hoplophoneus* (nimravid). 6B, *Eusmilus* (nimravid). 6D, *Barbourofelis* (nimravid). 6E, *Smilodon* (felid). [No Recent ecomorph.]

7. Pursuit carnivore: 7A, *Hyaenodon* (creodont). 7B, *Daphoenus* (amphicyonid). 7C, *Phoberocyon (hemicyonine ursid). 7E, Chasmoporthetes* (hyaenid). [Recent ecomorph is *Canis lupus* (the wolf, a canine canid).]

8. Bone-crushing carnivore: 8B, *Daphoenodon* (amphicyonid). 8C, *Amphicyon* (amphicyonid). 8D, *Osteoborus* (osteoborine canid). 8E, *Canis dirus* (canine canid). [No Recent ecomorph.]

size during the Tertiary appears to be typical only of primates, carnivores, ungulates and cetaceans. Even among and within these groups, increase in relative brain size occurred at different times and did not follow the same pattern or proceed to the same degree in all lineages.

For example, an increase in brain size in primates began early in their phylogenetic history (Martin, 1973), and this may in part have been related to a shift in life-history strategy (Shea, 1987). Nevertheless, relative brain sizes in the modern prosimian range were achieved by the Late Eocene (Radinsky, 1977), and prosimian brains have remained static throughout the Neogene and later (Jerison, 1985). Relative brain size in the earliest anthropoids is still poorly known, but there has been no pattern of increase (except among hominids) since the Oligocene, and there is no difference between apes and monkeys in encephalisation (Jerison, 1973; Radinsky, 1975). The dramatic increase in relative brain size within the hominids is a special adaptation of this lineage and began no earlier than 4–5 million years ago. Among ungulates and carnivores in both North and South America, modern levels of relative brain size had been achieved by Oligocene times (Radinsky, 1978; 1981).

The pattern observed for the mammals as a whole is thus one of inter-mittent increases in relative brain size in particular lines throughout the Tertiary radiation. This is an outstanding example of a trend that manifests itself as an increase in variance rather than as a general anagenetic trend (Gould, 1988; see also Chapter 1 of this volume). Since *decrease* in relative brain size would seem to offer few adaptive advantages, changes in a downward direction should be rare (but see Shea, 1983), so we would expect mammalian brain-size changes over time to generate a net increase in average encephalisation in the mammalian clade. However, this does not mean that increases in relative brain size were non-adaptive, nor that their timing does not reflect environmental change or expansion of adaptive diversity. The Tertiary increase in relative brain size in many mammalian lineages coincided with the beginning of Pulse 7, when the diversification resulting in the modern fauna began. Relative brain size is related, to a greater or lesser degree, to body size, diet, and habitat in a wide variety of extant mammal groups (Clutton-Brock and Harvey, 1980; Eisenberg and Wilson, 1978, 1981; Gittleman, 1986; Lemen, 1980; Mace *et al.*, 1981; Pagel and Harvey, 1989; but see Roth and Thorington, 1982).

Figure 13.3 Which is changing, the horse or the habitat?

Hyracotherium: Small size, three toes with food-pad, fairly short legs, low-crowned cheek teeth. Forest habitat.

Merychippus: Medium size, three toes but no foot pad, fairly long legs, moderately highly-crowned cheek teeth. Woodland to savanna habitat.

Equus: Large size, only one toe, very long legs, highly-crowned cheek teeth. Plains or prairie habitat.

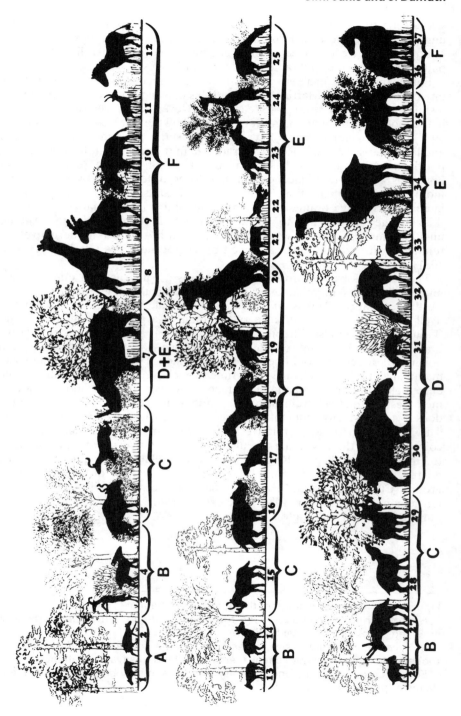

The evolution of larger body size

Cope's Rule is an empirical generalisation that animals tend to evolve toward larger body size over time. This is often interpreted as indicating widespread phyletic trends towards large size (see Chapter 4 herein). However, as noted above, preferential anagenetic trends toward larger size are not observed in detailed studies of Early Eocene species. Although there are many potential adaptive advantages to large size, there are potential disadvantages as well. The populations of mammal species of all body sizes have roughly equal potential to obtain trophic energy from their local environments, and to use this energy to grow and reproduce (Damuth, 1981; 1987). Stanley (1973) explained increase in average size of the members of a taxon as the result of the fact that the ancestors of most taxonomic groups tend to be relatively small and closer to the minimum size possible for the taxon than to the eventual upper limit. This, in turn, reflects the greater adaptive potential of generalised, small species; the fact that size changes at speciation should be

---

Figure 13.4 Differences in 'savanna-adapted' faunas.
Taken from Janis, 1982 (with permission, Cambridge University Press).
(I) Recent, East Africa
(II) Early Miocene, North America (Flint Hill Local Fauna, Batesland Formation, South Dakota; E. Hemingfordian).
(III) Late Miocene, North America (Love Bone Bed Local Fauna, Alachua Formation, Florida; L. Clarendonian).
Habitat types: A = forest; B = closed canopy woodland; C = open canopy woodland; D = woodland-savanna grade 1; E = woodland-savanna grade 2; F = open savanna.
Taxa:
(I): 1. *Hyemoschus aquaticus* (water chevrotain). 2. *Cephalophus nigrifrons* (duiker). 3. *Litocranius walleri* (gerunuk). 4. *Tragelaphus scriptus* (bushbuck). 5. *Tragelaphus strepsiceros* (greater kudu). 6. *Aepyceros melampus* (impala). 7. *Diceros bicornis* (rhino). 8. *Giraffa camelopardalis* (giraffe). 9. *Taurotragus oryx* (eland). 10. *Connochaetes taurinus* (wildebeest). 11. *Gazella granti* (Grant's gazelle). 12. *Equus burchelli* (Burchell's zebra).
(II): 13. *Parablastomeryx* (moschid). 14. *Barbouromeryx* (dromomerycid). 15. *Lambdoceras* (protoceratid). 16. *Diceratherium* (diceratherine rhino). 17. *Merycoidodon* (oreodont). 18. *Hypohippus* (browsing equid). 19. *Anchitherium* (browsing equid). 20. *Moropus* (chalicothere). 21. *Archaeohippus* (browsing equid). 22. *Merychyus* (oreodont). 23. *Parahippus* (mixed-feeding equid). 24. *Oxydactylus* (aepycameline camelid). 25. *Protolabis* (protolabine camelid).
(III): 26. *Pseudoceras* (gelocid). 27. *Pediomeryx* (dromomerycid). 28. *Tapirus* (tapir). 29. *Synthetoceras* (protoceratid). 30. *Aphelops* (aceratherine rhino). 31. *Merycodus* (merycodontine antilocaprid). 32. *Hemiauchenia* (cameline camel). 33. *Calippus* (dwarf tridactyl grazing equid). 34. *Aepycamelus* (aepycameline camelid). 35. *Neohipparion* (tridactyl grazing equid). 36. *Pliohippus* (monodactyl grazing equid). 37. *Astrohippus* (monodactyl grazing equid).

roughly proportional to the size of the parent species—yielding progressively greater average *absolute* size increases than decreases as the group diversifies —and the fact that large species are more susceptible to extinction than small ones, so there is a continual reradiation of new large species to replace them (as in the case of the initial Tertiary radiation of mammals after dinosaur extinction). Most size-increase trends in mammalian taxa are examples of change in clade variance (Gould, 1988). However, this is not to say that many of the trends towards size increase throughout time have not had a strong adaptive component. In particular, throughout the Tertiary changing climatic conditions have produced new kinds of habitat, the effective exploitation of which in many cases is favoured by large body size. A clear example is the evolution of larger body size in herbivores, as it is not possible to subsist on a fibrous diet such as grass at a small body size. Here the fossil record shows a coupling of the timing of the increase of body size in many lineages with climatic change: during the Paleocene and Eocene, when tropical forests were widespread, most mammalian species were of the small to moderate sizes typical of modern forest faunas. Although the mean size of mammals in the later part of the Caenozoic was probably greater than at the beginning, this largely reflects the·general change in climate and habitat rather than the unfolding of any type of 'adaptive potential'. The widely-observed 'island rule' (that large species decrease in size and small species increase in size when isolated on islands—see Van Valen, 1973a), and the dwarfing of many lineages in the post-Pleistocene (Marshall and Corruccini, 1978), though both imperfectly understood, imply that body size responds (sometimes rapidly) to environmental changes rather than being developmentally constrained.

The evolution of hypsodonty

Hypsodont (high-crowned) teeth are found in herbivores that eat abrasive herbage. Mammals may be somewhat hypsodont (mesodont), hypsodont, or hypselodont (with ever-growing cheek teeth). This is often seen as a logical 'progression series' with hypselodonty as the final outcome (Mones, 1982; Webb and Hulbert, 1986), but in fact there appear to be strong developmental constraints on the evolution of hypselodonty in large mammals, and hypsodonty is not the only mechanism by which teeth may be made more durable (Janis and Fortelius, 1988).

Hypsodonty is supposed to be an adaptation for eating grass, to counter the abrasive effect of siliceous particles in plant material (Van Valen, 1960; McNaughton et al., 1985). As such, the evolution of hypsodonty has been tied to the spread of grasslands in the Miocene (Webb, 1977) and so seen as related to climatic change in a fashion. However, even browsers and mixed feeders (which both graze and browse) have more hypsodont teeth if they feed in open habitats (Janis, 1988). Certain brachydont (low-crowned) fossil ungulates have dental microwear suggestive of a predominantly grass diet (Solunias et al., 1988); and grazing kangaroos are not especially hypsodont but make their dentition more durable by molar progression (Janis and Fortelius, 1988; Janis, 1990c). Grazers may lack highly hypsodont teeth if they eat grass that is in near-water environments and is not covered with dust and grit (e.g., waterbuck antelopes, hippos), or if they have a low metabolic rate so that their lifetime consumption of food is less than might be expected for their size (rock hyraxes) (see Janis and Fortelius, 1988, for review). Thus, the widespread trends in many mammalian lineages towards more hypsodont teeth (and other adaptations for durability of the dentition) reflect the Tertiary spread of open habitats and their exploitation by all herbivorous mammals, though grazers have evolved the most highly hypsodont teeth of any dietary group.

Evolution of cursoriality

Cursorial mammals are those with elongated distal limb segments, loss of lateral digits (side toes), change in foot posture from plantigrade to digitigrade or unguligrade, and restriction of the motion of the limb to the parasagittal plane (Hildebrand, 1974). Cursoriality is seen predominantly among carnivores (e.g., pack-hunting canids) and ungulates (equids and ruminants), and, as the name implies, has been assumed to reflect the coevolution of pursuit behaviour between aspirant predator and presumptive prey (Dawkins and Krebs, 1979; Bakker, 1983). However, there is no fossil evidence for coevolutionary coupling of the acquisition of cursoriality in carnivores and ungulates: ungulates obtain long legs at the start of the Miocene, but there are no truly cursorial carnivores (with limb proportions resembling those of present-day pursuit predators) until the Plio-Pleistocene. Thus, predator escape cannot have been the initial selection pressure favouring elongation of ungulate limbs. Among living cursorial ungulates and carnivores, relative limb length ('degree of cursoriality') is correlated with home range area but not with maximum recorded speed. Species with larger home ranges cover larger distances in their daily movements. Cursorial adaptations decrease locomotor costs at slow as well as fast gaits. It seems likely that the origin of cursoriality is related to efficient slow locomotion for daily traversal of home range area rather than for high-speed pursuit or avoidance. Since the size of home range area (and relative limb length) in living ungulates is also related to habitat type, with larger home ranges and longer legs characterising species living in open habitats, it is probably significant that the evolution of cursorial ungulates coincides with the opening up of vegetational habitats that occurred during the mid-Tertiary (see Janis, in press, for review).

PROCESSES INVOLVED IN TREND GENERATION AND DIRECTION

Extrinsic and intrinsic factors

Evolutionary trends are observed at all hierarchical levels in the palaeontological record. Simpson (1953, p. 146) expressed the traditional view when he concluded that 'all long- and most short-range trends consistent in direction are adaptively oriented'. That is, selection, an extrinsic and historically contingent factor, generates and directs observed trends within species and clades. The fate of a species or taxon, and its participation in a historical trend, cannot be explained by its characteristics alone, but requires explanation primarily in terms of environmental change (or opportunity) and the action of selection on the taxon's ecologically relevant subunits such as organisms and populations. In contrast, it is possible to view historical trends as having been controlled primarily by intrinsic factors (Gould, 1980), two of which will be discussed here: developmental constraints (especially, heterochrony), and characteristics possessed by different taxa that caused them to exhibit consistently different rates of speciation and/or extinction. It is important to note that control by neither extrinsic nor intrinsic factors implies that the later taxa or their members necessarily show higher levels of adaptedness (cf. Van Valen, 1973b).

Heterochrony and developmental constraints

If phenotypic variation,in mammals is highly constrained by developmental integration, characters will tend to covary, and only limited combinations of characters will be available for natural selection to choose among (Gould, 1980). If most heritable variation in adult characters is primarily the result of a small number of simple, early changes in developmental timing (heterochrony), variation might be further restricted to the coordinated character states exhibited at different ontogenetic stages. At the extreme, the most likely morphological evolution would be back and forth along a single ontogenetic pathway, or that which involved changing relationships among only a few major complexes of covarying traits (McNamara, 1982). By restricting variation this way, heterochrony (and intrinsic developmental constraints in general) could be said to be controlling or directing the path of evolutionary trends within and among species (see Chapter 3).

Where, however, the developmental covariance of traits can be relatively easily dissociated by selection, it is difficult to argue for a primary role for developmental constraint or heterochrony in *directing* trends, even though the sequence of morphological changes seen throughout the trend will necessarily still be redescribable in terms of specific developmental changes. Thus, the role of heterochrony in directing evolutionary trends is closely related to the more general issue of the ease with which characters can respond individually to selection.

The characters involved in most trends observed in mammalian history are quantitative phenotypic characters of the kind that show heritable variation in natural populations and that respond to artificial selection (Charlesworth *et al.*, 1982). However, other characters usually show correlated responses to selection on the target character(s), and such genetic correlations can significantly slow the rate of phenotypic evolution of the target character(s) (Lande 1979; Charlesworth *et al.*, 1982; Charlesworth, 1984; Maynard Smith *et al.*, 1985; Slatkin, 1987).

It is at the lowest levels, among trends involving a small number of closely related species, that the influence of correlated responses to selection has been most persuasively argued. Many closely related species differ from one another mainly in size, and many differences in form between closely related mammal species may represent correlated (allometric) responses to selection on size or its physiological correlates (Gould, 1974; 1975; Shea, 1983; 1985; 1988; Marshall and Corruccini, 1978; McKinney and Schoch, 1985; but see Prothero and Sereno, 1982). Such correlated responses are thought to be enhanced by particularly rapid size evolution (Shea, 1983). In these cases, simple heterochronic changes could be said to have been predominant in directing non-adaptive trends in the correlated characters. However, individual cases should be approached with caution. Apparently non-adaptive characters may reflect functions of which we are currently unaware. Observation of allometric relationships among characters of the adults of closely related species does not necessarily indicate that developmental processes constrained the evolution of the series (Gould, 1974; Lande, 1979; Maynard Smith *et al.*, 1985; Levinton, 1986). For studies of heterochrony, ontogenetic allometric series are difficult to obtain for most fossil mammal species; mammals do not exhibit the radical changes in individual form throughout postnatal growth that characterise many other groups, particularly those that exhibit complex life cycles. Also, there is seldom a major ecological separation between juveniles and adults that would allow us easily to assign adaptive significance to the characteristics of different developmental stages. These characteristics of the mammals relative to other taxa may partly explain the relatively small number of studies of heterochrony in fossil mammals that have been undertaken to date (McNamara, 1988).

In addition, there is some evidence that developmental constraints have played only a minor role in the direction of evolutionary trends in the mammals. The extensive examples discussed above (and presented in Table 13.1) show sometimes remarkable levels of convergence across species of all degrees of phylogenetic relatedness. This suggests that most such convergences among the mammals represent responses to selection for the solution of particular mechanical or functional problems posed by the environment (which recur in space and time), rather than being primarily the result of directed variation stemming from shared developmental programmes. For example, there are trends towards toe reduction and metapodial fusion in many ungulate groups whose members adopt digitigrade or unguligrade foot posture. Which toes are lost or reduced, and which bones are fused, bear a clear functional relationship in each group to the mechanical criteria that differ among the different weight-bearing configurations of the feet of each taxon (see, for example, Kent, 1969; Hildebrand, 1974). The mosaic nature of gradual evolution revealed by detailed studies of dental characters of early Caenozoic mammals (Rose and Bown, 1984; Bown and Rose, 1987) argues against tight developmental control of the phenotype on an evolutionary time-scale (see also Levinton, 1986). In interspecific studies adaptive dissociation of the allometric relationships of different characters or character complexes is frequently observed, arguing against unbreakable developmental constraints and supporting trend *direction* by extrinsic selective forces

(Shea, 1985; 1988; McKinney and Schoch, 1985). We note that the allometric scaling of antlers in the Cervidae presented by Gould (1974) in his classic paper on the Irish Elk *Megaloceros* has exceptions at all body sizes. For example, *Alces* has notably smaller antlers than expected for a cervid of its body size, and this deviation was plausibly interpreted by Gould in selective adaptive terms.

Cross-species trends

Phyletic trends are often observed at the lowest taxonomic levels, but trends at this scale do not appear to have been the primary determinant of large-scale macroevolutionary patterns in the mammals. Most large-scale trends observed in the Tertiary mammalian record that we have discussed here (such as overall increase in brain or body size, evolution of cursoriality, etc.) are associated with differential diversification and extinction, and/or production of new taxa. Trends in mean clade characteristics are due either to changes in clade variance or to other changes in relative taxonomic diversity, rather than being simply the higher-level effects of general anagenetic trends. In what sense could such cross-species trends be 'adaptively' driven or directed? There are two ways (not mutually exclusive) that differential speciation and extinction rates (and thus directional, non-anagenetic cross-species trends) can arise historically within a clade.

Intrinsic control: species sorting
One way, due primarily to intrinsic factors, is through differences in the basic ecology of different species, which cause them to exhibit different speciation and/or extinction rates. These differences in ecological characteristics will be passed on largely intact to the respective offspring species, which will in turn continue to exhibit different historical patterns of speciation and extinction. Environmental change is not required, or, if it is involved at any point, it is only to permit these intrinsic tendencies to be made manifest. The historical pattern is thus due to progressive sorting among species of the clade on the basis of these intrinsic properties, and trends appear in any features of the species or their members that covary with the intrinsic properties.

Such historical sorting has been called 'species selection' (Stanley, 1975), but, as traditionally formulated, it is not a process strictly analogous to natural selection among organisms (Damuth, 1985; Damuth and Heisler, 1988). However, this means that we can pursue the historical explanation of patterns of diversification within clades in terms of the historical causes of differences in speciation and extinction rates, without having to resolve issues in the units of selection controversy or hierarchical selection theory (which are formally irrelevant).

Demonstration that intrinsic factors are responsible for macroevolutionary trends in mammals has been limited and equivocal. The possibility that large species are more prone to extinction and/or speciate less readily has been mentioned (Stanley, 1973; Van Valen, 1975). Vrba (1984; 1987) argues that members of the African family Alcelaphinae (hartebeest and wildebeest) have had greater evolutionary turnover and exhibit greater diversity than their sister taxon, the Aepycerotini (impalas), because they are more likely to experience fragmentation and subdivision of their preferred habitat over evolutionary time, leading to geographic isolation of populations that will encourage speciation events. Vrba suggests that this is because the alcelaphines are specialist grazers and are thus restricted to certain kinds of habitat, whereas impalas are mixed feeders (grazers and browsers) and can thrive in a wide variety of habitats. We wish to point out that exactly the opposite pattern is exhibited by the radiation of horses (Equidae) in North America. Relatively little speciation took place among the Early Tertiary browsing equids—presumably dietary and habitat specialists. However, a massive splitting of clades took place among *mixed-feeding* (and presumably habitat-generalist) species in the Miocene (MacFadden and Hulbert, 1988). (These species are commonly described as grazers (see Simpson, 1951; Webb, 1983) but their levels of hypsodonty fall within the range of living ungulates that are mixed feeders rather than grazers—see Janis, 1984; 1988.) We do not claim that Vrba's explanation for the alcelaphine–aepycerotine example is incorrect but do question the generality of the association of habitat specialisation (stenotopy) with taxonomic turnover in the mammals.

Extrinsic control: 'adaptive' trends
Cross-species trends also arise, due primarily to extrinsic factors, if the reason for the divergence of daughter populations from the parental stock, which leads ultimately to speciation, is the adaptation of those daughter populations to new niches made available by environmental change (cf. the speciational trends of Grant, 1963; 1989; and Futuyma, 1987; 1989). Differential extinction (in the face of loss of adaptation due to disappearing environments/niches) may also be a factor in this case. Sustained or cyclical patterns of environmental change will result in sustained or iterative trends. However, in this case there are no particular species properties, passed from parent to daughter species, conferring higher speciation rates or resistance to extinction; rather, new species are taking advantage of new opportunities, and some established species are evolving or becoming extinct in the face of environmental change.

Both sources of differential speciation and extinction rates generate apparently progressive cross-species trends. However, we recognise only the latter, extrinsically directed, trends as being 'adaptive', because only here is changing adaptation to new environmental opportunities—at the level of individuals and populations—the primary factor involved in affecting speciation rates and directions. An analysis in terms of clade geometry (or observed differential speciation and extinction rates) alone is incapable of distinguishing between the two explanations. Support for either explanation is only possible given explicit knowledge about the palaeoecology of the species involved.

It should be clear at this point that we feel that the adaptive component has been dominant in historical trends observed in the Mammalia, largely because of the association of taxonomic evolution with directional climatic changes. Taxonomic evolution has resulted in new ecomorphs whose adaptations were consistent with contemporary environmental conditions.

CONCLUSIONS

At the highest levels of synapsid evolution, the timing and nature of the 'pulses' of major radiations were controlled primarily by extrinsic factors. Most large-scale within-clade trends observed in the Tertiary mammalian record involve diversification/extinction or changes in clade variance rather than simply reflecting the higher-level effects of general anagenetic trends. Few cross-species trends represent long-term 'progressive' perfection of adaptation within a single adaptive zone. However, there is ample evidence of specific climatic changes associated with many of these trends, and, where the functional significance of the observed changes is understood, many of these large-scale trends can be seen to be adaptive in nature. That is, the change in variance or mean across the species of a clade reflects radiation to exploit new adaptive opportunities or tracking of large-scale (extrinsic) environmental changes, and does not result from intrinsic factors affecting speciation and extinction rates independently of environmental selective factors. Convincing evidence for general intrinsic factors affecting speciation and extinction rates of mammalian taxa is limited. At lower levels, repeated parallel or convergent evolution of similar ecomorphs, separated by time or geography, indicates widespread adaptation to similar ecological conditions rather than orthogenetic direction by internal factors or developmental programmes. Simple heterochronic changes do not appear to have directed variation in large-scale trends. At the smallest scale, that of morphological trends involving a few closely related species, simple heterochronic changes (and correlated responses to selection in general) may plausibly underlie some of the observed trends. However, at this level it is difficult to distinguish developmental constraints and correlated responses to selection from selection on independent traits. The widespread evidence of mosaic evolution at all scales argues against a major role for developmental pathways in directing trends, though rates may be importantly affected. At all scales, examination of clade geometry alone is insufficient to arrive at causal explanations of historical patterns. Rather, a detailed knowledge of palaeoecology is required to interpret trends in the mammalian fossil record.

ACKNOWLEDGEMENTS

We thank M. Fortelius, S. J. Mazer, and J. J. Sepkoski, Jr, for comments on the manuscript. D. Jablonski kindly provided a copy of a manuscript in press. We also thank B. Regal for help with the illustrations.

REFERENCES

Allin, E.F., 1986, The auditory apparatus of advanced mammal-like reptiles and early mammals. In N. Hotton, III, P. D. MacLean, J. J. Roth, and E. C. Roth (eds), *The ecology and biology of mammal-like reptiles*, Smithsonian Institution Press, Washington DC: 283–94.

Anderson, J.M. and Cruickshank, A.R.I., 1978, The biostratigraphy of the Permian and the Triassic. Part 5: A review of the classification and distribution of Permo-Triassic tetrapods, *Palaeont, Afr.*, **21**: 15–44.

Archer, M., Flannery, T.F., Ritchie, A. and Molnar, R.E., 1985, First Mesozoic mammal from Australia — an early Cretaceous monotreme, *Nature*, **318**: 363–6.

Bakker, R.T., 1977, Tetrapod mass extinctions — a model of the regulation of speciation rates and immigration by cycles of topographic diversity. In A. Hallam (ed.), *Patterns of evolution, as illustrated by the fossil record*, Elsevier, New York: 439–68.

Bakker, R.T., 1983, The deer flees, the wolf pursues: incongruences in predator–prey coevolution. In D.J. Futuyma and M. Slatkin (eds), *Coevolution*, Sinauer, Sunderland, MA: 350–82.

Bennett, A.F., and Ruben, J. A., 1979, Endothermy and activity in vertebrates, *Science*, **206**: 649–54.

Bennett, A.F., and Ruben, J. A., 1986, The metabolic and thermoregulatory status of therapsids. In N. Hotton, III, P.D. MacLean, J. J. Roth, and E. C. Roth (eds), *The ecology and biology of mammal-like reptiles*, Smithsonian Institution Press, Washington DC: 207–18.

Benton, M.J., 1983a, Dinosaur success in the Triassic: a non-competitive ecological model, *Quart. Rev. Biol.*, **58**: 29–55.

Benton, M.J., 1983b, Progress and competition in macroevolution, *Biol. Rev.*, **62**: 305–38.

Bookstein, F.L., Gingerich, P.D. and Kluge, A.G., 1978, Hierarchical linear modeling of the tempo and mode of evolution, *Paleobiology*, **4**: 120–34.

Bown, T.M. and Rose, K.D., 1987, Patterns of dental evolution in early Eocene anaptomorphine primates (Omomyidae) from the Bighorn Basin, Wyoming, *J. Paleont*, (Supplement) **61**: 1–162.

Carrier, D.R., 1987, The evolution of locomotor stamina in tetrapods: circumventing a mechanical constraint, *Paleobiology*, **13**: 326–41.

Charlesworth, B., 1984, The cost of phenotypic evolution, *Paleobiology*, **10**: 319–27.

Charlesworth, B., Lande, R. and Slatkin, M., 1982, A neo-Darwinian commentary on macroevolution, *Evolution*, **36**: 474–98.

Cifelli, R.L., 1981, Patterns of evolution among Artiodactyla and Perissodactyla, *Evolution*, **35**: 433–40.

Clutton-Brock, T.H. and Harvey, P.H., 1980, Primates, brains and ecology, *J. Zool. Lond.* **190**: 309–23.

Collinson, M.E., and Hooker, J.J., 1987, Vegetational and mammalian faunal changes in the Early Tertiary of southern England. In E.M. Friis, W.G. Chaloner and P.R. Crane (eds), *The origins of angiosperms and their biological consequences*, Cambridge University Press, Cambridge: 259–304.

Coombs, M.C., 1983, Large mammalian clawed herbivores: a comparative study, *Trans. Am. Phil. Soc.*, **73**: 1–96.

Crompton, A.W., 1980, The biology of the earliest mammals. In K. Schmidt-Nielsen, L. Bolls and C.R. Taylor (eds), *Comparative physiology: primitive mammals*, Cambridge University Press, Cambridge: 1–12.

Crompton, A. W. and Hylander, W. L., 1986, Changes in mandibular function following the acquisition of a dentary-squamosal jaw articulation. In N. Hotton, III, P. D. MacLean, J. J. Roth, and E. C. Roth (eds), *The ecology and biology of mammal-like reptiles*, Smithsonian Institution Press, Washington DC: 263–82.

Damuth, J., 1981, Population density and body size in mammals, *Nature*, **290**: 699–700.

Damuth, J., 1985, Selection among 'species': a formulation in terms of natural functional units, *Evolution*, **39**, 1132–46.

Damuth, J., 1987, Interspecific allometry of population density in mammals and other animals: the independence of body mass and population energy-use, *Biol. J. Linn. Soc.*, **31**: 193–246.

Damuth, J. and Heisler, I. L., 1988, Alternative formulations of multilevel selection, *Biol. Phil.*, **3**: 407–30.

Dawkins, R. and Krebs, J. R., 1979, Arms races between and within species, *Proc. R. Soc. Lond.*, Series B, **205**: 489–511.

Eisenberg, J. F. and Wilson, D. E., 1978, Relative brain size and feeding strategies in the Chiroptera, *Evolution*, **32**: 740–51.

Eisenberg, J. F. and Wilson, D. E., 1981, Relative brain size and demographic strategies in didelphid marsupials, *Am. Nat.*, **118**: 1–15.

Emmons, L. H. and Gentry, A. H., 1983, Tropical forest structure and the distribution of gliding and prehensile-tailed vertebrates, *Am. Nat.*, **121**: 513–24.

Flynn, J. J., and Novacek, M. J., 1984, Early Eocene vertebrates from Baja California: evidence for intracontinental age correlations, *Science*, **224**: 151–3.

Futuyma, D. J., 1987, On the role of species in anagenesis, *Am. Nat.*, **130**: 465–73.

Futuyma, D. J., 1989, Speciational trends and the role of species in macroevolution, *Am. Nat.*, **134**: 318–21.

Gingerich, P. D., 1976, Paleontology and phylogeny: patterns of evolution at the species level in early Tertiary mammals, *Am. J. Sci.*, **276**: 1–28.

Gingerich, P. D., 1984, Mammalian diversity and structure, In P. D. Gingerich and C.E. Badgley (eds), *Mammals: notes for a short course, Univ. Tennessee Stud. Geol.*, **8**: 1–16.

Gingerich, P. D., 1985, Species in the fossil record: concepts, trends, and transitions, *Paleobiology*, **11**: 27–41.

Gingerich, P. D., 1987, Evolution and the fossil record: patterns, rates, and processes, *Can. J. Zool.*, **65**: 1053–60.

Gittleman, J. L., 1986, Carnivore brain size, behavioral ecology, and phylogeny, *J. Mammal.*, **67**: 23–36.

Gould, S. J., 1974, The origin and function of 'bizarre' structures: antler size and skull size in the 'Irish elk', *Megaloceros giganteus, Evolution*, **28**: 191–220.

Gould, S. J., 1975, On scaling of tooth size in mammals, *Am. Zool.*, **15**: 351–62.

Gould, S. J., 1980, Is a new and general theory of evolution emerging?, *Paleobiology*, **6**: 119–30.

Gould, S. J., 1988, Trends as changes in variance: a new slant on progress and directionality in evolution, *J. Paleont.*, **62**: 319–29.

Gould, S. J., Gilinsky, N. L. and German, R. Z., 1987, Assymetry of lineages and the direction of evolutionary time, *Science*, **236**: 1437–41.

Grant, V., 1963, *The origin of adaptations*, Columbia University Press, New York.

Grant, V., 1989, The theory of speciational trends, *Am. Nat.*, **133**: 604–12.

Hildebrand, M., 1974, *Analysis of vertebrate structure*, John Wiley and Sons, New York.

Jablonski, D. and Bottjer, D. J., in press, The ecology of evolutionary innovation: the fossil record. In M. Nitecki (ed.), *Evolutionary innovations*, University of Chicago Press, Chicago.

Janis, C. M., 1982, Evolution of horns in ungulates: ecology and paleoecology, *Biol. Rev.*, **57**: 261–318.

Janis, C. M., 1984, The use of fossil ungulate communities as indicators of climate and environment. In P. J. Brenchley (ed.), *Fossils and Climate*, Wiley and Sons, London: 85–103.

Janis, C. M., 1988, An estimation of tooth volume and hypsodonty indices in ungulate mammals, and the correlation of these factors with dietary preferences. In D. E. Russell, J.-P. Santoro and D. Sigogneau-Russell (eds), *Teeth revisited: Proceedings of the VIIth International Symposium on Dental Morphology, Paris, 1986,. Mém. Mus. Natn. Hist. Nat., Paris*, series C, **53**: 367–87.

Janis, C. M., 1990a, Correlation of cranial and dental variables with dietary preferences: a comparison of macropodoid and ungulate mammals, *Mem. Qld. Mus.* **28**: 349–66.

Janis, C. M., 1990b, A climatic explanation for patterns of evolutionary diversity in ungulate mammals, *Palaeontology*, **33** (in press.)

Janis, C. M., 1990c, Why kangaroos (Marsupialia: Macropodidae) are not as hypsodont as ungulates (Eutheria), *Aust. Mammal*, **13** (in press.)

Janis, C. M., in press, Do legs support the arms race in mammalian predator/prey relationships? In J.R. Horner and K. Carpenter (eds), *Vertebrate behavior as derived from the fossil record*, Columbia University Press, New York.

Janis, C. M. and Fortelius, M., 1988, On the means whereby mammals achieve increased functional durability of their dentitions with special reference to limiting factors, *Biol. Rev.*, **63**: 197–230.

Jenkins, F. A., Jr. and McLearn, D., 1984, Mechanisms of hind foot reversal in climbing mammals, *J. Morph.*, **182**: 197–219.

Jerison, H. J., 1973, *Evolution of the brain and intelligence*, Academic Press, New York.

Jerison, H. J., 1985, Issues in brain evolution, *Oxford Surv. Evol. Biol.*, **2**: 102–34.

Kappelman, J., 1988, Morphology and locomotor adaptations of the bovid femur in relation to habitat, *J. Morph.*, **198**: 119–38.

Kappelman, J., (in press), The paleoenvironment of *Kenyapithecus* at Fort Ternan, *J. Hum. Evol.*

Kauffman, S. A., 1989, Cambrian explosion and Permian quiescence: implications of rugged fitness landscapes, *Evol. Ecol.*, **3**: 274–281.

Kemp, T. S., 1982, *Mammal-like reptiles and the origin of mammals*, Academic Press, London.

Kent, G. C., Jr, 1969, *Comparative anatomy of the vertebrates*, 2nd edn, Mosby, Saint Louis, MO.

Kirsch, J. A., 1977, The six-percent solution. Second thoughts on the adaptedness of marsupials, *Am. Sci.*, **65**: 276–88.

von Koenigswald, W. and Schierning, H.-P., 1987, The ecological niche of an extinct group of mammals, the early Tertiary apatemyids, *Nature*, **326**: 595–7.

Krause, D. W., 1986, Competitive exclusion and taxonomic displacement in the fossil record: the case of rodents and multituberculates in North America. In K. M. Flanagan and J. Lillegraven (eds), *Vertebrates, phylogeny and philosophy*, *Contrib. Geol., Univ. Wyoming Spec. Pap.* **3**: 95–117.

Krause, D. W. and Jenkins, F. A., Jr, 1983, The postcranial skeleton of North American multituberculates, *Bull. Mus. Comp. Zool.*, **150**: 199–246.

Kurtén, B., 1973, Early Tertiary land mammals. In A. Hallam (ed.), *Atlas of palaeobiogeography*, Elsevier, Amsterdam: 437–42.

Lande, R., 1976, Natural selection and random genetic drift in phenotypic evolution, *Evolution*, **30**: 314–34.

Lande, R., 1979, Quantitative genetic analysis of multivariate evolution, applied to brain:body size allometry, *Evolution*, **33**: 402–16.

Langer, P., 1988, *The mammalian herbivore stomach: comparative anatomy, function and evolution*, Gustav Fischer Verlag, Stuttgart.

Lee, A. K. and Cockburn, A., 1985, *Evolutionary ecology of marsupials*, Cambridge University Press, Cambridge.

Lemen, C., 1980, Relationship between relative brain size and climbing ability in *Peromyscus, J. Mammal.*, **61**: 360–4.

Levinton, J. S., 1986, Developmental constraints and evolutionary saltations: a discussion and critique. In J. P. Gustafson, G. L. Stebbins, and F. J. Ayala (eds), *Genetics, development, and evolution. 17th Stadler Genetics Symposium*, Plenum, New York: 253–88.

Lillegraven, J. A., Kielan-Jaworowska, Z. and Clemens, W. A. (eds), 1979, *Mesozoic mammals: the first two-thirds of mammalian history*, University of California Press, Berkeley.

Lillegraven, J. A., Thompson, S. D., McNab, B. K. and Patton, J. L., 1987, The origin of eutherian mammals, *Biol. J. Linn. Soc.*, **32**: 281–336.

Lister, A. M., 1989, Rapid dwarfing of red deer on Jersey in the Last Interglacial, *Nature*, **342**: 539–42.

Maas, M. C., Krause, D. W. and Strait, S. G., 1988, The decline and extinction of the Plesiadapiformes (Mammalia: ?Primates) in North America: displacement or replacement? *Paleobiology*, **14**: 410–31.

MacArthur, R. H. and Wilson, E. O., 1967, *The theory of island biogeography*, Princeton University Press, Princeton, NJ.

McDonald, J. N., 1984, The reordered North American selection regime and late Quaternary megafaunal extinctions. In Martin, P. S., and Klein, R. G. (eds), *Quaternary extinctions*, University of Arizona Press, Tucson: 404–39.

Mace, G. M., Harvey, P. H. and Clutton-Brock, T. H., 1981, Brain size and ecology in small mammals, *J. Zool., Lond.*, **193**: 333–54.

MacFadden, B. J. and Hulbert, R. C., Jr., 1988, Explosive speciation at the base of the adaptive radiation of Miocene grazing horses, *Nature*, **336**: 466–8.

McKenna, M. C., 1980, Eocene paleolatitude, climate, and mammals of Ellesmere Island, *Palaeogeogr., Palaeoclimatol., Palaeoecol.*, **30**: 349–62.

McKinney, M. L. and Schoch, R. M., 1985, Titanothere allometry, heteochrony, and biomechanics: revising an evolutionary classic, *Evolution*, **39**: 1352–63.

McNab, B. K., 1983, Energetics, body size and the limits to endothermy, *J. Zool., Lond.*, **199**: 1–29.

McNab, B. K., 1986, Food habits, energetics, and the reproduction of marsupials, *J. Zool., Lond.*, **208**: 595–614.

McNamara, K. J., 1982, Heterochrony and phylogenetic trends, *Paleobiology*, **8**: 130–42.

McNamara, K. J., 1988, The abundance of heterochrony in the fossil record. In M. L. McKinney (ed.), *Heterochrony in evolution: a multidisciplinary approach*, Plenum, New York: 287–325.

McNaughton, S. J., Tarrants, J. L., McNaughton, M. M., and Davis, R. H., 1985, Silica as a defense against herbivory and as a growth promoter in African grasses, *Ecology*, **66**: 528–35.

Marshall, L. G. and Corruccini, R. S., 1978, Variability, evolutionary rates, and allometry in dwarfing lineages, *Paleobiology*, **4**: 101–19.

Marshall, L. G., Webb, S. D., Sepkoski, J. J., Jr and Raup, D. M., 1982, Mammalian evolution and the great American interchange, *Science*, **215**: 1351–7.

Martin, L. D., 1983, The origin and early radiation of birds. In A. H. Brush and G. A. Clark, Jr (eds), *Perspectives in ornithology*, Cambridge University Press, Cambridge: 291–338.

Martin, L. D., 1985, Tertiary extinction cycles and the Pliocene-Pleistocene boundary, *Institute for Tertiary–Quaternary Stud., TER-QUA Symp. Series*, **1**: 33–40.

Martin, L. D., 1989, Fossil history of the terrestrial Carnivora. In J. L. Gittleman (ed.), *Carnivore behavior, ecology, and evolution*, Cornell University Press, Ithaca, NY: 536–68.

Martin, P. S., 1984, Prehistoric overkill: the global model. In P. S. Martin and R. G. Klein (eds), *Quaternary extinctions*, University of Arizona Press, Tucson: 354–403.

Martin, R. D., 1973, Comparative anatomy and primate systematics, *Symp. Zool. Soc., Lond.*, **33**: 301–37.

Maynard Smith, J., Burian, R., Kauffman, S., Alberch, P., Campbell, J., Goodwin, B., Lande, R., Raup, D. and Wolpert, L., 1985, Developmental constraints and evolution, *Q. Rev. Biol.*, **60**: 265–87.

Mones, A., 1982, An equivocal nomenclature: what means hypsodonty? *Paläont. Zeitschr.*, **56**: 107–11.

Pagel, M. D. and Harvey, P. H., 1989, Taxonomic differences in the scaling of brain on body weight among mammals, *Science*, **244**: 1589–93.

Parker, P., 1977, An ecological comparison of marsupial and placental patterns of reproduction. In B. Stonehouse and D. Gilmore (eds), *The biology of marsupials*, Macmillan Press, London: 273–86.

Pascual, R., Vucetich, M. G., Scillato-Yane, G. J. and Bond, M., 1985, Main pathways of mammalian diversification in South America. In F. Stehli and S. D. Webb, (eds), *The great American biotic interchange*, Plenum, New York: 219–47.

Prothero, D. R., 1985, North American mammalian diversity and Eocene–Oligocene extinctions, *Paleobiology*, **11**: 389–405.

Prothero, D. R., 1989, Stepwise extinctions and climatic decline during the later Eocene and Oligocene. In S. K. Donovan (ed.), *Mass extinctions, processes and evidence*, Belhaven, London: 217–234.

Prothero, D. R. and Sereno, P. C., 1982, Allometry and paleoecology of medial Miocene dwarf rhinoceroses from the Texas Gulf Coastal Plain, *Paleobiology*, **8**: 16–30.

Radinsky, L. B., 1975, Anthropoid brain evolution, *Am. J. Phys. Anthrop.*, **41**: 15–28.

Radinsky, L. B., 1977, Early primate brains: facts and fiction, *J. Hum. Evol.*, **6**: 79–86.

Radinsky, L. B., 1978, Evolution of brain size in carnivores and ungulates, *Am. Nat.*, **112**: 815–831.

Radinsky, L. B., 1981, Brain evolution in extinct South American ungulates, *Brain Behav. Evol.*, **18**: 169–87.

Raup, D. M. and Sepkoski, J. J., Jr, 1984, Periodicity of extinctions in the geologic past, *Proc. Nat. Acad. Sci., USA*, **81**: 801–5.

Rose, K. D. and Bown, T. M., 1984, Gradual phyletic evolution at the generic level in early Eocene omomyid primates, *Nature*, **309**: 250–2.

Roth, V. L. and Thorington, R. W., Jr., 1982, Relative brain size among African squirrels, *J. Mammal.*, **63**: 168–73.

Scott, K. M., 1990, Postcranial dimensions of ungulates as predictors of body mass. In J. Damuth and B. J. MacFadden (eds), *Body size in mammalian paleobiology*, Cambridge University Press, Cambridge (in press).

Shea, B. T., 1983, Phyletic size change and brain/body allometry: a consideration based on the African pongids and other primates, *Int. J. Primatol.*, **4**: 33–62.

Shea, B. T., 1985, Ontogenetic allometry and scaling: a discussion based on the growth and form of the skull in African apes. In W. L. Jungers (ed.), *Size and scaling in primate biology*, Plenum, New York: 175–205.

Shea, B. T., 1987, Reproductive strategies, body size, and encephalization in primate evolution, *Int. J. Primatol.*, **8**, 139–56.

Shea, B. T., 1988, Heterochrony in primates. In M. L. McKinney (ed.), *Heterochrony in evolution: a multidisciplinary approach*, Plenum, New York: 237–66.

Simpson, G. G., 1951, *Horses—the story of the horse family in the modern world and through sixty million years of history*, Oxford University Press, Oxford.

Simpson, G. G., 1953, *The major features of evolution*, Columbia University Press, New York.

Slatkin, M., 1987, Quantiative genetics of heterochrony, *Evolution*, **41**: 799–811.

Solunias, N. and Dawson-Saunders, B., 1988, Dietary adaptations and palaoecology of the late Miocene ruminants from Pikermi and Samos in Greece, *Palaeogeogr., Palaeoclimatol., Palaeoecol.*, **65**: 149–72.

Solunias, N., Teaford, M. and Walker, A., 1988, Interpreting the diet of extinct ruminants: the case of the non-browsing giraffid, *Paleobiology*, **14**: 287–300.

Stanley, S. M., 1973, An explanation for Cope's Rule, *Evolution*, **27**: 1–26.

Stanley, S. M., 1975, A theory of evolution above the species level, *Proc. Nat. Acad. Sci., U.S.A.*, **72**: 646–50.

Sues, H.-D., 1986, Locomotion and body form in early therapsids (Dinocephalia, Gorgonopsia, and Therocephalia). In N. Hotton, III, P.D. MacLean, J. J. Roth and E.C. Roth (eds), *The ecology and biology of mammal-like reptiles*, Smithsonian Institution Press, Washington, DC: 61–70.

Szalay, F. S. and Decker, R. L., 1974, Origins, evolution and function of the tarsus in late Cretaceous eutherians and Paleocene primates. In F.A. Jenkins, Jr (ed.), *Primate locomotion*, Academic Press, New York: 223–59.

Upchurch, G. R., Jr and Wolfe, J. A., 1987, Mid-Cretaceous to Early Tertiary vegetation and climate: evidence from fossil leaves and woods. In E.M. Friis, W.G. Chaloner and P.R. Crane (eds), *The origins of angiosperms and their biological consequences*, Cambridge University Press, Cambridge: 75–105.

Van Valen, L., 1960, A functional index of hypsodonty, *Evolution*, **14**: 531–2.

Van Valen, L., 1971, Adaptive zones and the orders of mammals, *Evolution*, **25**: 420–8.

Van Valen, L., 1973a, Pattern and the balance of nature, *Evol. Theory*, **1**: 31–49.

Van Valen, L., 1973b, A new evolutionary law, *Evol. Theory*, **1**: 1–30.

Van Valen, L., 1975, Group selection, sex, and fossils, *Evolution*, **29**: 87–94.

Van Valkenburgh, B., 1987, Skeletal indicators of locomotor behavior in living and extinct carnivores, *J. Vert. Paleo.*, **7**: 162–82.

Van Valkenburgh, B., 1988, Trophic diversity in past and present guilds of large predatory mammals, *Paleobiology*, **14**, 155–73.

Vrba, E. S., 1984, Evolutionary pattern and process in the sister-group Alcelaphini-Aepycerotini (Mammalia: Bovidae). In N. Eldredge and S.M. Stanley (eds), *Living Fossils*, Springer-Verlag, New York: 62–79.

Vrba, E. S., 1987, Ecology in relation to speciation rates: some case histories of

Miocene–Recent mammal clades, *Evol. Ecol.*, **1**: 283–300.

Webb, S. D., 1977, A history of savanna vertebrates in the New World. Part I. North America, *Ann. Rev. Ecol. Syst.*, **8**: 355–80.

Webb, S. D., 1978, A history of savanna vertebrates in the New World. Part II. South America and the great interchange, *Ann. Rev. Ecol. Syst.*, **9**: 393–426.

Webb, S. D., 1983, The rise and fall of the late Miocene ungulate fauna of North America. In M. H. Nitecki (ed.), *Coevolution*, University of Chicago Press, Chicago: 267–306.

Webb, S. D., 1985, Late Cenozoic mammal dispersals between the Americas. In F.G. Stehli and S. David Webb (eds), *The Great American Biotic Interchange*, Plenum, New York.

Webb, S. D. and Hulbert, R. C., Jr., 1986, Systematics and evolution of *Pseudhipparion* (Mammalia, Equidae) from the late Neogene of the Gulf Coastal Plain and the Great Plains, *Contrib. Geol., Univ. Wyoming Spec. Pap.*, **3**: 237–72.

Wing, S. L. and B. H. Tiffney. 1987. Interactions of angiosperms and herbivorous tetrapods through time. In E.M. Friis, W.G. Chaloner, and P.R. Crane (eds), *The origins of angiosperms and their biological consequences*, Cambridge University Press, Cambridge: 203–24.

Wolfe, J. A., 1978, A paleobotanical interpretation of Tertiary climates in the Northern Hemisphere, *Am. Sci.*, **66**: 694–703.

Wolfe, J. A., 1985, Distribution of major vegetational types during the Tertiary. In *The carbon cycle and atmospheric CO^2: natural variations Archaean to present, Geophysical Monogr.*, **32**: 357–75.

EPILOGUE

Kenneth J. McNamara

Having posed a series of questions concerning the nature of evolutionary trends in the preface to this book, it is incumbent upon me to see what answers the chapters which followed provide to these questions. Because of the current 'state of the art' of evolutionary trends it is not feasible to propose too many profound or all-embracing statements about evolutionary trends in general. However, a number of interesting patterns have emerged. I shall also briefly assess how the examples and observations from the taxonomic chapters in Parts Two and Three tie in with some of the theoretical aspects discussed in Part One. It should be borne in mind, however, that the authors of chapters in Part One (apart from myself) did not see the chapters in Parts Two and Three before they wrote their contributions. And the same, in reverse, is the case for the authors of the taxonomic chapters.

Although in my preface I indicated how this book aimed to go some little way towards redressing the imbalance that exists between studies on rates of evolution and studies of causes of directionality, there is an inescapable link that connects the two. This point emerges in Steve Gould's opening chapter. Arguing strongly that most trends are speciational phenomena, and not the direct result of the accumulation of phyletic trends, the subsequent chapters largely support this viewpoint. The vast majority of chapters document evolutionary trends at the species level and above. Does this mean that phyletic gradualism is dead? There are many scientists who would argue vehemently that this is not so. Or is there a subconscious perception by the authors that only trends at the species level and above have any real influence in the large-scale patterns of evolution? Where phyletic trends are discussed, notably in Chapters 9 (echinoids) and 13 (mammals), the inference drawn is that the morphological changes arising from phyletic gradualism do not seem

347

to have a great bearing on trends at the species level and above. The exception to this is in bryozoans (Chapter 10), where it is suggested that there may be a relationship between intra- and interspecific trends.

One possible source of confusion in the debate concerning phyletic gradualism and directionality is in the use of the terms 'anagenesis' and 'phyletic gradualism'. There has been a tendency of late to conflate the two. This, I believe, should not be done. Phyletic gradulaism refers to *rate* of evolution, whereas anagenesis is concerned with *directionality*. As I explain in the Glossary, anagenesis can occur intraspecifically *or* interspecifically. Although it might merge with phyletic gradualism in intraspecific evolution, it need not, particularly if it can be demonstrated that the rates of intraspecific change are not gradual, but episodic. Stepped anagenesis can also be accommodated within any model of speciational trends. The examples that I have described in echinoids, and which are also described in this book, for instance in ammonites, trilobites and crinoids, seem to indicate that speciation, *without branching*, can occur within anagentic lineages. While this may stick in the craw of many cladists, whose only perception of speciation is one of cladogenesis, that is, the branching of one species from another, I believe that anagenetic speciation does occur. It can be considered as merely representing an extreme form of cladogenesis.

In the *Lovenia* lineage that I documented in Chapter 9, it could be argued that the evolution of one species from the next was a relatively rapid event and occurred when predation levels reached a critical level, beyond which the interrelationship between predator and prey species became chaotic. During this period, the old species and the new may have overlapped temporally, but the high levels of predation quickly extinguished the ancestral form. Once the descendant morph established a state of relative equilibrium with the predator species, the new species became established. Critics will probably argue that the three species of *Lovenia* merely represent a gradually evolving lineage, yet the earliest species persisted relatively unchanged for about 10 million years, before giving way to a morphologically quite distinct new species. Furthermore, being a living genus it is possible to use the same criteria in differentiating these fossil species from one another as can be used in differentiating between living species. Such speciating anagenetic trends may, therefore, just form branches on the greater cladogenetic tree. The branches of an asymmetric cladogenetic tree constitute a higher-level trend.

So, to what degree are the evolutionary changes in these single lineages reflected in the trends seen at higher levels? The evidence presented in the book suggests some differences between groups in this regard. There is a remarkable similarity in patterns of trends between echinoids and bivalves, and, to a lesser extent ammonoids and trilobites, in that trends at the species level are reflected in changes at high taxonomic levels. For instance, the overall trend to blindness in a number of groups of trilobites can also be tracked in single species-level lineages. Similarly, ordinal-level trends in echinoids of an onshore to offshore vector of diversification reflect a pattern that is evident in species-level evolutionary trends. In mammals and some fishes (lungfishes), on the other hand, large-scale trends appear to be more influenced by major extrinsic, often abiotic, factors.

At the species level, heterochrony appears to play a major role as one of the factors that channels evolution along particular morphological pathways in most groups, both invertebrate and vertebrate. This should not come as any great surprise, because organisms' own ontogenies are themselves morphological trends. Extending or contracting them provides a ready-made source of directionality both inter- and intraspecifically. The all-pervasive influence of heterochrony at this level arises from substantial phenotypic changes that can occur with a minimum of genetic change. However, as most authors argue, the availability of suitable niche axes is a prerequisite for the development of such heterochronically fuelled trends—although again in Chapter 13 Christine Janis and John Damuth argue that in mammals extrinsic factors, in the form of ecomorphs, play a more dominant role than heterochrony in controlling the direction of trend development at lower levels.

The frequency of parallel trends in many groups of both invertebrates and vertebrates (including mammals) would appear to arise from the twin constraining factors of heterochrony and niche-axis characteristics. Assessing whether these changes are adaptive or not is a moot point. Most authors still see most changes, particularly those at lower taxonomic levels, as being predominantly adaptive. But as Mike McKinney discusses in Chapter 4, lineages that show consistent changes in body size, either increases or decreases, argue for the need for stronger consideration to be given to non-adaptive explanations, such as selection for 'life-history strategies' or reproductive selection. Teasing such interpretations from fossil data will be a major challenge for palaeontologists in the years to come.

At lower taxonomic levels two driving forces for trends emerge: opportunistic speciation as environments changed (e.g., lungfishes), or as ecological niches became vacant; and predation. The direction of evolution in echinoids, bivalves and bryozoans appears to have been strongly influenced by predation pressure. However, as Steve Gould argues, the extent to which such 'arms race' struggles influence large-scale trends in many groups is hard to judge, and will only become apparent when more research has been carried out. Certainly, the current indication is that in some groups at least predation-driven trends at lower taxonomic levels have affected the overall patterns of evolution in the groups as a whole.

The proposal that many evolutionary trends are merely the reflection of increased variance through time is strongly supported in many of the chapters in the book. This may, perhaps, be taken to imply the common occurrence of asymmetric cladogenetic patterns. Still, a number of authors have taken pains to stress that a substantial number of trends are *not* only the result of such increase in variance. Such cases can be demonstrated when ancestral morphotypes are replaced by the derived forms (as in the predation-driven trends).

And lastly, to what extent do large-scale extrinsic effects, generally abiotic, such as mass extinctions or gross climatic changes, affect evolutionary trends? Here there appear to be some differences of opinion. In Chapter 6, Arnold Miller contends that mass extinctions had little or no effect on the course of trends in bivalve evolution. Yet according to Mike Benton (Chapter 12) they had a profound effect on the course of reptilian evolution, as they did on fishes—extinction of placoderms at the end of the Devonian was followed by diversification of hondrichthyan fishes into the same niches. In other groups, such as echinoids, that were almost wiped out by one such event at the Permo-Triassic boundary, all that happened was that the echinoid evolutionary clock was reset. Following this some patterns were repeated (such as the evolution of forms with flattened body shapes), but other, quite different, major trends emerged (most notably bilateral symmetry and a burrowing habit). Factors such as unidirectional climatic change appear to have strongly influenced trends in mammal evolution. It would seem likely that evolution in other terrestrial homoeotherms would similarly have been affected.

At the beginning of this book I quoted an observation made by Pierre Teilhard de Chardin: 'The past has revealed to me the structure of the future'. This was no mere self-indulgent whim. For what it suggests is an intriguing aspect to the study of evolutionary trends that seems rarely to have been considered: the potential for predicting the course of evolutionary trends. In such a complex system as biological evolution that depends on such a range of interacting factors, it might seem facile even to consider predicting the course of trends. Yet what emerges, I believe, from this book is that many evolutionary trends are the outcome not of random events, but of heavily constrained evolution, channelled both developmentally and ecologically along set pathways. The more we can identify the agents that constrain and direct trends, the more possible it will be to predict the course of evolutionary trends.

Let me conclude with an example. In Chapter 9, I described the changes that have occurred in a lineage of the pea-sized echinoid *Echinocyamus planissimus* over a period of about 2 million years from the Pliocene to the present day. There has been a steady decrease in the number of pore pairs in the ambulacra and the test has got bigger, as have the tubercles that cover the entire surface of the test. On the assumption that the rate of morphological change continues at much the same rate as it has over the last 2 million years, and that man's influence does not radically change the environment of the echinoids, then it is possible to suggest what the species might look like in, say, a million years' time. The adult specimens, that today barely reach 8 mm in length, would be larger, perhaps up to 12 mm long. They would have even fewer pore pairs, perhaps only four in the largest adults; and fewer, but larger, tubercles. Similarly, with an evolutionary lineage identified from the fossil record, predictions could perhaps be made of the likely appearance of an undiscovered extension of the lineage.

The study of evolutionary trends may well be a study of the history of the diversification of life in the past. But never forget that we, like all other life forms on this planet, are just part of a series of evolutionary trends currently sitting on top of the geological time-scale. While some of our activities in the last few hundred years may have prematurely terminated a number of trends, many will continue into the future.

GLOSSARY

The numbers in parentheses refer to the chapters in which the term is used.

Acceleration (3,4,7,9,12) – heterochronic process involving faster rate of development in the descendant; produces a peramorphic trait in the adult **phenotype**.

Adaptive zone (5,13) – a set of ecological niches. A population or taxon makes a transition from one zone to another across an **adaptive threshold**.

Allometry (3–9,11,13) – the study of size and shape; the change in size and shape observed. **Complex allometry** occurs when the ratio of the specific growth rates of the traits compared is not constant, a log-log plot comparing the traits not yielding a straight line (k in the allometric formula not remaining constant). When there is no change in shape with size increase, **isometry** is said to occur. In other words, a log-log plot yields a straight line with a slope of $k = 1$. **Positive allometry** occurs when trait x is increasing more slowly than trait y ($k > 1$). **Negative allometry** is the reverse ($k < 1$).

Allopatric speciation (5) – model of speciation in which establishment of a geographical barrier that severely restricts genetic exchange between populations leads to the evolution of a new species. Allopatric ancestral and descendant populations have geographical distributions that do not overlap.

Anagenesis (1–4,6,9,10,13) – a lineage in which there is directional morphological change. It may occur intraspecifically (e.g., the echinoid *Echinocyamus*, see Chapters 1,9) and interspecifically (e.g., the echinoid *Hemiaster*,

351

see Chapter 9), in which case there is no overlap between species. There will be no net increase in diversity. Morphological change in either of these need not be gradual. Periods of stasis, followed by rapid bursts of morphological change, produce a pattern of **stepped anagenesis**.

Apomorphy (5,12) – in cladistic analysis, a 'derived' or 'specialised' character.

Aptation (7) – a term to encompass both adaptation and **exaptation**.

Astogeny (10) – the growth and development of a colony.

Autocatakinetic (3,4) – self-generating; the mechanism that drives clado-genetic **heterochronoclines**.

Canalisation (3) –buffering of the developmental programme against per-turbations.

Cladogenetic asymmetry (2) – the 'nowhere to go but up' process, whereby a group (clade) originates at a character state (e.g., size) that is physically restricted to expansion mainly in one direction (e.g., a mammal can only get so small before encountering metabolic problems); the descendant radiation is therefore asymmetric in that character state (e.g., larger size). The resulting 'trend' is an increase in variance.

Competitive exclusion (5) – the principle that no two species can coexist in the same place if their ecological requirements are identical.

Competitive assymetry (4) – organisms which, by virtue of a particular trait or set of traits, have a disproportionate advantage over others. For example, organisms with larger body sizes often have a competitive advantage over smaller organisms because they can eat food that smaller ones cannot, consume more, and have foraging advantages.

Cope's Rule (2,4,8–10,12,13) – evolutionary tendency of many fossil lineages to increase in body size.

Developmental constraint (2,3,9,13) – the ontogenetic 'rules' which determine the course of evolution; the degree of 'intrinsic' control over evolutionary direction (in contrast to 'extrinsic' natural selection).

Dissociated heterochrony (3,9,11) – an ontogenetic change in rate or timing of a trait that does not occur in some other trait; a change that only occurs in a local growth field.

Ecomorph (13) – combination of morphological features reflecting functional adaptation to a particular ecological niche.

Effacement (5) – the tendency for dorsal furrows on trilobites to become obliterated.

Endotomy (8) – In crinoids, bifurcation in two main arms which give off branches only on their adradial side.

Evolutionary ratchet (2,4) – as organisms evolve, particularly those that become more complex, there is an accumulation of contingencies: an increasing interdependence among their components, making radical changes more difficult—that is, a 'hardening' occurs.

Evolutionary trends (1–13) – persistent directional evolutionary changes. These may be **anagenetic**, where the trends occur in single, non-branching lineages; or **cladogenetic**, where they occur in branching lineages. These parallel, respectively, Eldredge's **transformational** and **taxic** trends (see Chapters 2,3). Cladogenetic trend patterns may be **symmetrical** or **asymmetrical**, and each may be **accretive** or **non-accretive**, depending on whether older state variables (see below) are retained. In the former case **variance** (see Chapter 1) is increased, in the latter it is not.

Exaptation (1,3) – a feature of an organism that evolved for a reason other than its current utility, then was co-opted for a new use, or which evolved, but had no particular function, but became co-opted for a use at a later stage.

Global heterochrony (3) – ontogenetic change in rate or timing that affects the entire individual.

Heterochronocline (2–4,9) – an evolutionary sequence wherein **ontogenies** show regressive or progressive heterochronic changes, either **paedomorphosis** (**paedomorphocline (3,8–11)**, e.g., the trilobite *Olenellus*, see Chapter 5) or **peramorphosis** (**peramorphocline (3,7,9,10,12)**, e.g., the rhyncosaur reptiles, see Chapter 12). **Anagenetic** heterochronoclines occur when there is no temporal overlap between species in the sequence, and thus no increase in species number (e.g., the echinoid *Lovenia*, see Chapter 9). **Cladogenetic heterochronoclines** occur when there is evolutionary branching, a descendant heterochronic morphotype coexisting with the parent species (e.g., the brachiopod *Tegulorhynchia*, see Chapter 3). A **stepped** anagenetic or cladogenetic heterochronocline occurs when evolution is rapid and followed by a period of stasis. **Mosaic** heterochronoclines occur when some traits are affected by paedomorphic processes, and others by peramorphic ones (e.g., the echinoid *Hemiaster*, see Chapter 9). **Dissociated** heterochronoclines are the same, but the different patterns all produce the same overall end result, either paedomorphosis or peramorphosis (e.g., the echinoid *Schizaster*, see Chapter 9).

Heterochrony (1–13) – change in timing or rate of developmental events, relative to the same event in the ancestor.

Hierarchy (6,9) – any system of organisation based on levels or ranks.

Homoplasy (5,10) – parallel or convergent evolution of similar characters.

Hypermorphosis (2,3,4,7,9,12) – late cessation ('offset') of developmental events in the descendant; produces peramorphic traits when expressed in the adult phenotype. Late sexual maturation can produce global hypermorphosis; but late cessation in local growth fields can also produce hypermorphosis.

Iterative evolution (5,13) – a model of evolution in which an ancestral stock periodically gives rise to morphologically similar descendants.

K-selection (4,9) – selection that occurs in relatively constant, stable environments where competition and other density-dependent factors are important: favoured traits include large size, delayed reproduction, repeated reproduction, few offspring, and parental care of offspring.

Lazarus effect (5) – the disappearance and apparent extinction of taxa that reappear unchanged in later strata.

Macroecology (4) – the study of ecosystem-level body-size evolution.

Monophyletic group (12) – in cladistic analysis a group in which any species belonging to the group is more closely related to any other species, also in the group, than to any species which does not belong to it, by virtue of having at least one shared character which defines the group.

Neoteny (3,7–9) – slower rate of developmental events in the descendant; produces paedomorphic traits when expressed in the adult **phenotype**.

Niche (3,12) – the totality of environmental factors into which a species (or other taxon) fits: the outward projection of the needs of an organism, its specific way of utilising its environment.

Ontogenetic niche (3,4) – the ecological **niche** occupied by an organism at a given stage in its **ontogeny**. In many organisms this will change as the individual grows.

Ontogeny (2–4) – growth (usually size increase) and development (both size increase and differentiation of traits) of the individual.

Orthoselection (2) – a single dominant external force operating consistently in one direction.

Paedomorphocline – see **Heterochronocline**.

Paedomorphosis (2,3,5–7,9–11) – the retention of sub-adult ancestral traits in the descendant adult.

Parallel evolution (5,6,12,13) – evolution of similar characters separately in two or more lineages of common ancestry.

Parapatric speciation (5) – evolution of a taxon while it remains in contact with its ancestral population. Local selection pressures are strong enough to prevent homogenisation by interbreeding.

Paraphyletic group (12) – an assemblage of non-**monophyletic groups** (e.g., the Reptilia).

Peramorphocline – see **Heterochronocline**.

Peramorphosis (3,5,7–11) – development of traits beyond that of the ancestral adult.

Phenotype (3) – the totality of characteristics of an individual (its appearance) as a result of the interaction between genotype and environment.

Phyletic gradualism (1,2,5) – a slow, directional change of the gene pool producing a transformation of the entire population from one species to another through a series of graded intermediate forms.

Plesiomorphy (12) – in cladistic analysis, a 'primitive' or 'generalised' character.

Postdisplacement (3,9) – late initiation ('onset') of developmental events in the descendant; produces **paedomorphosis** when expressed in the adult phenotype.

Predisplacement (3,4,7,9,12) – early initiation ('onset') of developmental events in the descendant; produces **peramorphosis** when expressed in the adult **phenotype**.

Progenesis (3,5–9,12) – early cessation ('offset') of developmental events in the descendant; produces paedomorphic traits when expressed in the adult **phenotype**. Early sexual maturation will produce global progenesis, but early cessation in local growth fields can also produce progenesis.

Punctuated equilibrium (1–3,5,7) – following a prolonged period of morphological stasis, geologically abrupt change occurs by the influx of a new morphotype that has evolved at the periphery of the population (so-called **allopatric speciation**).

r-selection (4,9) – selection occurring in an unpredictable environment, where physical, density-dependent factors are important; traits selected for include: small size, early maturation, single reproduction, large numbers of offspring, little parental care.

Scaling (4) – changes external to an individual organism, such as in population density, predator or prey size, that occur with body-size increase.

Size (2,4,9,12) – a highly subjective term best measured by some multivariate method, such as principal component analysis, where size is defined as a general vector that best accounts for all the observed covariances. Deviations from this vector may be thought of as 'shape' changes, a similarly highly subjective term.

Species selection (1) – unlike natural selection, which acts at the level of the individual, species selection is selection acting on species. Whereas natural selection depends upon an individual's ability to survive, or least put off, death and on its rate of reproduction, species selection depends upon species' 'ability' to survive against extinction and on their rate of speciation.

State variable (2) – virtually any characteristic deemed to be of evolutionary interest. It may be specialised (e.g., tooth size) or general (e.g., body size).

Synapomorphy (11,12) – in cladistic analysis, a shared or 'derived' character. Only synapomorphies can define a sister-group relationship between two groups.

INDEX

357

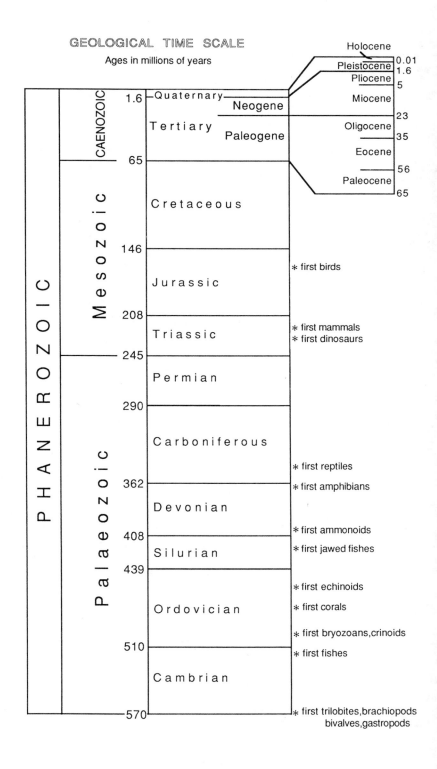

GEOLOGICAL TIME SCALE

Ages in millions of years

PHANEROZOIC

CAENOZOIC — 1.6 — Quaternary
Neogene
Tertiary — Paleogene
— 65 —

Holocene — 0.01
Pleistocene — 1.6
Pliocene — 5
Miocene
— 23
Oligocene — 35
Eocene
— 56
Paleocene — 65

Mesozoic

Cretaceous
146
Jurassic   * first birds
208
Triassic   * first mammals
                            * first dinosaurs
— 245 —

Palaeozoic

Permian
290
Carboniferous
                            * first reptiles
362   * first amphibians
Devonian
                            * first ammonoids
408   * first jawed fishes
Silurian
439
                            * first echinoids
Ordovician   * first corals
                            * first bryozoans,crinoids
510   * first fishes
Cambrian
— 570 —   * first trilobites,brachiopods
                            bivalves,gastropods

THE PERMANENCE OF WAVES

THE PERMANENCE OF WAVES

C. J. CLARK

LANGMARC
PUBLISHING
AUSTIN, TEXAS

The Permanence of Waves
Fable and Drawings
by C. J. Clark

The cover photograph of Betty, Margorie and
Fowler D. Brooks was taken at Stony Point on
Battle Lake in Minnesota around 1935.

PUBLISHED BY
LANGMARC PUBLISHING
P.O. Box 90488
AUSTIN, TEXAS 78709-0488
www.langmarc.com

Library in Congress Cataloging: 2006926371
ISBN: 1880292-815

DEDICATION

For Grandpa, who brought us here,
for Grandma, Mother, Dad, Pam, Joe, Jennifer,
and Catherine who believed in it,
and for Scout, who discovered the secret stones

Do you not see
the shimmering stones you took
from beneath sun-bleached docks
lost their dance of light
when placed in drawers forgotten?

Did you not sense
the blue heron lake wore those speckled
stones like brilliant buttons
To hold together
her glistening garment of waves?

Couldn't you instead store
treasured green and maroon flecked stones
somewhere inside you
to dance on your light, in your memories,
leaving the lake
Unbroken?

From "Whispers in the Wind" by Pam Trogolo

TABLE OF CONTENTS

- 1 -

BLUE WATER STONES

———————◁▷———————

Otter Tail County Road 16 was sprinkled with just enough cold rainwater to make it slick for travelers. Especially cyclists. And even worse for an elderly man peddling uphill. The lone cyclist braced himself against the steady pinpricks of water, tipping his cap just over his eyes. He peered out at the bleak landscape, marveling at how May in Minnesota could always manage to hurl the last remnants of winter back at spring. His red flannel shirt and blue jeans were a shock contrast to the gray of the morning. His bones ached, and he was chilled and shivering from the dampness, but he kept peddling. Painstakingly, he climbed the hill passing the town cemetery, which had once been rolling prairie before grave markers began to appear. The graves had tripled in number since he'd first traveled this highway, when it was still a dirt road. But he did not look in that direction. His eyes were fixed on the road ahead of him.

A car sped past him, splashing water. He could see that the car was filled with laughing teenagers.

Thunderous music blared from their open windows, sounding like a car wreck in progress. It dissipated like a muffled scream as they moved further down the road. Sighing, he welcomed the quiet of his bicycle tires on the wet pavement and listened intently to the singing meadow birds. He knew the birds by their individual songs. He knew things that he felt no one cared about anymore.

On both sides of the road there was open land descending into woods that led to a small lake. He would always pause beside the lake because the water was such a brilliant blue. He felt healed just looking at it. He could breathe more deeply. Except for an occasional fishing boat, the little lake was uninhabited by cabins on the shoreline. Even on this dreary May morning, it looked more blue than gray to him.

Taking a sharp breath, he noticed a white-tailed deer lying by the side of the road. He stopped. Slowly he got off his bicycle and moved toward the deer. It was a doe. She had been killed by a car probably earlier that morning. What a waste, he thought, shaking his head and stooping down to run his gnarled hand along the doe's smooth coat. Her coat had already begun to turn to the more reddish color of summer.

The man's legs were stiff, making the task of getting back onto his bicycle almost unbearable. Gradually he resumed his peddling, turning sharply down a gravel road that was so bumpy he considered walking his bike the rest of the way. The Main Camp Road. We used to call it the Main Camp Road, he thought. It was because we came here not to intrude; we came here just to set up camp. We came to build just a few noninvasive cabins by the lake and coexist with the land and with every creature from the weakest to the strongest. He thought of the dead doe with sadness.

He was the last of his generation now. He looked down at the package he carried with him. It contained a loaf of brown bread, a slab of butter wrapped in wax paper, a tin of peaches in heavy syrup, and the day's mail. He'd almost forgotten why he'd made this trip to town. His pace was excruciatingly slow. He gazed with appreciation at the thick deciduous woods on both sides of the road. Patches of early spring green dotted the woods. He became preoccupied with identifying which trees were yet to emerge with emerald leaves that would fill in the gray spaces left by winter.

Eventually the road narrowed and divided into a fork. It still amazed him that a sign had been installed here by the county. Stony Point Trail. Decades had passed without that sign. He turned and applied the brake pedal against the steep downward bend. A rush of rain-scented woods filled his nostrils. It was his favorite aroma in the world. What was distressing was that the scent had somehow become diluted over the years. The trail had once been a tiny grass lane with towering trees and underbrush, crowding it from both sides into a narrow passageway. As he approached the cabin, his pedaling had slowed to nearly a standstill. He was greeted by a boisterous yellow Labrador retriever.

"Hello there, poochie dog!" He patted the joyful dog on the head and looked into her liquid brown eyes. This was a good dog. Folks wondered aloud if he was perhaps too old to own such an energetic dog. But Scout was different. She was gentle, and she seemed even more perceptive than a lot of people he knew. She never chased the chipmunks or squirrels. She just didn't do dog things. And her devotion to him was touching. What would become of her if they had to leave here, he wondered. She followed him into the cabin.

He sat in his old rocker that smelled of wood smoke and moth balls and began to thumb through the stack of mail from town. There were bills and announcements and a letter from his granddaughter, Olive, who lived in Kansas. He held it for a moment, feeling almost too drowsy to read, but he opened it anyway. Her handwriting was childlike for a thirty-year-old woman. He smiled at this, then began to read aloud to Scout since her face was plunked down onto his knee, waiting.

Dearest Grandpa,

By the time you read this letter, Lottie and I will be on our way there to see you. I hope you are well. I've been worried because I've not heard much from you. I called the neighbors and they told me you are getting along fairly well, but they are so far from you that they see you only once in a while.

I hope you are considering my offer to come back with us to Kansas. Our house isn't big but you would have your own room across from the bathroom. There are lots of good doctors in Kansas City and there are interesting things to do. You might like it. And of course, Scout is welcome! Lottie loves her dearly. We can talk about this when we get there.

Well, I have to finish packing and get this mailed. See you soon.

Love, Olive

He folded the letter and closed his eyes. It would be good to see them. Lottie was eight now and her world was changing rapidly, caught up in friends and activities. Her parents had been killed in a car accident leaving Olive, her mother's sister, to care for her, something she had done admirably. His own daughter, Olive's mother, had died years ago from an incurable heart ailment. Olive and Lottie were his only heirs. One day they would inherit this cozy cabin, the beach filled with its large, colorful stones leading to the clear lake, the blue sky, the expanse of wooded land, and every other beloved thing. Would they ever truly appreciate the natural peace and brilliance of it?

Would they ever learn, as he had, to gaze into the silent lake's reflection at sunset and see the world as pure and perfect? Or maybe he saw these things because he was old and had walked these stony shores for so very long. It was a part of him. Like another layer of skin. It just couldn't be described in words. Maybe because it was too simple. You either loved the natural world or you didn't. You either saw the goodness of it or you didn't. You either respected it or you didn't.

He stared out the window at the morning sky. The gray was beginning to break apart into sapphire patches and wisps of lavender clouds. A blue heron glided by. He watched the reflection of it in the water, thinking how unspoiled the image was. It was graceful and serene but when he looked back at the sky, he felt there was something peculiar about the way the heron was flying. Herons rarely seemed uncertain, but this one did. Perhaps it was because of the construction of new cabins and widening of roads at the far end of the lake. The heron's habitat had likely been disrupted.

It was inevitable that, in spite of everything, people would come and go, and in their attempts to extract recreation from the lake and woods they would somehow disturb it. They would disturb a whole world they couldn't even see. Well, maybe he had become a sentimental old fool after all. He chuckled. Scout's tail thumped vigorously against the cold linoleum floor.

The man pried himself up and out of the old rocker, still feeling the ache from his bicycle ride. He shuffled his way to the kitchen and made a mug of instant coffee, black and strong. He blew on the steaming coffee, then took big slurps of it, not caring how loud he sounded. Noticing that the fireplace still had a few orange embers, he stirred them into dancing little flames and added wood. Good solid slow burning ironwood. For a moment he stood mesmerized by the spit and dance of the rekindled fire. Some of the ache in his bones melted away.

In the distance he heard a flock of Canadian geese honking. The sound became louder as the geese got closer, drawing his attention to the lakeshore. The light breeze of morning had begun to develop into a gusty wind causing the lake to become choppy. He watched waves spill onto shore, washing the beach stones into brilliant colors.

Scout loved the stones. It was an odd fascination, really. Sometimes he saw Scout stare at the stones as though she were willing them to acknowledge her, and at other times she would step into the shallow water, dip her nose into the lake, gingerly pick up a stone, then drop it carefully onto the beach with the other stones. The funny thing was, she did this only if she thought he wasn't watching her.

The two of them walked out into the day. Scout bounded ahead, stopping ever so often while he navigated the steep bank leading to the stone beach. A noisy cawing crow announced their presence. Leaning heavily against the oldest stooped cottonwood tree, he decided that he would need to spend the day tidying the cabin for the arrival of his granddaughter and great granddaughter. Not a small task. His housekeeping skills had been rudimentary at best. There were endless stacks of books, winter jackets thrown across the backs of chairs, and clean dishes crowding the countertops ready for his use. There were Scout's dog-haired furniture covers and wood chips to be swept away. Well, perhaps it wasn't all that horrible. He gazed into the old cottonwood's branches, then at the ancient beach stones, momentarily envious of their simplicity.

"No walk today, Scout. We're expecting company," he told the dog who had been anticipating her daily trek to the creek. But the dog bristled and stood rigid, looking away from him towards the cabin. They heard a car door slam.

"Grandpa! We're here! Grandpa!" He followed the dog's gaze to see Lottie racing down the bank towards him. Scout slowly began to wag her tail. Behind Lottie was Olive, who was sprinting towards him. Their momentum was so great that if he hadn't been leaning against the old cottonwood tree, he might have been sent sprawling into the water by their joyful hugs.

"Welcome! Welcome to our Stony Point!" he exclaimed happily. It was what he always said when they came to visit. Our Stony Point. Olive and Lottie had come to expect this greeting.

"You made good time. I just read your letter this morning. The cabin is pretty cluttered, I'm afraid. Not all that much in the way of food either."

"Oh, now don't worry about that," Olive said, winking at Lottie. "We stopped and bought groceries on the way here. Really, we're just glad to be here."

"How long can you stay?"

"Not too long this time, Grandpa, but we can talk about all that later. You look tired. Are you okay?" Olive asked, standing back to study him. She thought he looked frail and drained of color. A sudden gust of wind threatened to throw him off balance. Gently she took his arm and led him away from the stony beach. For an instant, he struggled to keep his footing on the large stones. Olive tightened her grip on his arm.

"Lottie, how do you like school? Are you glad to be almost finished for this year?" he asked in attempt to shift their focus from his momentary weakness.

As they walked together back to the cabin, Olive listened to her grandfather and niece talk enthusiastically about her school work, her plans for the summer, her friends, and her passion for art. He had always had such a keen interest in kids. He'd been a brilliant psychologist and had retired to write books on various topics. The writing had not been overly successful, but he had been very content living here at the lake cabin he and his brother had built over forty years ago. It was a good place.

Olive had been coming here every summer since her childhood, as had her mother. It was somewhat remote, very quaint, and nothing at all like the life she was used to in Kansas City. But it always happened that once she was here, she felt a soothing sort of peace envelop her. She felt as though each breath she took somehow cleansed her soul, like waves washing over stones.

Getting settled took some doing. Olive had to clear counter space, restock the refrigerator, make the beds with clean sheets, and straighten up some of the clutter. She was captivated by the old photographs on the fireplace mantle. A photo of her mother and grandmother were predominantly placed. There were school pictures of Lottie and one of Lottie's mother and father. And there was one of herself taken after her divorce. She thought she looked gaunt and preoccupied. The photo was taken long before Lottie had come to live with her.

Olive spent some time dusting and sweeping up wood chips and returning books to the bookshelves. When she was finished cleaning, she realized it was late enough for supper, so she busied herself preparing a meal for the three of them. It was then that she silently rehearsed what she planned to say to her grandfather. She dreaded the conversation. It would be such a turning point. She had always idolized her grandfather. It didn't seem natural to tell him what she thought would be best for him.

As daylight slipped into the cold darkness of night, a brisk lake wind began to gust around the cabin. Olive said nothing until Lottie went to bed. Her grandfather sat in his old worn rocker, drumming his fingers on the big wooden arms of the chair. A roaring fire in the fireplace warmed the room. They listened in silence to the crackle and hiss of it. Finally she spoke.

"Grandpa, have you given any thought to coming with us to Kansas?"

She moved closer to the edge of her chair.

"No, Olive. I can't say that I have. Although I do appreciate the offer." He smiled at her as though that would be the end of the discussion.

"But listen, Grandpa, I fixed a room for you. It's the big corner room with the south and east windows. There's that beautiful old mimosa tree on the south side. Remember all those little wren houses you built? I hung them in that tree. And during the day you'll have the whole house to yourself. I think you'll find that Kansas City is a good place to live. There's a church down the street from us and a branch library, too. And we're really close to doctors and a hospital. Plus, there are a lot of things to do. If you want to get out and walk, there's a park nearby. It's very peaceful." She realized that she was babbling like a salesperson and stopped talking.

"It sounds very nice, Olive. I know you're happy there. And Lottie, too. But this is my home. I'm not sure I have the words to describe how I feel about this land. It's no ordinary lake cabin in the woods. Besides, Olive, I'm very set in my ways. You would find me quite annoying after a while." He chuckled, reaching down to pet Scout who stared at him adoringly.

Olive didn't laugh. She felt frustrated and decided to try a different approach. "The thing is, I worry about you up here all by yourself. Oh, I know you have neighbors who check on you but, Grandpa, do you think it's a good idea to ride your bicycle all the way to town? I mean, you have the car. And what about staying warm or just getting around? What if you fall or get sick? Your phone doesn't even have an answering machine. And…"

For a moment the two of them sat listening to the wind howling outside. They could hear waves crashing against the shore, while inside, the cabin glowed with warmth and memories reaching across generations. Olive watched her grandfather rest his head against the back of the rocking chair.

"Grandpa, I'm just saying, you're not getting any younger." Her tone was becoming more pleading. "I think it would be best if you lived where someone could make sure you were safe."

He leaned forward and patted her hand the way he had when she was a child.

"Olive, do you remember that drawer in your room here, the one where you put all the pretty stones and other things you collected from the beach when you were little?"

"Yes, I guess so, but…" She didn't understand the shift in conversation. It irritated her. She would not look at him.

"All those things were treasures to you, each special in its own way, worthy of collecting and worthy of putting away for safekeeping. But in time you forgot about them, and so, they still sit in the drawer, useless and waiting, out of their element." He waited for her to look at him.

"But what does that have to do with what I'm saying?" His words angered her.

"Olive, don't you see? That's how I would feel moving to Kansas City. Just like one of those treasured but forgotten things. Waiting, and out of my element. Useless, and serving no purpose." With some effort, he stood up, signaling the end of their conversation. Finally he added, "Let's talk some more tomorrow. I'm all tuckered out."

Olive pictured herself jumping to her feet and shouting at him that this was not okay, that he had to start being realistic, to think about all the things that could go wrong here, that she was offering him a comfortable, easy life. But then she pictured the disbelief that would cloud his thoughtful face at her harshness and waved away the image. She had never spoken to him like that. Olive thought he looked weary, but strong, like he had always been. It was hard to think of him as old. Turning to fix her gaze on the fire in the fireplace, she said quietly, "Okay. Fair enough. Good night, Grandpa."

"Good night, Olive."

She watched him leave with Scout trotting dutifully behind him. Maybe he would be more reasonable in the morning after a good night's sleep, she thought. She felt discouraged though, bothered by what she concluded was his stubbornness. That was it. He was stubborn and set in his ways. He didn't know what was best for him. He just wasn't being practical. People become that way when they get older. Olive's annoyance grew. She was determined not to leave him alone at the cabin.

Before her mother died, she'd made Olive promise that she would look after her grandfather, but he was making it very difficult. He was just impossible. He had practically ignored her letters asking him to consider coming to live with her. When they did talk about it, he would invariably change the subject or speak in riddles.

When she was a child, she'd always enjoyed his fables. He told his fables to explain things she couldn't understand. It was how she'd learned many lessons. But now, she had to be the adult, the responsible one. Olive remembered her mother's exact words. "Olive, your Grandpa is the wisest, kindest man alive. You make sure his days are comfortable and contented, that he has whatever he needs. He's always looked after us, you know. You must do this for me, Olive, you must promise." And Olive had promised.

Olive looked around her. Her mother's life had ended here in this cabin with her grandfather by her side. Oddly, the image of it didn't distress her because it was what her mother wanted, and in her heart, Olive believed it was a good place to live and to die. She stood up, walked to the front window and looked out into the blackness of night.

The wind still raged. She could see that the moon was beginning to rise above the lake, creating a choppy path of white light in the water. The light was just enough that she could see the waves slam into shore, one after another after another. Each wave was different from the one before it. She was mesmerized by the rhythmic repetitiveness of it. It's like all of us, she thought, we come and go, leaving our inscription on the stones of the earth. That's all the permanence we get, really. Life goes on. Fleeting and predictable.

She placed another log on the fire. It rolled to the back of the fireplace, creating a brief crash and spark, blue flames licking it uncertainly. She watched this wondering how it was that her grandfather managed to ride his bike to town, stack firewood, read seven books a week, and thrive on so much reclusion. He did thrive in a way. He still had his glow and was as sharp as ever. But, no, she knew he couldn't stay here. What about the promise she had made to her mother? Besides, it just didn't make sense. He was old and moved slowly, slept more soundly. It's what people do when they get older; they move to safer places where someone can look after them.

Olive was restless. It would be impossible to sleep. The clock on the mantle ticked loudly. She began to pace without realizing it. Her pacing led into her grandfather's study. It was a small room, but cozy and inviting. She switched on the light and began to look through his collection of books. History, psychology, nature, novels, every topic imaginable was here.

Her eyes searched the rest of the room. The massive bookshelves, file cabinets, and stacks of periodicals made the little room appear vital, almost as though it had its own pulse. Maybe there was something here, just one thing, that would serve as evidence to her that he really was declining. Then she would be able to gently persuade him with a specific reason why he should return with her to Kansas. It would be that simple.

On his desk sat his old manual Royal typewriter, which he loved dearly. She smiled, thinking of him pounding out letters to congressmen, newspaper editors, and family on this obsolete thing. It was in immaculate shape, almost as though he'd polished it regularly. Behind the typewriter was a small cardboard box labeled "pencils—red, blue and #2B" and next to it a smaller box that contained several pink, white, and gray erasers.

Stacked in front of the boxes were three wooden rulers, a brass letter opener, and a large magnifying glass. Olive noticed that her grandfather had inscribed his initials on the rulers. Beside the three rulers were two field guidebooks for identifying northwestern songbirds. The books were earmarked with so many small scribbled notes that she wondered how he was ever able to successfully turn the pages.

Beneath the books, there was a folder filled with newspaper clippings. All of the clippings pertained to either local or national politics. He had written comments in red pencil on some of the clippings, as if he'd intended to write letters to the editor. Most of his comments were dated.

Under this folder was another folder he'd entitled "Erasable Bond," which contained a small stack of slightly yellowed typing paper. Still another folder held short grocery lists and partially completed letters he'd composed. The grocery lists were all similar: apples, pumpernickel bread, powdered milk, plum jelly, green beans, saltine crackers, and peppermints. The letters appeared to be to no one in particular. Most were written in pencil, and behind them were little sketches of birds and geese. Although the pencil was smudged in spots, Olive marveled at how detailed the sketches were.

At the bottom of the stack of books and folders, she noticed a package. It was some sort of book wrapped in oilcloth and tied with a piece of twine. Was it an intended gift? He had obviously taken great care in wrapping it, although the oilcloth looked worn. Very worn.

Olive decided it was not a gift. Slowly, she began to unwrap it. Halfway through, she stopped, feeling intrusive. She listened to the silence around her.

Overcome with the weight of responsibility she felt for her grandfather and Lottie, Olive sat down, pulling the chair close to the desk. Minutes passed as she listened to the ticking clock on the mantle. She hugged the oilcloth package to her as though it were a small breathing thing.

Eventually she placed the package on the desk and opened it tentatively, smoothing the oilcloth flat. It was a book. A sketchbook. No, a journal, some sort of journal. She began to thumb through it. There was an unsealed envelop just inside the journal's pages. It was not addressed to her, but to Lottie.

For a moment she just stared at it. She examined it, then placed it back inside the journal exactly as she had found it. She wrapped the journal back in the oilcloth and tied the twine. So it had been intended as a gift after all, she thought. And yet, it appeared to have been wrapped quite some time ago. Had he forgotten to give it to Lottie? Was it supposed to be a secret? But why? And what sort of book would he have wrapped so long ago? Lottie was barely eight years old.

Olive closed her eyes, inhaled deeply, then reached again for the package. She took a deep breath, then reached for the letter, opened it, and began to read.

My Dear Great Granddaughter Lottie,

Some rainy day, I hope you will find this letter and read it. I am very proud of what a bright young lady you have become. You have been a big help to your Aunt Olive and to me as well. Your mother and dad would be very proud of you. They loved you very much.

Someday, Lottie, your great Grandpa will not be here anymore. It's not a sad thing really, just the natural order of everything. I do wish I could tell you about what I have learned about life over the years on this land, which will someday belong to you. Truly, I've discovered a world I'd never imagined in my earlier years, which were spent obtaining degrees and titles.

There is a timeless wisdom that awaits you here unlike any other. You only need to be willing to listen and to see, to really see, what is around you. To be sensitive to nature, respect every creature, every puffy cloud, every breeze that blows, every tree that sways, and every frog that croaks.

Olive tells me you like to read, so I have written a story for you to read someday after I am gone. Perhaps this story will inspire you in some way to walk a gentle path through these north woods. And when you see your reflection in the lake, you will know you have not intruded. You will know that you belong.

Love,
Great Grandpa

Olive carefully folded the letter and tucked it back in the envelop. She opened the journal to the first page, placing the letter there. The first page appeared to be the beginning of a neatly typed story titled *The Legend of Blue Water Stones*. She smiled, remembering how her grandfather's stories had always been so imaginative. Sometimes the stories were full of adventure and other times full of comfort. She remembered sitting spellbound beside him when she was a young girl while he spun inventive tales that seemed to be just for her. He could make the most commonplace things enchanting.

Perhaps he knew he would not have enough time left in his life to do the same for Lottie. She thought about his letter. Lotttie was too young to truly understand the words, but she adored her great grandfather so much that she would have tried to understand. She would have tried to do exactly as he had said. Lottie would have willed herself to listen, to see, to appreciate every tiny thing, to walk a gentle path through the north woods.

Olive sighed. She pushed the chair back from the desk and resumed her pacing. There was no hope for sleep. The clock on mantle chimed quietly, then resumed ticking. The fire had died down to popping embers. She walked slowly to the kitchen and fixed a mug of steaming hot cocoa. While she sipped on the sweet chocolate, she gazed at her grandfather's empty rocking chair by the fireplace. For a moment she imagined him never again sitting there in its tapestry upholstered softness and wide wooden arms. She took a sharp breath.

Returning to his desk, she picked up the story meant for Lottie. She hesitated, setting it carefully back on the desk, patting the pages into place. Tentatively, she picked it up again. Reading a few pages would help lull her to sleep, she thought. She felt in need of a bit of comfort anyway.

Olive walked back to her grandfather's rocking chair, balancing the hot cocoa mug in her other hand. She lowered herself into the chair, sinking into the rocker's cushions, covering herself with a soft puffy quilt. Immediately she felt small, but safe. Very safe. She closed her eyes for a moment, then opening them wide began to read.

© C.J. Clark

📝 For my great granddaughter Lottie,
from Grandpa

The Legend of the Blue Water Stones

My dearest Lottie, one day when you travel far, far to the north, past gray fields of slumbering farmlands, you will discover a most magical place where the earth turns blue when it meets the sky. You will see brilliant blue pouring gently from the horizon onto open land, creating a beautiful and mysterious lake. From the heart of the lake, soft breezes form waves that continually wash blue onto a stone-covered beach, causing the stones to be more richly colorful than any other known to the world.

It will be clear to you that these are not ordinary stones. There is something very curious about them and something unexplainable about this place. There is something strange in the way the old cottonwood trees bend so easily to touch the blue water. And something stranger still is the way the birds, animals, and other creatures cling to it. Almost no one knows the origin of the unusual stones. Almost no one knows this, the true legend of the Blue Water Stones...

You see, very few know that the blue water stones came to be along the lakeshore ages ago at the time when the moon broke free from Mother Earth to dangle in the darkest night sky. In the very instant that the young moon was alone, legend tells us that he became filled with a loneliness bigger than the whole universe. He could not yet understand his importance to the earth. Nor did he know anything about the rising and setting sun.

At first, the young moon stayed in the sky too long at dawn while the sun rose, just so he could be a part of the light. He became distraught when the great sun, who colored the sky with pinks and lavenders, chased him away into the darkness.

"Who goes there?" exclaimed the sun in a booming voice, as the sky became lighter and lighter.

"It is me, the moon. I am from Earth, and I don't know what I'm to do!" the moon answered quietly. "I'm not used to such darkness."

"Well, Moon from Earth, you cannot stay here. There is only room enough for me and a few clouds in this sky. You must go and find your own way. Look for the North Star, she can help you. But you must leave now! We cannot both be here. Now, go! I must get back to work!"

So the moon sadly went in search of the North Star. And long before he discovered her and his new and dear friends the stars, even before he learned how bright he could shine in the night, he shed great solid tears of misery that fell to the earth.

His tears plunged to the small northern lakeshore, forming stones that scattered all along the blue water's edge. As each teardrop touched the ground, it became a new stone, taking on warmth and color and substance and character.

The moon blinked in astonishment as he gazed down at Earth and saw that his tears had acquired finely sculpted features, each one different from the next. They were like tiny planets, and when he looked hard enough he could see that they had big expressive faces with almost no bodies except for rounded bellies and knobby feet. Every one of them could close their eyes, become perfectly still and then ever so quietly retreat into themselves so that they looked as ordinary as any other common stone of the world. That was what would keep them safe. Henceforth, they would be the guardians of the sparkling blue water.

Anyone coming to the lake would have to pass over the stones. It was quite clear they were not adapted for travel, but then, there was no reason for them to ever leave the blue water shores.

As he marveled at the stones looking up at him in the night, the sky around him became a little brighter. And just then, the moon heard a soft voice call, "Do you see? Your teardrop stones will stay beside the blue water as a reminder to you that despite your earlier sadness, there will always be good things that follow. And from this time on, for those who have eyes to see and the hearts to follow, the stones will share secrets of long life. Now, come along, let me show you the night sky."

And at that very moment he knew. The young moon of Earth had found the North Star. He looked back one more time at the blue water stones.

- 2 -

WHISPERLEAF

Now, even in this age, you will see that high above the blue water's shore, along the bank, old cottonwood trees still watch over the stones night and day. One of the eldest of all the cottonwood trees was called Whisperleaf because she spoke so softly that she could barely be heard above the unceasing waves. The harsh winds of winter had twisted her trunk so close to the shore that she could gently touch the blue water stones when the breezes were just right.

She knew each stone by name. The moon's teardrop stones were her friends. Whisperleaf knew these stones were far more ancient and far wiser than she was. She had learned many things of great importance from them. Even in times of great bleakness, they'd told her, "Whisperleaf... look for the good...you must try to look for the good."

When she was young, and winters were long and sharp with ice, or summers were hot and waterless, she did not know what they meant. But as her trunk began to bend with age and she saw the timeless moon come and go, she began to realize what it was to look for the good.

Even so, Whisperleaf could not forget that day long ago when one of the stones was taken away from the lakeshore by a human child. The stone was called OldSkogmo, and he was the only stone Whisperleaf could ever remember actually leaving the blue water shores. She was a very young tree then, and it made her sad. And yet the other stones assured her that OldSkogmo would return someday with tales of other lands, and it would be good.

But to Whisperleaf, he was just gone. He had been her closest friend. She would never forget his kind and knowing face, his orange and black specked roundness, and the secrets of other times he shared with her. He had told her of the times before humans came when the cottonwoods were tall and plentiful. Yes, she missed that. Sometimes she wondered if he had somehow returned to the moon. She was very old now, and he had never come back.

And now, it was spring once again. All of them had survived another winter, and on this day Whisperleaf took courage in the early dawn's yellow sun rising up and over the woods behind her. She watched as a white-tailed deer arrived for a drink of the blue water. The young deer stepped lightly along the shore, never disturbing a thing. She was always extra careful when crossing over the little seashells.

Whisperleaf noticed a blue heron wading in the shallow water. As usual, the heron was stern and silent, like a statue, and seemed to be admiring his own reflection in the lake. At times he would bend his neck towards her and lift his giant wings for no apparent reason. Then suddenly, just as he had done so many times before, he lifted himself out of the water, gliding silently away to some secret place. Whisperleaf knew he would return to the same spot that evening as though he had never left.

As she watched the heron fly away, a giant Canadian goose landed noisily in the water. Two more geese landed after him. The first goose created a hubbub and was loud when he came to the blue water. He blurted out all his thoughts despite the stones' reminders that many things are better left unsaid. The goose tried hard to learn from the stones. He listened to all he was told and was better for it, but he still managed to be cumbersome and silly.

And yet, thought Whisperleaf, when the three huge geese returned to the sky, they were amazingly graceful, as though flight was the most effortless thing in the world.

Whisperleaf was gazing upward at the geese as they glided above the lake in a miniature V formation when there was a sudden crash and fluttering in her upper branches.

"Cau,,,cau...cau...caution! A storm, a dangerous storm...look to the north....cau... cau... caution!" screeched a large black crow who had landed heavily on her branch. Then all of a sudden he was gone. And there was nothing more. Only the bright chill of a spring morning.

It was so calm, yet Solosong, the black crow, was never wrong. As the day unfolded, darkness overtook the skies. A biting north wind began to whip through Whisperleaf's leaves. The deer lifted her head, alert, motionless, water dripping from the corners of her mouth, alarmed by the sudden change in the air. Then suddenly she bounded into the woods without a glance back. She did not see the lazy lapping waves turn into crashing white-caps slamming into the stones.

The wind whirled around faster and faster, lifting bits of sand and small shells off the shore. The stones began one by one to close their eyes and retreat into themselves.

Whisperleaf braced herself against the onslaught of icy rain. Her new spring leaves were twisted, torn, and blown away. The sky became darker, the wind even stronger. She felt her trunk bend to the point of almost breaking. The smaller cottonwoods shivered and tried to resist, only to be tossed around like twigs. Stabbing ice pellets mixed with snow, frozen clouds dropped to the earth and swirled into drifts of white on the ground.

The wind became a green black hissing beast devouring everything in its path. Lake debris swirled through the air. Immense cracking sounds overtook the screaming wind. It became colder and colder as if winter had returned in a savage fury against spring. The wind raged on, spitting ice, which turned to sleet, then sheets of endless rain. The sky seemed to roar with gray water rolling out of black clouds.

Then there was just rain. Somewhere in the distance, Whisperleaf thought she heard the black crow calling, but the downpour drowned out the sound. The wind had stopped. After a long while, the rain became a pitter-patter sound and a streak of pink light appeared in the western sky. It was over. Only then did Whisperleaf realize that one of her largest branches had been partially ripped and was hanging close to the ground.

"Whisperleaf?" Large round eyes were fixed on her. It was Skeedad who spoke. He was a portly black stone with bold orange stripes outlining his perfectly sculpted face. He had a spiral of narrow stripes wrapped around his belly and stubby feet planted firmly in the sand. A small wave washed over him. His tiny nose wrinkled slightly as he gazed up at her. He frowned thoughtfully, then smiled. His smile was nearly as big as his face. "Are you alright?"

Whisperleaf saw that he was filled with curiosity and warmth, causing her to forget the storm and answer, "Why, yes, of course, except for this branch..."

Skeedad tottered closer to her on his knobby feet. He tumbled forward, then righted himself. Looking up, he surveyed the injured branch, blinked his giant eyes at her and sighed, "Ah, yes...I can hear you better now...look how very close you are to us."

She had not realized this and just as she began to marvel at their closeness, the giant Canadian goose and his two friends landed in the lake with a great splash, causing Skeedad to roll backward into the shallow water.

"Gaak! How ghastly, how hideous! Look at that ugly branch! And look how your leaves are torn away! Whisperleaf, the storm has absolutely ravaged you to bits. Gaak, how horrid! How very hid—"exclaimed the goose who stopped himself immediately when he saw the stones turn scornful eyes to him, just as Skeedad was rolling himself back onto the sand.

"Um...yes...well, it was a fierce storm and you survived it quite well enough, good Whisperleaf," the second goose said. The third goose, stood tall and flapped his great wings in agreement, while the first goose looked shamefully at the ground.

"How very excellent it will be to have your shade in the heat of summer. Don't you think so, too, Silverdew?" Skeedad asked, casting a warning glance at the first goose. A tiny droplet of water trickled down Skeedad's face.

Whisperleaf considered Skeedad's words and did not mention that she felt a bit weaker and older.

In the days and weeks that followed, spring turned slowly to summer. There was real warmth from the sun. The sky became a powdery blue with friendly clouds dancing here and there. The white-tailed deer and the blue heron came each dawn and each sunset, and the three geese spent whole days wandering lazily along the shore. Skeedad and the other stones traded tales of other springs and summers and absorbed the goodness of each day with Whisperleaf close by. Her tender spring leaves were full now. Even the ripped branch had shiny leaves that touched the damp sand.

Once on a clear and peaceful night, when the moon was full and reflected on the water, one of the youngest cottonwoods asked Whisperleaf, "What are we to fear, Whisperleaf?"

"Should we fear storms?" another asked.

"And, will there be other dangers?" a very tiny cottonwood asked.

"Storms are not to be feared. Yes, some are very powerful but the black crow, Solosong, lets us know before they are upon us so we are ready. It is just the way of nature. Humans are another thing altogether. You cannot ever be ready for humans. They are to be feared because you cannot know what they might do," Whisperleaf said in her quiet careful way.

"Yes, and although that is true, there are humans who are good and would not harm us," Skeedad said. "Once long, long ago before Whisperleaf's time, when the cottonwoods were tall and plentiful, there were humans who came here at the break of every day. As they aged, their footsteps became shaky and slow. Sometimes they would sit in complete silence under the cottonwoods. Their skin had lines deeper than bark and their hair was the color of a moonless night. They never caused us any harm."

"I remember them well," a smaller red stone said. She had an earnest face with big inquisitive eyes, a rounded nose and mouth. She looked upward when she spoke as though she was snatching words right out of the sky. Her red color, unusual for a blue water stone, was speckled all over with white and black dots, which made her glitter even in the earliest moonlight.

"They wore feathers in their hair and their steps were light as any deer," the red stone continued, looking at each tiny cottonwood. "They came here from the water in a little boat made of trees."

At the mention of trees, the young cottonwoods began to whisper to each other. They seemed worried.

"A boat made of trees?" one of them finally asked.

"Yes, but the trees' spirits had already gone to be above the sky," she answered. Sensing their relief, she continued, "There were other humans, too, at that time, but they stayed away because they believed rival spirits danced on the shores of this land."

"Which, in fact, is true," Skeedad said. Skeedad knew all about spirits. He had an ability to perceive all the ghosts that had ever been along the blue water shores.

"Those humans were very courageous, indeed," the small red stone concluded. Skeedad nodded in agreement, momentarily tipping too far forward, his face touching the beach.

"And, there was one human who came here alone many moon seasons later. We called him Brookdweller. You remember him," a plain orange stone said looking from Whisperleaf to Skeedad.

The orange stone had an elongated face with large droopy eyes under an angular brow that made him appear to be frowning all the time. He was larger than the other stones and a bit clumsy. Because of his odd shape, he had to lean slightly against the small red stone to keep from teetering forward. Despite his gawky appearance, they listened to him intently because he was always able to recall any detail of any time and jog their memory of it. He'd once told them that he could even remember as far back as when he had been a tear falling from the moon.

"He built his house over there beyond the grassy bank at the edge of the woods without disturbing the land. Very uncommon for a human," he said. Whisperleaf followed his gaze towards the bank and waved her upper branches slightly.

"Yes, yes, we remember Brookdweller," the stones replied in unison.

"Brookdweller's spirit will always be here," Skeedad said. "He was kind and gentle. He loved the earth and respected every creature, a rare quality for a human. He understood about the order of creation."

Whisperleaf knew this had always been important to Skeedad. She felt the same way. Creatures needed to be respected for who they were. It was only right.

She swayed a little in the evening breeze, remembering Brookdweller. The first time she saw him, he was a young man. He wandered so frequently along the lakeshore to and from the little brook, and he became known as Brookdweller. Every day he walked in the same direction the heron had flown in the morning. It was as though he might be searching for the heron. And yet, Brookdweller kept a respectful distance from creatures who might be wary of him. He seemed to know he was a stranger to this place.

"Brookdweller would take a blue water stone in his hand and turn it over and over, studying the stone as if he were searching for something," Skeedad said. "Many nights he would stand at the water's edge, gazing up at the moon. He spoke no words. Then, at times, he would speak directly to the stones as though he actually saw us. Yes. actually saw us for who we are! How could that be, with our eyes closed and looking like any other common stone of the world?"

Skeedad closed his eyes for a moment, then opened them wide and added, "It was quite odd, of course, but not frightening and certainly not a danger. He's been gone a long time now, but we have not forgotten him."

"Not forgotten him...not a danger," several of the stones repeated quietly.

Skeedad looked at the young cottonwoods with great fondness and said, "Humans are not always to be feared. Just be cautious and be prepared for the unpredictable."

"Be prepared for the unpredictable," Whisperleaf murmured.

The young cottonwoods gazed appreciatively at her and bowed slightly.

"Be prepared for the unpredictable..." they whispered.

"Be cautious," the small red stone added, looking upward at the night's first twinkling stars. She watched the enormous yellow moon rise above the lake into the black velvet sky. It created a path of warm light in the dark water.

All of them followed her gaze and felt sheltered in the moon's light.

"Be cautious, be prepared....and always remember to look for the good," Skeedad concluded thoughtfully.

And so it went. There were other stories of times gone by told by the stones. As days grew longer, the sunsets became more orange and the warmth of each day stayed with them into the night. Whisperleaf considered all they spoke of and wondered what had become of OldSkogmo who told the best tales of all.

In fact, Whisperleaf was beginning to wonder where the black crow Solosong had gone. It was not like him to be away for so long. She had not heard his voice since the strong storm of spring, which had torn her branch. She mentioned his absence to the stones, but they seemed unconcerned. They never fretted about things like that. Time moved differently for them.

©C. J. Clark

- 3 -

SOLOSONG

---○---

The small black crow gazed at the moonlit lake from where he was perched in an old pine tree. The water was so still, it could have been ice. Even the slightest ripple could be heard. Owls hooted into the darkness and night animals scurried through underbrush. There was a plodding sound of deer hooves following narrow trails though the woods to the lake, then pounding away after a cool drink of water. The sound seemed to go on forever. Far in the distance there was the yip and howl of a coyote's song blended into melody. Then silence. The quiet was like a vast space not yet filled. For the small crow it was not like any other night. He waited. He knew how to wait.

When dawn finally arrived in the eastern sky, the little crow stretched his wings to greet the new day. All was well on North Point. All was well and yet as the sun began to warm the sky, he began to practice his warning call, "Cau...Cau...Cau...!"

He kept this up for a time until the morning was filled with twittering songbirds. The raucous call of a blue heron and the laugh of a loon resounded across the lake. A formation of Canadian geese appeared in the sky above him. The geese seemed to fill the entire sky. Their honking sounds were deafening. His call was lost, and he fell silent.

He soared upward, then down, following the tree line along the lake and landing in a young birch tree. This caught the attention of the creatures along the shore which pleased him. Since he had nothing to warn them about, he flew away, returning to the old pine tree, last night's haven. All was well on North Point. He dozed.

Suddenly he was seized with panic as a dark shadow fell over him. Startled, he looked up only to be blinded by the sun. He looked down and saw it again. A looming darkness on the land seemed to grow into a giant shape. There was a crash in the branch beside him.

"All is well on North Point?" It was Solosong from Stony Point. He had returned as promised. The younger crow was in complete awe of the large older crow and was unable to utter a sound. Solosong seemed to understand the silence. He waited.

It was Solosong's patience that made it possible for this young crow to learn his duties. Solosong had patrolled North Point and Stony Point all spring and early summer since the Elder CauSong of North Point had gone to live above the sky. Long, long ago, Elder CauSong had taught Solosong the ways of the lake creatures and what he must do as an honorable crow to alert them to danger.

And now, here was this young crow, willing to learn the ways. It was not a usual life for a crow, but the little crow was bright and capable of carrying on the path of Elder CauSong. Solosong was sure of this. It was already summer and little crow had spent his first day and night alone. It had gone well. He was ready. Solosong could return to Stony Point.

"Solosong, there are some things I still need to know," the young crow began tentatively.

Solosong looked ahead, scanning the horizon for anything different. He waited for the young crow to finish.

"How long am I to stay with the lake creatures after I have sounded the warning alarm?" He wasn't sure, and it seemed important.

Solosong turned to look at the smaller crow and said, "You do not stay...you alert them to what is coming and then you go. You fly away. When the trouble has passed, you alert them from where you have taken cover. They will hear you. You are a crow. It is your song. It is unlike any other."

Together they listened to the sounds of the day, the lake breeze rustling through leaves, the steady slapping sound of waves on stones, a woodpecker hammering a hollow tree. There were sounds of finches flitting about, of chattering chipmunks, and the sweep of pine needles against the breeze.

Solosong knew that the young crow would follow a solitary path all his days. He watched a small white cloud drift by and said nothing more for a time.

Finally the little crow interrupted the silence. "What will they call me, Solosong?"

"Lonesong of North Point," Solosong said thoughtfully.

Lonesong puffed himself up as large as he could and settled into the name. It was good.

"Of all the lake creatures," Solosong said, "the Blue Water Stones will always tell you what you need to know. They have been here the longest, even here on this North Point. Their wisdom comes from an endless time of absorbing the goodness of spirits that have gone on and from being intense observers. They will teach you how to look for the good. You will come to understand if you just listen to the stones." With this, he soared upward and was gone.

Solosong dipped and soared and sliced through clouds and deep blue sky, believing himself to be an eagle or a hawk. He called into the air as loud as he could, his voice lost in the wind. He flew towards the sun, the warm summer sun. It was red and low in the sky. He soared upwards over treetops and creeks of babbling water. Onward he flew, hurled through space as though he were just one feather, light and swift.

Finally he landed in a stand of poplars where he heard the sound of another crow calling.

"All is well on the West Point?" Solosong bellowed.

"All is well. You are returning to Stony Point, Solosong?" the calling crow asked.

"Yes. It is time. The sun is warm. The North Point is at peace." Barely did he finish the words when he shot up and away and on to Stony Point. It was nearly nightfall.

When he returned to Stony Point, he landed with unusual softness in the upper branches of his friend, Whisperleaf. He made no sound. He could see that Whisperleaf had been damaged by the last violent storm of winter, and he felt sad. Still, he said nothing. The stones, shells, and cottonwoods, the white-tailed deer, the blue heron, and even the three silly geese looked up at him in anticipation. But he had no warning alarm for them this time. The stones were the first to understand this.

"Welcome back, Solosong." Skeedad was always the first to speak.

Making no reply, Solosong launched himself upward with all the grace he could muster. He glided above the shore, turned sharply, and flew higher, above the trees, the ironwoods, the elms, the pines, and spruce trees. He continued on until there was just open meadow below him. Then he turned once more, back toward the lakeshore where he stopped a fair distance away, settling in a giant maple. All was well on Stony Point.

From his perch in the maple tree, he could hear the stones telling tales of ages ago when there were large fierce animals that came to the lakeshore to drink. Even the peaceful animals needed a loud crow to warn them of the presence of such beasts so they could flee the danger. The white-tailed deer had never had to fear those animals. They had long since gone to be above the sky. Skeedad spoke of them with great fondness and seemed to miss them, although there had been many crow lifetimes since then.

Solosong was certain that the fierce animals had been replaced by humans. Skeedad probably knew this, too, but it was said that he did not speak often of humans after OldSkogmo had been taken away by the child.

Skeedad spoke of ice storms and winter blizzards that covered their entire world in slumber. The stones would withdraw into their own warmth and the only trees still green and watching were the pine, fir and spruce.

One quiet summer morning, Solosong awoke from his rest with an overpowering feeling of sheer dread. He looked to the skies and saw nothing unusual. There was nothing strange in the south breeze. He could see that the stones, Whisperleaf, and all the creatures were content in their places. Still, he felt a sense of darkness that would not go away. It stayed with him throughout the day.

As he became increasingly nervous, he decided to fly to the North Point by the longest route. This meant he would travel above fields and great open meadows south before switching back north. It just seemed right. He didn't know why. It did not matter why. He was never wrong about his feelings of danger.

With a quick glance back at Stony Point, he set off on his journey. He flew higher and further than he ever did this time of year. There were cheery puffs of white clouds hanging low in the sky that greeted him along the way. Southward he flew, over a small town and over farms with enormous fields of young corn, cows plodding through pastures, solitary human dwellings, tiny lakes, and pine woods. The wind was warm and strong. Solosong saw everything. And yet there was nothing of danger. It was just the feeling of something not right.

He circled back to the north, hovering, soaring, darting to the east, then west. He lingered in the sky, watching.

Then he saw it. There was no mistaking the slow methodical movement of a car on the road below. It was a big shiny intrusion. It stirred the gravel road into great clouds of brown dust. It was loud. And it was ugly against the landscape. Humans, he thought.

The car traveled past the pines, the tiny lakes, the small dwellings, the cows, the fields of corn, and was almost to the small town when Solosong, sure of its path, flew fast like a bullet taking the crow's route back to Stony Point.

As he streaked across the sky, the wind zipping through his wings, stinging him with fatigue, he thought he remembered Skeedad saying, "...not all humans are to be feared." Still, he had to warn them. "Just be cautious," Skylo, the red stone, had said.

So he flew with every bit of strength he had, thinking only of his friends Whisperleaf, the stones, the sea shells, the white-tailed deer, the heron, and the geese. He thought of their world, peaceful and good. The storm was one thing, humans were something else. It was his duty to warn them. He was Solosong of Stony Point.

When he spotted Whisperleaf, he landed this time with a crash and screech, darting from uppermost branch to branch, screaming, "Cau...Cau.. Caution!"

"What is it, Solosong?" Skeedad asked, alert, curious.

"Solosong?" Whisperleaf was shaken.

"Humans coming from the south. Cau... Cau...Caution!"

Solosong repeated the alarm just once until he could feel all the lake creatures become tense and wary. It was all he could do. He was exhausted. His task was complete. He shot upward and was gone.

- 4 -

goodScout

One more turn to go and they would be there. The big yellow dog pressed her nose against the window and whined until it was opened for her. Crisp northern air rushed in and filled the car. It was cool and vaguely familiar to the dog. As the car rounded the last corner and made its way down the hill, the strong scent of rain-washed woods filled her nostrils. There it was. The lake. Yes. She could see it now. She almost leapt out of the car window. There was a crow's loud cawing, and looking up, she spotted three geese climbing higher into the blue sky. She was almost certain she had been here before. Yes, she was sure she had.

As the car pulled into the narrow drive, the sun was setting in an explosion of bright colors reflected perfectly in the lake. The door nearest the dog opened and she streaked out, barely touching the ground. Splash! She was shoulder deep in the cool blue water.

"Wait for me! Wait!"

A brown-haired child raced after the dog, picked up a small branch and hurled it into the water. She ripped her sneakers off, tossed them onto the bank, and pushing stray curls off her face, she rolled up her jeans. Timidly, she stepped into the lake just as a frothy wave washed over her toes. The water was so cold, she shivered a little.

"Go, go get the stick, go on—it's over there!" she squealed.

The big yellow dog splashed happily through the shallow water, pretending to search for the clumsily thrown branch. The child bounded into the water after her yelling, "No, look! See? It's there! Over there!"

The dog swam toward the branch, snatched it out of the water, and brought it to the child.

"You found it! Good Scout! Good dog!" The child was drenched and laughing. The dog adored this game.

Scout charged up the bank and rolled onto her back in the tall grass. Panting, she gazed up at the cloudless dusky sky, then suddenly sprung to her feet and dashed back to the lake. Dipping her nose into the clear blue water, she took a drink. The water was cool and refreshing, not like anything she had tasted before.

A woman's voice sounded from the cabin. "Don't let Scout go in the water. It's almost time to come inside."

Well, it was too late for that. The girl and the dog continued frolicking in and out of the water until the sun dipped out of sight. Then out came the big towel, and Scout had to endure being dried off for what seemed like forever.

"Good, Scout. C'mon, let's go in before we get in trouble."

But the dog did not follow. Instead she stopped and tuned her senses to everything around her. Somewhere in the distance she could hear the eerie laugh of a loon calling from the liquid darkness of the lake. The sound hung there in the stillness of the new nightfall.

Then there was that other sound. Not a dog. It was more like a howl, no, a yip, but not familiar to her. It was hypnotic. She took a few steps away from the cabin into the dark. An owl hooted in the tree above her. It startled her, and she jumped backward, overturning a bucket. The bucket made an unbearably loud clanking sound that broke the stillness. From the lakeshore, a deer bounded past her, crashing through underbrush into the depths of the woods.

Then it was quiet. Looking back at the lake, Scout was puzzled to see the huge old cottonwood tree sway slightly, although there was no breeze or anything to cause it to move like that. Just as she was about to investigate, she heard her name called.

"Scout! Where are you?" It was the woman looking for her in the first bit of moonlight. The moon was bright and full, creating a large path of white light on the water.

Starting towards the woman, Scout thought she heard tiny voices coming from the lake. She froze with uncertainty. She barked, just once. It was a dull and flat sound that did not belong here. An immediate hush fell over the lake. The little voices vanished.

"There you are! C'mon Scout, what are you doing out here?"

This time Scout followed along. Inside the cabin she found a fleece dog bed by the fireplace where a friendly crackling fire warmed the cabin. Suddenly overwhelmed with weariness, she curled up and slept.

Early the next morning, Scout awoke to the sound of geese honking. Rising from her comfy bed, she stretched, trotted to the window, and looked out. Three Canadian geese had landed in the water where she'd played the evening before. They were massive creatures. She had a vague recollection of chasing geese like these far into the sky when she was just a puppy. It had been a ridiculous game. There had been a whole flock of them and they had waited until she was right up on them, when suddenly, with a massive flapping of wings, all of them took off at once, honking loud protests all the way. The rush of wing beats was staggering. And there she was, left alone on land, with nothing but her own shadow.

She watched the three geese as they swam in the shallow water close to the cabin. They glided leisurely, creating long ripples in the lake. The water was so still, like liquid glass, reflecting the lavender morning. One of the geese waddled up on to the shore and lingered by the old cotton-wood tree, making occasional loud cackling sounds. Scout watched the lakeshore activity from the window as she waited for the woman and child to awaken. At that moment, it seemed as though time did not move.

But eventually it did. Soon the woman and child were up and about starting their day. After the usual bustle of morning household activities, they pulled on hiking boots and grabbed Scout's leash as they headed for the door.

"Scout! Want to go for a walk?" the woman called. With a hasty pat on the big dog's head, a leash was snapped to her collar. The woman was dressed warmly in jeans and a flannel shirt with a baseball cap to shade her eyes. A small good-natured person, she looked ready for anything.

"Scout, we're walking to the creek," the child said. Her curly hair was pulled back away from her face, showing off big blue eyes and rosy cheeks. The red flannel shirt she wore was too large for her and likely belonged to the woman. It made her look tiny.

"Does she have to have this leash? Scout won't run away."

"She was just a puppy when she was here last, Lottie. She could get lost. Besides the woods by the creek is probably really overgrown. Who knows what we'll find! We'll have to blaze our own trail. Exciting, huh?" the woman said, laughing lightly.

The creek, thought Scout, now that was something. It was all coming back to her even though she had been such a pup last time. The creek held all sorts of wonders. Leading to it they found narrow paths made by deer and other smaller animals. The trees grew thick, making it almost impassable. The creek itself was shallow and clear as air. It curved and nestled itself between banks of tall trees with shimmering leaves. At one end there were cattails, so tall and dense that it was difficult to see where the creek met the lake.

Part of the creek tumbled over stones and reeds, forming a little brook of constantly moving water. It became wider and deeper as it moved further away from the brook, as though it gained power and purpose in its journey to the next lake.

Scout wanted to follow the creek, lake to lake, chasing the scents that enveloped her at every turn. She kept her nose to the earth, breathing in and memorizing all the unfamiliar smells.

"Look, Lottie, this is where Grandpa used to hike when he first came here," the woman said, pointing to a pile of stones that made a natural bridge across the brook to a huge old tree stump. It was a perfect sitting stump.

All of a sudden the child gasped. "Aunt Olive! Look!"

Scout, suddenly alert, jerked her head up in time to see a large blue heron glance sternly in their direction, then lift off like a slow plane out of the cattails. It had been so close, they could see the stark yellow of its eyes and feel the movement of air from its wing beats. It didn't occur to Scout to bark, or chase, or to do anything but stare in disbelief. It had been so close.

"Did you see that, Scout?" Lottie asked, as if the dog might answer.

Scout wagged her tail in response. What was it about that heron? Something odd and beckoning about that stern look. The heron knew something. Something important. She shook off the feeling and bounded ahead.

They took their time hiking back to the cabin. The path they took through the woods was so thick that the sun could only break through in tiny patches of dancing light. The earth was black and rich and had the strong aroma of decaying leaves.

They hiked through an open meadow towards the lake, which led them to a stony beach. Waves lapped over the shore, spraying a fine mist of lake water over the stones. They stepped carefully over slippery stones. Scout slowed her pace for the woman and child who kept stumbling over the larger stones, laughing and talking and hanging on to each other to keep from falling.

They just weren't paying attention to the same things the dog noticed. Didn't they see that there was something peculiar about the stones? Scout had an odd feeling. She sniffed at the stones. What was it? She heard a loud cawing sound close by. She looked up. The sound was persistent and unsettling. The woman and child didn't seem to be aware of it. They kept talking about how beautiful the stones were. Scout didn't find the stones beautiful. There was just something curious about them.

"What's that?" Lottie asked suddenly, hearing a new sound in the distance.

Scout cocked her head to one side, listening. There was a buzzing noise coming from the grassy bank in front of the cabin. It was becoming louder and louder as they approached.

"It sounds like...oh, no, surely not...!" Aunt Olive exclaimed, quickening her pace.

She tripped over a pile of larger stones, caught herself, then scrambled ahead in the direction of the sound. She was almost running.

"Mr. Peterson!" Aunt Olive shouted, waving her arms, trying to be heard over the buzzing saw. "What are you doing?"

The noise stopped. A lanky blonde man flashed a wide smile at her.

"Hello, Olive." He stood beside the old cotton-wood with the saw running on low at his side. A few branches lay on the ground next to the tree. "Good to see you. When did you get up here?"

"Mr. Peterson, what are you doing?" Aunt Olive asked again, ignoring his greeting.

"Oh! Well, this tree. I figured you'd want it cut down soon. I was just getting a start on it. We had a bad storm early spring. High winds, ice, a lot of rain. This old cottonwood looks like it got beat up

real bad in that spring storm. I could stack up
some firewood for you." Hesitantly, he added,
"Your grandfather always had me go ahead and
take care of the trees."

"My grandfather. Well, you know Grandpa's
not here anymore. It's just Lottie and me now, and
I don't want you to touch that old tree."

"You must be joking. Look at it. It's so old.
How many years could it have left anyway?" His
voice trailed off as Lottie and Scout climbed onto
the bank.

Scout trotted over to the tree and sat staring at
Mr. Peterson. She considered growling at him, but
he already looked uneasy.

"No. I'm not joking. Please. Maybe later." Aunt
Olive didn't know why she felt so strongly about
the old cottonwood at that moment. But as she
looked from Scout to Lottie, she became strong in
her resolve.

"Well...okay. It's your choice," Mr. Peterson
said as he shrugged his shoulders. He ran his
fingers through his blonde hair and gave her a
friendly grin.

"Hope you folks have a good vacation. Let me
know if you need anything." With that he walked
off without looking back. They could hear him
climb into his truck and drive away.

Aunt Olive looked back at Scout and Lottie.
She really didn't know why she felt the way she
did about the old tree. It was obvious that it might
not live through too many more winters. Maybe
she should have let Mr. Peterson cut it down.

Just as she began to doubt her decision, Scout
jumped up and slapped her big paws onto the
cottonwood, looking straight up into its branches,
wagging her tail. The sight was so comical that she

and Lottie began to laugh, forgetting about Mr. Peterson's visit.

The following weeks at the cabin were long and peaceful. Aunt Olive and Lottie took Scout with them everywhere they went. She went with them on their trips to town, trips in the boat, trips to visit old friends, and even trips to the county flea market. Mostly though, they stayed at Stony Point where each day was a new adventure. They were more relaxed than they had ever been.

It was the nights that made her uneasy. There was just too much activity outside the cabin after dark. The moon was brighter and more imposing than she had ever seen. On several occasions she heard those small hushed voices coming from the lakeshore. And the cottonwoods seemed to have an ability to rustle their leaves without the slightest breeze. When Scout tried to alert Aunt Olive and Lottie, they ignored her and told her to come inside.

Every dawn, without fail, a crow would call from the old cottonwood tree. It was that same persistent, disturbing call she'd heard before. It always followed that the three geese who were there at daybreak would fly away, not to be seen again until the next morning.

One evening, Lottie picked up an orange stone and tossed it into the shallow water. Immediately Scout felt compelled to retrieve the stone. She ran to it, dunked her head under the water and carefully scooped it up into her mouth, dropping it back on the lakeshore exactly where it had been.

Considering this to be an amazing feat, Lottie picked it up to throw again, but this time Scout caught it midair. She gently carried it in her mouth, then dropped it back on the shore again. Lottie lost interest in the game when Aunt Olive called her from the cabin.

As Lottie dashed away, Scout glanced back at the bright orange stone. It was large and angular, not particularly unusual. Just as she started to turn away, she thought she saw a little face looking back at her. She was startled to see that the stone had eyes, a tiny nose, and a mouth that formed a wide smile. And the little face was gazing steadily back at her with giant unblinking eyes. She shook her head in disbelief. A face? And it was looking right at her. How is that possible? When she looked again, the face was not there. The orange stone was nestled amongst the other stones just as it had been. Puzzled, she stood there for a long time, watching.

After that, she spent even more time by the water's edge, wandering this way and that, sniffing the breezes. Sometimes she would paw at the stones, turning them over trying to find the orange stone's face, but she never did.

She rested on the sand in the shade of the huge cottonwood, listening to the water wash over the stones. The cottonwood had a low lying branch that made a perfect hiding place. It was good that they'd not allowed Mr. Peterson to cut down the old tree.

Aunt Olive and Lottie began to spend more time on the lakeshore, too. Lottie made little castle-like structures out of the sand and collected stones and seashells in a small pail. When she wasn't looking, Scout would pick the stones and shells out of the pail one by one and deposit them back on the shore.

The woman and child splashed and swam in the lake and paddled the canoe a short distance into the water. Sometimes Scout would swim alongside the canoe, and they would cheer her on, as though it were a race and the dog was winning.

Late one afternoon, close to the end of their vacation, Lottie decided to take a walk with Scout, leaving Aunt Olive behind at the cabin. They were going to walk along the shore only a short distance. Aunt Olive told her not to go too far because it looked as though it might storm.

Off they went into the sunless afternoon, stepping over stones and wading in the shallow water where minnows nipped playfully at their toes. They charged together up onto the soft grass of the bank and paused there while Lottie fed cracked corn to the greedy chipmunks. Racing back to the shore, they chased and splashed and made a game of it all.

The wind began to blow in strong gusts, but they didn't notice it until Lottie's cap was blown straight up off her head. It flew away as though it had wings, and together they chased it along its flight path.

The cap was carried by the wind so far that they almost gave up running after it. By the time they did catch up to it, it had lodged in a tree deep in the woods, and Lottie had to climb up to retrieve it. Only then did they realize how dark the sky had become. It would be night soon and the oncoming storm made the daylight disappear faster. The sky turned from gray to dark blue to black. They started back towards the lakeshore when it began to sprinkle and then rain in a soft steady shower.

All at once Lottie stopped. Scout sat beside her searching the horizon. They had gone far. Somehow, they were all turned around and had lost their direction. The wind and rain and darkness made things worse.

"It's this way, Scout..." she said, turning the opposite direction. She began to walk faster, wiping rain water off her face. It was so dark. There was no moonlight to guide them, not a speck of any kind of light.

"No, wait, this can't be right." Now Lottie was frightened. She could not see ahead of her.

"Scout, can you find the way?"

Scout sniffed the air, but all the scents mixed together in the rain.

She pawed at the ground, digging up more scents but they were unfamiliar. Everything was beginning to smell like damp sand or wet wood. The wind howled just enough to bully the waves into whitecaps. In their fury to crash into shore, they drowned out any sounds familiar to Scout.

As it began to rain harder, they took cover under a huge fallen tree, but it was not much shelter. They were soaked, and Lottie began to shiver.

"We're lost, Scout," she whimpered, her voice carried away in the wind. She began to sob.

Scout licked the little girl's face and tried to comfort her, but she knew they had to move away from this spot and find their way back to the cabin. She listened. What was it? Something out there. There was a brief lull in the storm.

"GoodScout, over here, this is the way...come back this way...follow the shore..."

Startled, the big yellow dog stood up, her hair bristled with alarm. The voice was calling her from far away. Who was that? It was such an unusual voice, small, but strong and sure. She shook her head as though to clear the sound from her ears. Her tension and disbelief only served to scare Lottie, who crawled even further under the tree trunk.

"GoodScout...follow our voices, it's this way," another voice beckoned.

For some reason that she would not understand for a long, long time, Scout felt calm and certain of the path ahead. She nudged the child, who at first refused to budge until she sensed the dog's urgency. Lottie did not seem to hear the voices. She got up only at Scout's insistence.

Instantly, the two were blinded and drenched with sheets of rainwater. A tree limb snapped and crashed to the earth beside them. The darkness was like a black veil, and the wind tossed leaves and twigs around them in frantic swirls. Lottie took hold of Scout's collar but could barely keep up with the dog as she forged ahead.

"GoodScout...this way...not far...keep on..." said the voices, closer now. And the big yellow dog kept on, slogging through water, tripping over unseen obstacles, on and on she plodded. Every time she would open her eyes in an attempt to see ahead, rainwater would blind her. She just had to follow the voices.

"Go up on the bank, stay out of the water! This way! GoodScout, almost..." The dog and child climbed out of the water onto the bank just as thunder clapped and a lightening flash brightened the sky.

In the momentary light, Scout could see that they had narrowly escaped a deep drop off in the water. The pouring rain never let up as they continued trudging along the bank, losing their footing every few steps. The voices were closer and closer, and the dog moved faster and faster with Lottie hanging on to her collar.

"One more bend, follow the shore, goodScout."

Just then a small dot of light parted the rainwater and they heard Aunt Olive shouting, "Lottie! Scout!" as she shined a flashlight in their direction. She ran to them, snatching Lottie in her arms and hugging the dog at the same time.

"Lottie, I told you not to go far," she scolded.

"But my cap blew away…we had to chase it." Lottie whimpered, then began to laugh. She looked at the dog. "Scout brought us back. I don't know how she knew the way. We were really lost. It was so dark."

Together they hugged the dog, then trudged inside, shaking off water as they went. Scout paused at the door, looking back at the lakeshore. The voices were gone now. There was only the plink of raindrops on leaves, earth, and stones.

The next day was cool and bright. It would be autumn before long. Aunt Olive and Lottie began packing to return south, while Scout wandered on the beach. She sniffed and nudged the stones, searching for the little stone face she'd seen a few weeks before.

After a while, one of the stones held her interest more than the others because she caught a glimpse of it looking at her when she sat very, very still. It had a kindly face and did not seem at all afraid of her. It watched her intently. Did this stone help her find the way out of the storm? Did the voices she heard sometimes at night belong to the stones? It was just all so odd.

And why had they called her goodScout instead of Scout, which was the name given to her by humans. Even so, she felt protected and at ease here, and strangely she felt like she belonged.

From this time on, she would be goodScout, no matter what humans called her. After all, humans missed so many obvious things. She stretched her front legs so that the tips of her paws touched the water's edge just as a wave washed into shore. Lake mist lightly sprayed her face. She closed her eyes and shook briskly, her ears flapping over her head. Yes, it made perfect sense to her; she had become goodScout.

The old cottonwood's leaves brushed lightly against her face as if to confirm the kinship to the lakeshore. It would be time for the dog to leave soon.

"Scout, let's go!" Aunt Olive called.

Lottie appeared on the bank and called to her, "C'mon, Scout time to go in the car." But the dog didn't come to her.

Scout sniffed for one last moment at the kindly-faced stone .

"Aunt Olive, look at this beautiful stone Scout found," Lottie exclaimed, now standing beside the dog on the lakeshore. She reached for the black stone with the bold orange stripes.

Aunt Olive bent to look at the stone and agreed that it was very beautiful indeed and picked it up. Scout gently took the stone from her hand and deliberately dropped it back on the shore.

"Good, Scout. Do you like that stone?" asked Lottie, misunderstanding the dog's intent.

Aunt Olive picked it up again and this time Lottie took it from her saying, "Let's take it along. It's so pretty. Look how shiny and colorful it is."

And so, she carried the stone away from Stony Point with Scout close behind her. As they pulled away to leave, they could hear the loud cawing of a crow from a distant meadow.

©C.J. Clark

- 5 -

Deweye

The last thing he remembered was Whisperleaf telling him to look for the good. She didn't say final words like farewell or good-bye. Even so, she had faltered with those few words she'd murmured. In the early autumn breeze, her leaves fluttered like paper butterflies. Soon they would turn lemon yellow and fall softly to the earth. Perhaps she thought she would not see him again, that he would not return in her lifetime. All he had ever known was Stony Point, and he was filled with gloom at going away.

It didn't lessen his sadness all that much when goodScout retrieved him from the floor of the car and placed him beside her on the seat. The child would exclaim again how beautiful the stone was and the woman would look over and agree in a preoccupied sort of way. The child noticed everything, it seemed.

"Aunt Olive, look at those three geese up there. It almost looks like they are following us."

The geese did fly along for a time. Swift and mighty creatures that they were, they could not follow for long. He recognized the voices of Silverdew and his two friends.

They sounded so forlorn, calling, "good-bye, good-bye, come back soon, Skeedad." After a while, their voices trailed off, and all that could be heard was the heavy drone of the car.

Onward they drove, past the small town, past the farm fields of harvested corn, and cows plodding through pastures, past solitary human dwellings and tiny lakes surrounded by pine woods. Puffs of jolly clouds winked at them along the way.

And so began the long journey south for Skeedad. There was not a touch of water or a hint of anything green and growing. They traveled on and on as the day aged from dawn to dusk. The woman and child spoke of so many things that their conversations began to sound like the babbling brook of the inner creek. Skeedad tried to keep up with it. He thought if he could just listen long enough he would come to understand why they took him away, but he began to feel a slumbering numbness overtake him like winter, and he dozed.

By nightfall, the woman took him from goodScout, who had never let the stone out of her sight. GoodScout watched her intently as the woman casually dropped Skeedad into a small suitcase, where he spent the remainder of the journey in smooth darkness with dozens of unfamiliar objects.

"Hello? Is anyone there?" he whispered to the soft textured things around him. He was nestled in so snuggly he could barely see around the interior.

"Anyone at all?" he tried again, louder. "I am Skeedad of Stony Point, and I'm not supposed to be here. I belong beside the Blue Water."

Thunk! All of a sudden, the suitcase was moved, and he was rolled to the side with several other objects. The soft textured things stayed where they were, but suddenly he was face to face with a square ticking creature. It was the same size as he was, but it was cold and mechanical. Even its shape was peculiar to Skeedad.

"Tick, tock, tick tock, tick tock," it replied.

"Hello. I was just saying, I belong at the lakeshore. Do you know where they are taking us?" Skeedad asked hopefully.

"Tick tock, tick tock, tick tock, tick tock," came the answer.

"Well, I don't mean to offend you, but that really doesn't help. I mean, where exactly is Tick Tock? Oh dear, I most definitely do not belong here," Skeedad sighed. He knew his words were lost on this tick tock thing, so he kept quiet and tried harder to listen to his surroundings.

He heard raindrops on the car once and maybe even a bird singing in the far distance, but the sounds were muffled. The woman and child continued talking and laughing. There was no sound from goodScout, although Skeedad was sure she was still there.

Despite the boxed-in darkness, he knew that there had been at least two sunrises and sunsets and that the air was different. It was heavy, like thick warm mud. The car grumbled onward. They had already traveled very far and did not seem close to stopping.

He had so much time to think that he relived many things in great detail. The further the car traveled, the further he traveled into his own road of memories.

He thought about Whisperleaf as a young tree, inquisitive and attentive. She was the wisest and the most gentle of all the cottonwoods that had ever lived on Stony Point. She was also the eldest.

Skeedad thought fondly of Meadowsteps, the white-tailed deer, when she'd first appeared as a wobbly fawn with her mother and brother. Her mother taught the eager young fawns how to step lightly along the shore. Meadowsteps was not even a yearling when she arrived without her mother or brother one dark autumn day. She was solemn and barely drank from the lake. Something unspeakable had happened to her family. From then on, the stones were her dearest friends. She lingered by the water's edge well into the day. The stones knew better than to ask her what might have happened, and thankfully, Silverdew, who would have, was not there at the time.

Silverdew could not quite learn to think before speaking his mind. It was just his way. After all, his life had not been easy. He'd lost his mate long before, and he lived solely as a sentry goose. His life was dedicated to warning his two companion geese of any approaching trouble. He had to be alert at all times. There was not usually time to be courteous. More than once his boisterous voice had saved them from danger. They were truly grateful and quietly tolerated his bluntness. Skeedad missed the three geese.

And Solosong. Skeedad believed that the crow was more devoted to Stony Point than any crow before him. True, he was powerless to save them from danger, but he could warn them of it, and he always did. He appeared aloof and at times seemed quite arrogant. But the stones knew better. It was just Solosong's way. This, too, Skeedad missed.

Yes, he could imagine all of them. Blue Heron wading in the shallow water, pretending to ignore the ancient tales told most likely by other stones. Maybe it would be Skylo, the small red stone, who would be the one to weave a lively story based on days gone by. And everyone would listen attentively.

Everyone loved the story of Stony Point because it had no beginning and no end. Their world was simple and true, and there was good in everything when one learned to look for it. Things changed, and yet somehow things remained the same. The sun always set in the evening on the western shore and rose up over the treetops from the east the following dawn. Winter would always blanket them in snow, then melt into deep greens again in the spring. For some, there would be many more sunrises than for others. All the lake creatures knew and accepted that. The stones had seen more sunrises and more moonlight than anyone. Now, as the big car rumbled along, for the first time ever, Skeedad wondered about the good and where he would find it.

Just as he was considering this, the car jolted to a stop and goodScout barked happily. There was a flurry of activity, and Skeedad felt the suitcase being carried into some sort of human dwelling. He could hear voices and movements that made no sense to him. Time passed slowly. The tick tock thing next to him had finally stopped ticking. Was it listening, too? Skeedad waited in the darkness, wondering.

All at once, there was a rush of light as the suitcase opened wide. He saw goodScout and the child peering in to look. Then another child appeared beside them.

"Look at this beautiful stone Scout found!"

"This? What's so beautiful about some old rock?"

Rock? Skeedad had heard that word once before. It was a human word. He thought it meant a type of stone made by humans, although the very idea was strange to him. He knew that he was not a rock.

The child glanced appreciatively at Skeedad and said, "Well, it was beautiful. It came from the lake. There are millions of them. Scout liked this one the best."

"Beautiful? It's just some old rock. Put it away. Let's go," the other child said, turning away.

"I think it's a pretty stone. I don't care what you think!" the first child exclaimed reaching into the suitcase and picking up Skeedad. GoodScout sniffed the stone and began to whine.

"It's okay, Scout. I'll put your stone in the desk drawer where it will be safe," she said as Skeedad felt himself being placed into a wooden drawer. Before the drawer was shut, he looked up to see a large gray bird glaring down at him with piercing light-colored eyes. The bird had a curved black beak and bright red tail feathers. It was not nearly as large as Blue Heron, but it had that same look of stern reproach. It had the air of Solosong's haughtiness but seemed more hawk-like than anything. Oddly, it was caged in some sort of wire enclosure.

"Put it away, put it away. Old rock, old rock!" Skeedad was shocked to hear the bird speak human words.

"Don't be mean, Mr. Littlejohn." The first child spoke directly to the colorful bird.

How peculiar, Skeedad thought. Humans did not ordinarily speak with creatures. In fact, they never seemed to hear creatures speak. Never once did a human gaze into the face of a stone. They just couldn't see or hear the same things. Of course, the stones knew to close their eyes and retreat into themselves when humans were near, but that was just an extra precaution learned many ages ago. This conversing bird was another matter. It was something too unusual for him to understand yet.

As the drawer closed him into darkness, he heard the bird say mockingly, "Mean Mr. Littlejohn, mean Mr. Littlejohn," then, "old rock, put it away, put it away!"

Inside the drawer it was blacker than a moonless night and Skeedad felt he was lost forever. How would he ever get home to Stony Point? Would he ever see his friends again? Was goodScout trustworthy? He believed she did have a gentle and kind spirit. This was her home. He heard her trotting from the room with the children. It was quiet then.

It was so quiet that it unsettled him. For a very long time he listened to the emptiness. He wondered again why they had taken him away, claiming that he was so beautiful, only to put him here in this dark musty place. The stones never thought of themselves as beautiful. They were stones, wise and solid. They had knowledge that grew with the ages. That was their beauty.

What good could he find here? He listened so long and so intently to the silence that he entered a state of near slumber and began to hear the lake waves in his mind. Then, the voices of Whisperleaf and Solosong and even Silverdew filled the emptiness. It was autumn there now. How would he ever know the change of seasons, the approach of storms, the rising and setting of the sun? What could he learn from this? There wasn't anything but silent darkness. It was just a wooden drawer that smelled of old bark. He closed his eyes and began to retreat into his deepest memories to comfort himself. He would need to discover a path home.

"Hullo? Skeedad of Stony Point?" whispered a light melodic voice, interrupting the stillness.

Startled, Skeedad peered into the darkness and saw the outline of a huge feather. It was a giant goose feather. Next to the feather he could make out the shape of a seashell.

"Halloo. Skeedad of Stony Point. I am Deweye and this is Daysee of Stony Point," the feather announced, moving closer to the stone, who did not speak at first because he thought he was dreaming.

"Hello, Deweye and Daysee, I am glad to see you," he said hesitantly.

The feather, Deweye, had fluffed himself to an impressive size, causing the seashell to appear even more fragile and delicate. What an unusual pair, thought Skeedad, still trying to adjust to these surprise companions.

"Yes, yes, yes, you think you are dreaming, but you are not. We are short timers of Stony Point. We know you because you are a stone, the bravest and wisest of all," Deweye continued, bending dramatically to one side as if to bow.

"Well, the wisest and bravest, I don't know..." Skeedad said. Daysee interrupted him. Her words were like carefully measured bits of sand.

"Skeedad, sir...we welcome you."

Deweye chuckled. He lowered his voice to a meaningful whisper and hissed, "Yes, we welcome you...to the Drawer of Forgotten Things!"

- 6 -

DRAWER OF FORGOTTEN THINGS

A tiny streak of light broke through the entrance to the Drawer of Forgotten Things. It was just enough light to illuminate Skeedad's new companions and their surroundings. He was becoming accustomed to the dimness of this strange world. Deweye dramatized all his words, while Daysee spoke with polite softness. Skeedad knew the two were short-timers of Stony Point; they would have followed a fleeting path for only a few seasons. There was good in that. But here, in this dark dry place, they would be long-timers. They were not meant to be long-timers. It was not natural. How could any of this be good?

As he was pondering this, Deweye and Daysee studied him intently. After a moment, Daysee spoke.

"We want to go home. We want to find the way north." She sounded so forlorn. It was as though she had given up hope of returning home and now just recited the words.

"I want to warm myself in the summer sun by the water's edge," she added dreamily.

Skeedad sensed her hopelessness. How long had they been here?

They could only tell him that it had been a very long time, that they had been brought here by humans for their natural beauty, but then had been left and forgotten. They wanted to hear stories of Stony Point, but Skeedad wanted to learn more about the Drawer of Forgotten Things. Most of all, he wanted to figure out how they might escape. He had to think. There had to be a way. There was always a way, but he had to know more.

Deweye became impatient with the momentary silence. He fluffed and preened. When he could wait no more, he began to talk about himself.

"It's true, of course, I am only a feather, yes, a short-timer. You would think that I would not have feelings on this subject of returning home, but I do!"

All of a sudden he began to sing.

> *"I am Dew-Eye*
> *Light and swift*
> *I spin, then fly,*
> *And dream to drift*
> *In the summer sky,*
> *The breeze will lift*
> *Me up so high*
> *What an honor*
> *To be Dew-Eye..."*

Daysee joined in and together they sang louder and louder, *"What an honor to be Dew-Eye, what an honor to be Dew-Eye..."*

Deweye paused, while Daysee continued humming the little tune. He eyed his companions and spoke in a booming voice.

"Listen, all I ever wanted was to be just what I am. A feather. Yes, that's right. Just a feather. A goose feather meant to be swept away in the first strong breezes of summer, to twirl and turn and fly high above Whisperleaf, to touch the sky, and travel on and on, forevermore."

While Skeedad was considering Deweye's words, the feather suddenly whirled around and asked him, "May I show you around the Drawer of Forgotten Things, Skeedad of the Blue Water Stones?"

"Yes, please. I must see," Skeedad said.

"Well then, this way sir," the feather said with a sweep and a bow. He moved slowly, allowing the stone time to roll along. Daysee stayed by Deweye's side.

"First, there is the Box of Forgotten Places. Go ahead, look inside. It will surprise you."

The Box of Forgotten Places was not much taller than Skeedad. It was plain and made of something foreign to him. The top was open slightly, and he peered inside. He was astonished to see rainbows and sunshine, deep blue skies with huge white clouds and sunsets on the lake. There were red, yellow, and orange leafed trees and a blue heron frozen in flight above the cattails. Behind each miniature forgotten place was another and another.

Each of the forgotten places was completely still. Not one thing moved. How could that be possible? He saw the small town, the farm fields and plodding cows. He saw the tiny lakes and the

pine woods. He was awestruck. The forgotten
places seemed to go on forever. Skeedad wanted
to look away. It was too much to understand.

"Next is the Box of Forgotten Humans,"
Deweye said, moving ahead while Skeedad
continued staring into the first box.

"Ahem! The Box of Forgotten Humans is next,"
Deweye repeated. He flitted back to where Skeedad
was. Skeedad followed along reluctantly, still
looking back at the first box.

The Box of Forgotten Humans was the same
size as the Box of Forgotten Places, and again
Skeedad found it slightly open. He peered inside.
He was shocked to see the smiling face of
Brookdweller staring at him. Skeedad quickly
backed away.

"How is it possible that—" he started, searching
Deweye for an answer.

"Go ahead. Take your time. You will find it
fascinating," Deweye said. He watched Skeedad
move closer to the box, then slowly peek inside
again.

Yes, it was Brookdweller. He was a young
man, and he was surrounded by other humans.
All seemed happy and were laughing. They were
so tiny. One of the tiny humans was holding up a
miniature fish from the water on some sort of
string thing. It was startling to see it. And the most
eerie thing was that no one actually moved. They,
too, were frozen in place. Behind them were even
more humans, all smiling and all gaping at him.
And there was another Brookdweller, except older,
and not smiling. And then, still another Brookdweller sitting, a Brookdweller walking, and a
Brookdweller standing. He was shorter than
Skeedad. How could that be?

Just as he thought he had seen enough, Daysee tapped at the next box, which she explained was the Box of Forgotten Sounds. There was a slight hiss, then a melody unlike any he had heard from Blue Water songbirds or loons or even the distant coyotes. He listened, mesmerized. It was a peaceful sound, but it stopped almost as suddenly as it had begun.

"Again?" Daysee asked as she tapped against the Box of Forgotten Sounds, which started the tune once again. This time it played a bit longer, and they fell silent in reverence to its song.

"There's more..." Deweye interrupted.

"I want to see everything," Skeedad said, rolling along behind the feather who made dramatic sweeps this way and that.

"This," Deweye said, absently dusting a rectangular object, "is the Box of Forgotten Thoughts."

Skeedad stared at the box and could see that it had a front and a back bound together with a thick middle, which Deweye explained contained pages of human words that no one cared about anymore. This made no sense to Skeedad at all. He listened to Deweye tell him that the words were a human's thoughts written onto pages and pages and then forgotten. This made even less sense to Skeedad.

"Now, over here is the Box of Forgotten Time." Deweye was showing him a tick tock creature, just like the one Skeedad had traveled with, only it wasn't making a sound. It just sat there, dusty and unmoving.

Deweye said it was called a clock and that the Box of Forgotten Thoughts was called a journal. The Box of Forgotten Places and the Box of Forgotten Humans contained something called photographs. He was not able to explain the Box of Forgotten Sounds. Deweye was about to show Skeedad more when Skeedad stopped him.

"Deweye and Daysee, may I ask, how do you know all this?" Skeedad asked amazed.

"Oh. Um, well, we learned it all from OldSkogmo," Deweye replied casually.

Skeedad gasped. How could they have learned from OldSkogmo? OldSkogmo was taken away by humans long ago when Whisperleaf was still a young tree. No one had seen or heard about him since.

As though reading his thoughts, Daysee whispered, "OldSkogmo is here…in the Drawer of Forgotten Things, Skeedad, sir."

"What? Where? And how?" Skeedad asked. This was unbelievable!

"There," she whispered, looking to the back of the drawer.

Skeedad followed her gaze. What he saw was both alarming and a source of joy. It was indeed OldSkogmo. He was all the way at the back of the drawer in the darkest corner. Yes, it was OldSkogmo, alright. There was no mistaking the orange and black speckles and the smooth round shape. Skeedad was filled with hopefulness. And yet, he saw that OldSkogmo was dull and lifeless. His eyes had been closed a very long time, and he did not respond in any way as the three moved closer to him.

"OldSkogmo? It's me, Skeedad," he called to his friend.

"Oh, poo, he can't hear you. He's become a rock!" Deweye announced.

"Deweye! Don't say that!" Daysee hissed. "OldSkogmo is just resting!"

"Yes, forevermore," Deweye chortled, lowering his voice.

"A rock? A rock! He cannot be a rock because he is a stone. A Blue Water stone," Skeedad exclaimed. "And that is forevermore!"

"Well, I don't know, Skeedad, sir. He used to talk with us and tell us incredible tales, all true, and all first-hand accounts. He wasn't always in the Drawer of Forgotten Things. I don't know when he came to be here exactly," Deweye said.

He looked towards OldSkogmo and added, "Before he was left in this drawer, he discovered a lot about humans from being in their world. He once said that it was such a tiresome thing to be a paperweight, but he learned all he could from listening to Brookdweller. Somehow, he was able to find good in everything around him. He was put here after Brookdweller went to be above the sky."

Brookdweller above the sky? Was that possible, Skeedad wondered. He could hardly imagine humans above the sky. They were always so unable to rest long enough to truly absorb goodness. It just didn't seem likely. But then, Brookdweller was exceptional.

"When we came to be here, OldSkogmo was very kind and helped us to find good. It wasn't easy," Daysee said. "I believe that as so very many sunrises came and went, he drifted into a state of winter's deepest slumber and has not returned. But," she added, with a stern look in Deweye's direction, "he did not become a rock!"

Ashamed, Deweye spoke more softly to Skeedad. "Do you think you can bring him back? He's been silent for a long, long time. Ages."

Skeedad studied OldSkogmo. It was still OldSkogmo; he was still in there. "I will stay with him and remind him of the lakeshore. I'll try."

It was a somber sight. Skeedad stayed beside OldSkogmo well into the sunrise without a word spoken. Deweye and Daysee reluctantly left the two stones at the back of the Drawer of Forgotten Things and waited. They knew that any hope for OldSkogmo's return would come from what Skeedad might do.

Several more sunrises and sunsets passed, and still Skeedad stayed with his silent friend. Stones are very patient creatures, Daysee reminded Deweye.

One day though, he did leave OldSkogmo. He had to. There was a distracting commotion at the entrance to the Drawer of Forgotten Things. When he got there, he discovered Deweye and Daysee singing along to the melody of the Box of Forgotten Sounds. They were belting out the song as loud as they could.

"I am Dew-Eye
Light and swift,
I spin, then fly
And dream to drift—"

Suddenly there was a loud tapping outside the drawer, which became more persistent. The pounding was so loud that the two stopped singing at once.

"Uh oh," Deweye gulped.

The drawer creaked open slightly, allowing a shaft of blinding light to enter, as a voice snapped, "Be quiet! That is a stupid song, and I am sick of it! You will not be allowed to sing it again. Do you understand? I have spoken!"

It took Skeedad a moment to realize who was speaking to them. It was the large gray talking bird he had seen when he'd first entered the Drawer of Forgotten Things. Deweye and Daysee were cringing with fear and said nothing.

"It's not a stupid song. It's a good song. Why do you not want them to sing it?" Skeedad asked politely.

"Are you speaking to me, old rock?" the bird shrieked.

"Yes. Why, yes I am. You are Mr. Littlejohn? I am Skeedad of the Blue Water Stones. I am a stone. I am not a rock," Skeedad replied.

"How dare you speak out of turn! I am Lord Littlejohn to you, you boring old rock. Rocks are boring. You are boring me! Now, I want quiet in this Drawer of Forgotten Things this instant!" And with that he began to hammer on the drawer with his powerful beak until it closed.

Skeedad was aghast. As the light in the drawer was reduced to the dimness he'd become accustomed to, he turned to his companions.

"Deweye, Daysee, you must not be afraid. Mr. Littlejohn is a very troubled bird," Skeedad said. "Something is not right about him. Go ahead, sing your song, and I will go sit with OldSkogmo for a bit."

Deweye and Daysee didn't budge. They stared after Skeedad in awe of his bravery. He'd stood up to the mean Mr. Littlejohn. Just as they began to feel safe from the bird, the drawer flew open wide. There was the child, Lottie, peering in at them.

"There you are pretty stone," she said, picking up Skeedad. She admired the stone, holding him to the light, and then to everyone's amazement she tucked him into her jacket pocket.

Uh oh, this cannot be good, Skeedad thought.

Off they went. It was soft and dark in the pocket. He was bounced around as the child skipped and ran. For a split second, he sensed that goodScout was at the child's side but then she said, "No, you have to stay here." And the dog stayed.

After a while she was joined by other children and still more children until all Skeedad could hear was a multitude of children talking and giggling. Then an adult voice sounded above them, and the children became quiet. The adult voice droned on and on until the children began whispering. Skeedad had no idea what to think. He waited.

After a short time, a loud bell rang and the children began shrieking and laughing. The child lifted him out of her pocket and held him in her outstretched hand for all the children to see.

"See? It's a beautiful stone from way up north," she said.

"Can I have it, Lottie?" one child asked as he grabbed Skeedad from her.

"No, give it back," Lottie said.

Another larger child pushed forward and took the stone from the first child and said, "What's so special about it?"

"It's just an old rock," another child said.

The children all looked at Lottie. She snatched Skeedad back and hesitated, looking down at the stone. Skeedad could tell that Lottie was growing uncomfortable when all of a sudden she blurted, "Well, it's a magic stone!"

Magic? Skeedad wondered what that could possibly mean.

"Oh yeah? What does it do then?" sneered the large child.

Again, the children pressed forward and wanted to touch the magic stone. They gazed from the stone to Lottie, then back to the stone. They looked back at her again, expectantly.

"Well..." she paused, thinking, then said triumphantly, "it shows you the way north!"

"Big deal! I'll show it north," one child snatching the stone and tossing it high above him. Skeedad soared upwards, feeling oddly buoyant and weightless. He sailed higher and higher toward blue sky. He was a bird. He was Solosong or Silverdew ascending to the clouds above treetops. It was wonderful. Except that just when he thought he might make it to the moon, he began to fall. Down, down, faster and faster. He became dizzy, bracing himself for his crash to earth.

"How exactly does it show you the way north?" challenged the larger child, catching Skeedad just before he landed.

Lottie shifted and grew increasingly nervous. She thrust her chin forward and said, "It just does! You just have to know what to look for. It can show you the way north even if you are in total darkness or even if you are in a box, or well, anything like that."

She had no idea what she meant and neither did Skeedad. Although he did know the way north.

"It's not that great. It's just an orange and black stone," a small boy said. "It's not even a good skipper rock." He took the stone from the large child and held it up for everyone to agree.

In one quick movement, Lottie grabbed Skeedad, dropped him into her jacket pocket and bolted away from the other children.

"I don't care what you think. It was my dog who found the stone anyway!"

Skeedad wondered why Lottie became so glum after the others had scoffed at her. He was amazed by her sensitivity. Lottie had claimed he was beautiful because she saw him as a Stony Point stone. She still pictured him surrounded by other lake creatures under the shade of Whisperleaf as the blue water swirled around him like liquid laughter. To her, each stone had sparkled in its own unique way.

Skeedad knew the peace of Stony Point was soothing and constant and that the stones had absorbed it all. Lottie had tried to share a glimpse of this with the other children, but they could just not feel it. Skeedad recognized that, even if he had looked directly at the children and boldly announced himself, the children would not have been able to see him for who he was because they did not have the eyes or heart for it. Lottie couldn't change that.

Later that afternoon when Lottie returned home, GoodScout greeted her as she approached, but she barely spoke to the dog. When she finally arrived back at the old desk, she opened the Drawer of Forgotten Things and looked up at Mr. Littlejohn.

"It is an ugly old rock. Just an ugly old rock," she said.

Skeedad was dropped into the drawer with a thud. Lottie closed the drawer and he was left in the dark. As she walked away, he heard Mr. Littlejohn whisper to him, "Ha! Now, you really belong in the Drawer of Forgotten Things!"

- 7 -

MR. LITTLEJOHN

As autumn gave way to winter, the north wind howled furiously. Deweye said that the wind, which could be heard despite being inside the Drawer of Forgotten Things, called his name. He knew it. He could hear "Deweye" clearly. He even answered. But his reply was lost as the glass window panes clattered from the icy air whistling through tiny cracks. The north was far away and dreamlike. Skeedad and Daysee would listen as Deweye would sing a song to the north wind.

The problem was that he would sing louder and louder until he was nearly shouting. Daysee could not stop herself from joining in, and it always happened that Mr. Littlejohn's voice would boom above them as he worked the drawer open with his beak.

"Stop it, Stop it, Stop it! How many times do I have to tell you?" he shouted impatiently. He peered in at them with his sharp eyes.

It was always the same. Skeedad would try to talk with the bird, but Mr. Littlejohn would become even more irritable, telling them that he was the ruler of the Drawer of Forgotten Things, and they had better do as he said, or else.

"Or else what?" Skeedad asked one day, wondering what could be worse than being in the dark drawer forever.

"I do not have to answer that," the large bird replied angrily and using his beak, pushed the drawer shut one jab at a time.

They never saw Lottie. She had not returned since the time she called Skeedad an ugly old rock. GoodScout's pattering footsteps were heard sometimes in the distance. OldSkogmo remained in a state of slumber while Skeedad stayed by him, certain that someday he could reach his friend.

"Skeedad, sir...I believe that OldSkogmo could no longer find good in anything. That's why he is, well...gone," Deweye explained one day. He said this in his kindest voice. Daysee felt differently and said so.

"Gone? No, he still has good inside. He's just resting. It's what stones do, right?" She turned to Skeedad.

"You are both right," Skeedad answered sadly. "OldSkogmo has been missed at Stony Point for a long, long time. He'll never be forgotten. He was so very sensible. He used to tell us all sorts of things from old legends that only he could remember. There was this lively way he had of storytelling what could hurl you back in time. I think it was because he could see and hear more than anyone else, and he remembered everything. I mean everything. He didn't miss even one falling leaf! Brilliant, he was just brilliant," he added.

For a while they did not speak. Finally Daysee said, "He still is a brilliant stone. Anyone can see that. He told us that the child who took him from the lakeshore used to speak to him as though she could actually see him. Can you imagine? As if she could see him! When she grew up, she gave him to her grandfather, Brookdweller, to keep on this desk as a paperweight."

"A paper what? What is *that*?" Skeedad asked.

"No, no, a *paper weight*!" Daysee said.

"Paper wait? What were the papers waiting for?" Skeedad was puzzled.

"No, no, not that kind of waiting. Well, sort of that, but also to keep the papers from getting away, I guess," Daysee replied, now a little puzzled herself.

"Paper wait? Now, that's just strange. Why would papers try to leave? What did he have to do to get them to stay? Just sit there?" Skeedad asked, curious to know the fate of his friend.

"No. Well yes, to hold down the Forgotten Papers whenever the north wind blew through the open window, I presume. The Forgotten Papers are here in the Drawer of Forgotten Things. I could show them to you. Maybe they might tell us the way north. OldSkogmo knew what the papers were. I know he told us. I just don't remember. Daysee, do you remember?" Deweye said.

"Papers and papers and papers of forgotten things. I'm not sure, Deweye. OldSkogmo said the papers were very important to Brookdweller," Daysee said.

"Forgotten papers..." Skeedad repeated thoughtfully. He just could not understand why there were so many forgotten things. Forgotten places, forgotten humans, forgotten time, forgotten sounds, forgotten thoughts, forgotten papers.

At Stony Point every single thing, tiny or huge, short-timers or long-timers, rainstorms or sunshine, had significance and value. It was all important. It all mattered. The stones knew this better than anyone because they had been there so long that they understood how, over time, things worked together to become good. It was really quite simple.

Deweye and Daysee were looking at him.

"So, what about Mr. Littlejohn? What do you know about him," Skeedad asked, changing the subject, "Don't you think it's a bit peculiar that a large bird lives in such a place?"

"He's not like any other bird we've ever known," Daysee replied.

"He doesn't fly away," Deweye added. "What good are his wings if he never uses them to fly away? He's quite mean and…"

"Lonely?" Skeedad added.

"Yes! Lonely! He must be very lonely," Daysee exclaimed.

"No one comes to see him?" Skeedad asked.

"Nope! He's too mean. Doesn't have one nice thing to say. Ever," Deweye said.

"No one? Ever?" Skeedad was stunned.

"Maybe no one visits him because he always says such horrible things. The little girl used to talk to him. Sometimes he just ignored her and sometimes he would say hateful things. She really did try though," Daysee said.

"Maybe he misses Brookdweller. OldSkogmo told us that he really only liked Brookdweller," Deweye said.

"Brookdweller? Hmmm…Mr. Littlejohn must be very old then. Almost a human lifetime." Skeedad pondered this.

"OldSkogmo also told us that Mr. Littlejohn wasn't always so mean. You know, you're right. He must be lonely! Don't you agree, Deweye," Daysee asked, turning to her friend.

"Well, I suppose…" Deweye said reluctantly, then added, "What does it matter? He's mean now!"

Daysee and Deweye looked glum. Mr. Littlejohn had tormented them for as long as they could remember. The worst part was that they never really knew why. It didn't matter if they were quiet or loud, polite or rude. The bird was always mean to them.

Suddenly Skeedad's mood brightened. "I have an idea. Come with me. We are going to talk to him."

"Oh, no! Oh no, no, no, no, no, no…he could hurt us," Deweye exclaimed, backing away from Skeedad.

"Deweye! Do you want to find the way north?"

"No! Well, okay, I meant…yes, yes, I do," Deweye said cautiously.

"Then, don't you see? We have to try talking to Mr. Littlejohn. He may know something important. In any case, he likely has a lot of good inside that needs to come out," Skeedad said.

"I sincerely doubt that," Deweye muttered.

Ignoring Deweye, Skeedad said in a gentle voice, "I believe that Mr. Littlejohn has forgotten his own good."

The three made their way to the front of the drawer with Skeedad in the lead. Deweye was so hesitant that Daysee had to push him with all her might.

"Hullo? Mr. Littlejohn?" Skeedad said.

There was no reply. They could hear the big bird moving about, but he did not answer.

"Hello? We would like to speak with you, please, Mr. Littlejohn," Skeedad shouted. Still the bird refused to answer.

"Okay! This is what we must do." Skeedad turned to his friends and said, "We need to sing that song. You know, the one you sing the most. Look, I will sing it with you."

"Ahem...*I am Dew-Eye, light and swift...*" Skeedad began. At first he was the only one singing but Deweye and Daysee finally chimed in, unable to resist singing their favorite song. Skeedad's voice was strong and deep. The three sang louder and louder. Deweye twirled as he sang and Daysee harmonized in her high-pitched voice. Louder and louder they sang. It wasn't long before they were belting out the song at the top of their voices.

Bam! Bam! Bam! Mr. Littlejohn pried the drawer open with his powerful beak and feet. He backed off a little when he saw that the three were at the very front of the drawer.

"Stop it! Stop the stupid song! Stop it now!" He screeched so loud that Deweye was flipped backwards into the drawer.

Skeedad ignored this. Instead he asked, "Mr. Littlejohn, what kind of bird are you? We have never seen such a colorful bird who can talk directly to humans."

Mr. Littlejohn eyed Skeedad suspiciously. "Why do you want to know?"

"Just do," Skeedad replied.

"Just do, just do....just do," Mr. Littlejohn repeated. He was searching for something hateful to say, but instead he found himself saying, "Just do, eh? I am an African Grey parrot!"

"Where are you from?" Skeedad asked, relieved that the parrot had not said anything mean.

"South. South of the world. I am an African Grey parrot from the deepest forests of Africa. Very, very, very far south. South of the world." He glared at Skeedad as though the question was meant as a challenge.

Deweye was shocked that the parrot was answering questions so easily. He moved forward, bowed slightly and asked the parrot in a croaking voice, "Um, well, then, why don't you just fly away to south of the world?"

Mr. Littlejohn jerked his head in Deweye's direction and snapped, "Because I am here and I am the ruler of the Drawer of Forgotten Things. You will go back and leave me alone now, this instant!"

Deweye was horrified. He had not intended to anger the parrot, but he had. Before Skeedad had a chance to ask anything more, Mr. Littlejohn slammed the drawer shut.

It was a very long time before they spoke with the parrot again. He would scold them from outside the drawer but refused to open it when they asked.

The blackness of winter passed into spring, and they could hear birds twittering somewhere outside, far away. A time of great rain came and all they heard was water driving down from the sky. It was then that Skeedad decided to try again to befriend the lonely old parrot. He had a new plan.

Early one morning, Skeedad began calling for goodScout. He called and called. His voice was low and determined. He believed that eventually the dog would hear him and would come to investigate.

"GoodScout! Over here, it's this way," he called softly.

After what seemed like a very long time, goodScout trotted into the room and sat by the desk with her head cocked to one side listening to her name being called.

Mr. Littlejohn screamed at her to go away. He told her she was a bad, horrible dog and that she had no business sitting by the old desk. GoodScout whined, not understanding. But, there it was again. Someone was calling her.

"GoodScout! Come here...can you help us?"

How unusual, thought the dog. She glanced up at Mr. Littlejohn who glared down at her.

"GoodScout, it's us. You remember, from the Blue Water."

GoodScout did remember the lake. She remembered the stone from the lakeshore. Yes, yes. That was it. It was the stone's little voice calling her from within the desk drawer. Then she remembered the time when she and Lottie were lost in the storm and those same little voices helped her find the way back. It had been the stones who spoke to her that day.

All of a sudden the old parrot decided to work the drawer open with his beak. He was furious!

"Why, why, why must you persist in annoying me? Don't you get it? You are in this Drawer of Forgotten Things because you are forgotten. That means you are nothing, and you need to behave like nothing. It means you must say nothing, you must do nothing, and you must sing nothing. You are nothing! Nothing, nothing, nothing but forgotten things," Mr. Littlejohn shouted.

GoodScout looked on in amazement. She did not budge.

"Please, we just want to ask you some questions. Please don't shut the drawer, Mr. Littlejohn," Skeedad pleaded.

"Well, for pity's sake, what is it? What, what, what do you want to know? Be brief!" The parrot was exasperated.

"Um, for one thing, we want to know—" Deweye, who was hiding behind Skeedad, began.

Skeedad interrupted Deweye, "—we want to know if you will assist us in finding the way back north."

"What? Are you seeds and nuts? Why in Africa's name would I do such a thing," the parrot screeched while goodScout continued to watch in disbelief.

"Well, I mean, don't you ever want to go home? To the south? To the deep forests of Africa? Don't you have friends and family there?" Skeedad asked cautiously.

Unexpectedly, the parrot looked away for a moment. He gazed out the window at the steady warm rain of late spring and said nothing. When he did speak, his tone changed slightly.

"They have forgotten me by now," he replied quietly.

"Oh, surely not! A regal parrot such as yourself? How long have you been gone?" Daysee asked.

Mr. Littlejohn continued to stare out the window and muttered, "A lifetime."

"What's it like? The south," asked Deweye carefully, afraid he might say the wrong thing again and make Mr. Littlejohn angry.

"Ah, it's paradise. Warm and abundant. It rains the most fragrant rain you could imagine. And colorful. Bright and so much green." The parrot's mood had become pensive, almost forlorn.

"Yes, yes, but what about Lottie? She is the great granddaughter of Brookdweller, your friend. Why are you so mean to her? You're just always so—" Deweye said without thinking. He stopped himself and cringed slightly, waiting for a reproach.

But it was as though the old parrot could only hear and see memories of his youth at that moment because he did not answer Deweye. Instead, he spoke almost to himself, "Humans came to the forest one day. They had giant brooms and began beating the trees. When I tried to fly away, I was captured in a huge net and stolen away from my homeland. So young. I was so young. The other Greys got away. Then, I was smuggled north of the world for a huge sum of money. It was horrid. Just horrid. I lost my territory, my family, my paradise. I would have lived many, many ages there."

Slowly he turned and looked directly at Deweye. He seemed to finally hear his questions. Deweye hid behind Skeedad, waiting for the on-slaught of scolding.

"Brookdweller, you call him? Hmmm. Yes, he was a friend," Mr. Littlejohn said. "A human, yes, but a friend, nevertheless. He was old, as I am now. He rescued me from that ghastly pet store where it was said that I was too mean to touch. He didn't pay one penny for me. He walked in and said, 'you're a long way from home, fella,' which was true. He saw a speck of good in me from my life before and ignored the CAUTION sign on my cage. He let me be myself and treated me with great respect. He told the shop clerk that it was a tragedy that I was not left in my natural world. That pet store was glad to get rid of me. He brought me here. To his library. This was the place he spent his days. We became the best of friends. But then, he died, and I had nothing. Just nothing."

A dark gloom settled on him as he turned to look out the window again. Rain water trickled down the window pane in chaotic streaks. The sky was heavy and colorless.

"But what about Lottie?" Deweye asked. "She seems to want to be your friend. She's quite tender hearted and bright for a human."

"Oh, yes, her. I don't know," Mr. Littlejohn replied.

"And there's goodScout, the kindest of all dogs," Skeedad said.

The big yellow dog wagged her tail when she heard this. The parrot gazed at her as though seeing her for the first time. The dog had never caused him any harm.

"Mr. Littlejohn, if you sing a song and act as though you feel cheery, before you know it…well, you will be," Daysee exclaimed, adding, "That's what we do."

"Yes, yes, I know. The song. The stupid song," he groaned, growing restless. He didn't want to talk any more. He just wanted to watch the rain fall.

He ended the conversation without another word, shutting the drawer more slowly than usual.

When Deweye tried to protest, Skeedad hushed him to silence. He knew Mr. Littlejohn needed to retreat to himself and ponder many things.

GoodScout remained beside the old parrot, sensing his need for her quiet presence. She settled on a small rug for a nap. She would do this many times before summer. Mr. Littlejohn became accustomed to her company, taking comfort in the dog's goodness.

Many days and nights passed and still Mr. Littlejohn stared out the window. He thought of his youth, of his home. Never could he return. So much time had passed. Was it true what OldSkogmo had told him after Brookdweller died? He had said to look for the good, no matter what. If you stopped looking for good, you would die inside. The seashell, Daysee, told him to act cheery, to sing a song, and that he would begin to feel cheery. It seemed absurd. Ridiculous, really. He looked at goodScout napping on the floor below him. She was a human's pet, and yet...

He thought of Lottie. The little girl had spent countless hours talking to him, bringing him special treats and toys. She had the same eyes as Brookdweller. Eyes that could look into a creature's tortured soul and heal the hurt. She had not been the one to bring him here. She would not take him back south. He belonged here now. It had been his home for a long, long time. He had just stopped seeing any good in it.

He stretched his wings and lifted his head high. The rain had stopped and there was a yellow sun illuminating the fresh green leaves of early summer.

"La, La, La, La, La, La, La, La, La, La La-Lottie, Lala Lottie!" He sang so loud, he startled himself. GoodScout jolted awake and stood at attention.

"Oh lalalalalalalaLottie," he sang out again and again. Soon there was a muffled chorus coming from the Drawer of Forgotten Things and even goodScout yipped along.

"Lalalalalalalala Lottie, lalalalalala Lottie, lalalalalala Lottie, la la la la Lottie!" It wasn't much of a song, but it was a start. Sure enough, Lottie appeared in the room, her eyes wide and her mouth open in astonishment.

"Mr. Littlejohn?" she exclaimed. "Are you calling me?"

"Yes! Yes, I am! La, la, la, la Lottie! Gooooood morning!" he replied. He was elated that she had appeared so quickly.

"Gosh, well okay. Good morning to you, too, Mr. Littlejohn," she said, still surprised. She bounced over to him and stroked the top of his head, something she had not done in years.

It was as though he had never said one hateful thing to her. She stretched out her arm for him to climb onto, which he did gladly. He was very proud and continued to sing, "La, la, la, Lottie, my friend Lottie. My friend, Scout." This felt great! The dog trotted along as they pranced through the room. Mr. Littlejohn could not remember when he had felt so jubilant.

During the day, Lottie dashed in and out of the room bringing him treats and new silly toys, but mostly her company. She talked to him earnestly, the way one would talk to a good friend who had been away for years. It was unbelievable to him. At the days's end, she covered his cage with a blanket the way she had long ago. The cage door was left open, as usual, but she had covered the cage. It was a wonderful thing.

After dark, he thought once again of his home, far, far, south of the world. No, he would never go back. And although he would always miss it and always have that sadness, he would be fine here with Lottie and goodScout.

"Thank you, OldSkogmo, Thank you, Skeedad, Deweye, and Daysee," he whispered, "Thank you for teaching me to look for the good."

"You're welcome," came the immediate hushed reply from the closed drawer.

"I will never return south, but for the sake of all that is good and right, you will find your way north. And I will help you," he exclaimed in a voice strong with resolve. He, too, had a plan.

"Yippee! We are going home! Did you hear that? Home," shouted Deweye. "But how? How? How? I just want to know how. You must tell us. Because the way north is far, far away—"

"Shhhh—hush! You must not let the humans hear you. They will come in here thinking it is me chatting in the night. I do have a plan. It will work," replied the parrot. "But everyone must be very careful and do exactly as I say."

In his most quiet voice, Mr. Littlejohn called goodScout into the room. After a moment, she came in and sat by the desk, looking at him expectantly.

"Scout, there is something you must do," Mr. Littlejohn said. He spoke easily and expertly as though he had thought this plan over for a long time.

"Go to Lottie's room. Find the small travel case. The one she always packs for trips. Bring it to me. But, don't let anyone see you. Be very, very quiet. You must not awaken the child. Understand?" He spoke simply and clearly so there would be no mistake what he wanted goodScout to do.

GoodScout wagged her tail, spun around and pattered softly out of the room. The instant she was gone, Mr. Littlejohn began prying the Drawer of Forgotten Things open. He worked hard at opening the drawer as wide as he could. He looked at Deweye and Daysee. They were at the front eagerly anticipating the way out.

"Where is Skeedad?" the parrot asked. He saw that the stone had stayed to the back.

"We cannot leave OldSkogmo," Skeedad said firmly. "He must return to north, too. It's his only chance. I will not leave him."

"Yes, I know. I thought of that, too. If I could just get this drawer to open wider…" Mr. Littlejohn struggled with the drawer. It had been one thing to pry it open enough to scold them, but opening it wide enough to get to OldSkogmo was almost impossible. Halfway open, the drawer stuck. He could not get it to budge another inch.

"Can you push him to the front?" Mr. Littlejohn wondered aloud. OldSkogmo was at the darkest back corner.

"We can try," Skeedad said. They could. They could try, he thought, knowing that OldSkogmo was considerably bigger than any of them. Skeedad, Deweye, and Daysee exchanged looks. Without saying a thing, they began to make their way to the back of the drawer.

Mr. Littlejohn watched as they disappeared into the darkness of the Drawer of Forgotten

Things. He was doubtful but listened intently as they tugged and pushed and shoved and prodded. He heard grunts and groans and pleading.

They were getting nowhere, but there was no stopping them. Their determination was so powerful, Mr. Littlejohn started to believe they could do anything.

"It's no use," Deweye wailed from the back of the drawer. "We can't push him!"

"Now we will never find the way north," Daysee cried.

"Just hold on, you two. There must be something," Skeedad said.

He was interrupted by goodScout who entered the room dragging a travel case. She lifted it every few steps in an attempt to make less noise. Mr. Littlejohn was pleased to see that it was the right bag.

The dog froze when all of a sudden a light was turned on somewhere in the house. They heard footsteps coming toward the room. They all held their breath.

"Drop! Pretend to be asleep," Mr. Littlejohn hissed.

The footsteps came closer and closer. Suddenly a switch snapped on flooding the room with light.

"Anyone in here," Olive asked uncertainly. She clutched her robe to her and peered around the room, her gaze falling on goodScout. Rubbing her eyes, she reached down to pet the dog just as Mr. Littlejohn started to shriek.

Startled, Olive turned off the light and stepped back. After a moment, she switched on the light again and said, "What is wrong with you? You old coot. You scared me to death!"

"Bird sleeping! Bird sleeping! Bird sleeping!" Mr. Littlejohn screamed, trying to sound indignant.

"Oh goodness! Okay, Mr. Fussy Bird, I'm leaving." She glanced around the room again. For a moment she stared at the suitcase next to good-Scout, then shrugged. She patted the dog's head, turned off the light and left.

Skeedad, Deweye, Daysee, and goodScout waited for Mr. Littlejohn to speak.

"Excellent job, Scout!" he finally whispered. "Now, Scout, can you jump up and put your front feet here?" He pointed his beak to the drawer entrance, which was still halfway open.

GoodScout tipped her head to one side as though trying to understand why she would be asked to do such a thing. She wanted to help her friends. After all, hadn't they shown her the way out of the storm at the lakeshore?

Her hesitation caused Mr. Littlejohn to become anxious. Did she understand what he was asking? He tapped the drawer in repeated motion with his beak and even placed one foot inside the front of the drawer.

"Scout, please. Just slap your front paws right here. Nothing to it. You know, like when you were a puppy and humans told you not to jump on things," Mr. Littlejohn said.

Of course she understood. And, as the dog jumped up landing her big front paws on the drawer, it tipped slightly, hurling everyone to the front. Skeedad, Deweye, Daysee, and even OldSkogmo flipped and rolled altogether to the entrance.

"It's a miracle!" Deweye shouted. "Yippee! We are going—"

"Shhhhhhh! Deweye! Please...you must be very, very quiet. In fact, you cannot speak again until you arrive north," the old parrot cautioned. "Scout, listen. You see these four? Take each one, one by one, very, very carefully and place them in the travel case." Thankfully, the case was open and partially packed with vacation clothes.

GoodScout did as she was told. She began with OldSkogmo, the largest, then Skeedad. She was ever so gentle and placed each stone deep into the bag, hiding them under the clothing. She picked up Deweye with her little front teeth, barely touching him. For a split second she let go, and he drifted softly to the inside of the bag. With her nose she rooted around, lightly pushing him beneath the clothes.

"Okay, Scout, now you must be the most gentle you have ever been in your entire life. Daysee is strong of heart, but so fragile you could crush her," Mr. Littlejohn said.

The dog hesitated. She opened her mouth and scooped up the seashell with her tongue. It was easy. Hiding her in the travel bag was a bit tricky, but the dog did as she was told and lifted a small corner of clothing so the little seashell could hide securely in the softest compartment.

Done. It was done! Mr. Littlejohn hopped down from his perch onto goodScout's head and tapped her nose lightly with his beak. "You have done a very good thing, Scout. Now, you must take the bag back where you found it. Very, very quietly."

The dog gazed up at the old parrot sitting on her head and wagged her tail happily. She wanted to bark but knew not to. Everything had gone smoothly.

As the parrot climbed back up on the desktop, he stopped for a moment at the Drawer of Forgotten Things, shutting it one last time. Then looking back at goodScout he said, "See you again in the autumn, my friend."

For a moment he felt a little sad and whispered to the others, "Farewell brave lake creatures. Farewell OldSkogmo, may you awaken one day and know that there was good here after all. Farewell my dear friends, Skeedad, Deweye, and Daysee."

©C.J. Clark

- 8 -

THE WAY NORTH

 "Whisperleaf?"

"Yes, Skylo?" came the faint reply.

"Were you sleeping?"

"No, just resting a bit." The great cottonwood sighed, gazing down at the stones. The small red stone looked up at her inquisitively. All the stones seemed to be looking at her.

Her leaves were sparse and dry, her branches brittle. It had been a hard winter of deep snow and fierce blizzards, but she had survived. She had survived to greet the soft blue and lush green of summer, a shadow of her former self.

Whisperleaf watched as Meadowsteps, the white-tailed deer, drank deeply from the lake with her new fawn by her side. The deer looked up, blinking her giant brown eyes. She was teaching her fawn the ways of the lake creatures.

In the silence of the shallow water stood the ever solitary blue heron. He was motionless except for an occasional glance back at the shore to make certain that his presence was appreciated.

Whisperleaf looked to the east. Early dawn had slipped away. The new day's sun rose into the sky, sprinkling shimmering dots on the lake, which was very still. Every sound was magnified in the quiet of morning. The cottonwoods on either side of her had grown to an impressive size and could rustle their leaves without the slightest breeze. She felt content to hold on to the good earth with the younger cottonwoods while birds, insects, and little animals darted by. Each day unfolded differently from the one before. And yet all the days were the same. Whisperleaf rested.

"Whisperleaf?"

"Yes, Skylo?"

"What is it like to be so tall? To be able to see so far and feel the breezes so high?"

"Good. It is good. There is so much. The blue of the sky vanishes when I touch it. The clouds are just out of reach. The sun rises up over the woods in beams of broken light, bit by bit until it is full in the sky. Birds fly so close that I could tickle them. And the stones. There are stones as far as I can see. Big stones, little stones. Stones of a lakeshore, a lakeshore of stones."

Together they listened to the hushed morning. In the far distance they could hear geese calling. Whisperleaf swept her lowest leafed branch softly across the stones.

"Skylo?"

"Yes, Whisperleaf?"

"What is it like to be so ancient? To be, and to be, and to be, for so very long…to slumber under ice in winter and to be washed with warm water in summer, to see and to see and to see so very much for so very long."

"It is good, Whisperleaf. The sun always rises in the east and sets in the west. The moonlight shelters us at night. Every good thing is within us forevermore. The caw of the crows, the call of the geese, the loons and herons, the breath of every breeze that blows, the change of leaves in the spring and autumn... to be old with the elders, and new again with the young. All at the same time."

"Look over there, Whisperleaf," Skylo continued, looking to the lake as a flock of Canadian geese landed in the water. Slowly, the geese began to drift towards the shore. "It's Silverdew, with his two friends and their young goslings. Silverdew has many charges now. As their protector, the extra responsibility will change him some. He has truly become himself, no longer a daft young goose. It is the course of things. It is good."

The small flock of geese swam leisurely into shore, then waddled onto the sandy bank making cackling sounds. The little goslings awkwardly followed their parents to the shade under the cottonwoods. Silverdew stayed slightly behind, scanning the sky and horizon.

"Gaak back. Gaak, back home," he bellowed and stood tall, flapping his wings. He was pleased to bask in the peacefulness of Stony Point. He greeted everyone.

It was one of those days that lingered softly, like sleep. A northern summer day was like floating on a gentle wave, warming away the last chill of winter. There was balance. There was harmony.

But suddenly the peacefulness was shattered when Solosong landed lightly in Whisperleaf's uppermost bare branches. He did his best not to disturb her. He began calling a persistent and familiar warning.

"Cau—cau—caution! Humans coming from the south! Cau—cau—cau—caution! Soon, very soon, close, very close, caution! Cau—cau—cau—caution!"

Then he was gone. All the lake creatures were startled. The blue heron lifted his giant wings. Gliding close to the water, he made his way to the creek without looking back. Meadowsteps and her fawn sniffed the air for a moment, then kicked up their heels and bolted back into the forest. Silverdew and the other geese waddled back to the lake and began to swim slowly away. There was a shrill call of a loon from somewhere near the creek. Then all was still.

It wasn't long after Solosong's warning that the big noisy car rumbled over the grassy driveway, coming to a halt beside the old cabin. The humans had returned.

"Wait! Wait for me!" Lottie called to goodScout, who was bounding ahead to the lakeshore. The big yellow dog began nudging and sniffing the stones. She wagged her tail and darted up the bank back to the child, coaxing her to come along.

Lottie charged ahead then abruptly stopped. "Look at that! Oh no, look at the old cottonwood tree!"

Aunt Olive joined them by the lake and together they gazed up at Whisperleaf. Finally Aunt Olive said, "This old tree has been here all my life and probably longer. I just hate to see it die. Mr. Peterson will want to cut it down, you know."

Lottie placed her arm around Whisperleaf's trunk and said, "Why? Why can't it just be left alone? If we weren't here, it wouldn't have to be cut down because there would be no one to see it. It would just die where it lived and in its own way." To Lottie, the tree was still more alive than dead and still grand and still beautiful.

Aunt Olive looked from Lottie to goodScout to Whisperleaf and sighed, "You know what, Lottie? You're right. I won't let anyone talk us into cutting it down while it still has a bit of life in it. It's a fine and dignified old tree."

GoodScout was still rooting around in the stones as Lottie and Aunt Olive went to unpack. It was a chilly day with hardly any breeze, causing their voices to echo across the lake as they laughed and talked, carrying luggage from the car to the cabin.

And then it happened.

"This is strange. Aunt Olive, how do you suppose these things got into my suitcase?" Lottie stood over the small travel bag, which she had just unpacked.

"What is it?" Aunt Olive asked, coming into the room.

"Well, here's Scout's stone and a little seashell and this huge pretty feather and look, another stone, even larger."

"You must have left them there from last year's trip, don't you think?"

Lottie knew that couldn't be. In fact, she remembered taking the smaller stone to school and telling the other children that it was a magic stone. It shows you the way north, she'd told them. They hadn't believed her, and she'd put the stone back in her grandfather's desk drawer. She didn't know what to think.

"It's a magic stone. It shows you the way north..." Lottie mumbled softly.

"What? What did you say?"

Just then goodScout trotted with a sense of purpose into the room. Remembering Mr. Littlejohn's instructions, she took the stone from Lottie's hand and strode to the screen door. She pawed at the door, which opened easily. Then she made her way down the grassy bank to the lakeshore. Lowering her head, she placed Skeedad gently by the water's edge near Whisperleaf, next to the other stones. Then, relieved that she had completed her mission, she rolled on her back next to the stones, kicking her feet in the air.

Lottie and Aunt Olive followed goodScout. They watched her in amazement.

"Let's go get the other things and bring them down here, too," Lottie said. It shows you the way north, she kept thinking. She ran back to the cabin with Aunt Olive behind her.

Aunt Olive picked up the larger stone, turning it over and over in her hand and said, "This stone looks familiar somehow. How could that be? Isn't that odd?"

But Lottie was still thinking of the magic stone. It shows you the way north. She pondered the wonder of it. How? How did it do that? She would never know. She only knew that she had tossed the stone back in the drawer calling it an ugly old rock and now, somehow, it was here. It was here, up north, where it belonged.

Carefully she picked up the feather and seashell and walked back to the water's edge with Aunt Olive. Together they placed Daysee, Deweye, and OldSkogmo beside Skeedad just below Whisperleaf on the lakeshore.

GoodScout looked up at Olive and Lottie and yipped in approval. The echoing sound momentarily broke the silence of the morning. She sat beside them then, gazing out at the lake. The water was smooth glass. Tufts of snow white clouds hung motionless in the sky. Lottie and Olive stood so close to the lake's edge that they could see their reflections mirrored clearly in the still water. The immense quiet made time stand still. For a long moment, the woman and child contemplated their reflections in the lake.

Aunt Olive saw herself reflected as a child again, taking a stone from the lake, wanting to make the pretty stone her own possession. Lottie saw herself as an adult, placing the stone back where it belonged. Neither spoke.

"Aunt Olive," Lottie finally said, "do you see?"

Aunt Olive nodded, understanding.

They watched as a tiny ripple of blue water circled the stones. Instantly their vibrant colors were restored. The blue water stones shone like precious jewels. The seashell, too, glistened as bright as the north star. And the goose feather, even with no breeze, seemed to flutter lightly, its tones distinct and soft.

"The stones, the seashell, and even the feather are all beautiful once again," Lottie exclaimed. It really was magic! They had found the way north.

- 9 -

BEYOND FOREVERMORE

Olive awoke to early daylight timidly peeking through the cabin's windows. The raucous calling of a blue heron echoed somewhere on the lake. Shivering slightly, she noticed the previous night's warm roaring fire had become cold ash. She stared at her empty cocoa mug resting on the arm of the rocking chair.

What time was it? Had she been sitting here all night? Her mind was muddled. She'd had the oddest dream. Or no, not a dream exactly. It must have been her grandfather's story, which now lay scattered in her lap. She must have drifted off to sleep. That was it. She'd been reading and fell sound asleep.

She rubbed her eyes. Why had she been so upset the evening before? All at once, a certain peace she had never known before enveloped her. She felt sheltered by it as though it were a protective cocoon.

Blue water stones in the drawer of forgotten things, she remembered suddenly. This image stayed with her as she stretched and yawned. The Drawer of Forgotten Things. Her grandfather had told her that he didn't want to become like a stone in her childhood drawer of forgotten things, that he would feel useless and out of his element. Yes, it would crush his spirit, she thought.

Olive stood up, carefully tucking the story pages under her arm, she walked to her bedroom. She opened the top drawer of her dresser and took a long deep breath. Inside the drawer were stones, seashells, a peacock feather, the top of a cattail reed, a piece of driftwood, and what appeared to be a Canadian goose feather. Everything looked dull and worthless. Why had she kept all this? She gently removed the entire drawer from the dresser, being careful not to disturb anything. Quietly closing the cabin door behind her, she carried her drawer of forgotten things to the water's edge .

By the time she reached the lake, her grandfather had awakened and silently watched her from the cabin window. He watched as she sat on the beach and one by one, delicately placed each stone, seashell, feather, driftwood piece, and cattail reed by the water's edge. After she had done this, she stood up and studied the horizon.

The wind had shifted to the south, causing the waves to change direction. They sloshed into shore one after another after another, washing over the objects Olive had placed there. She bent towards the water, extending her hand to the waves as though to stop them, if only for a moment. For a long time she stood motionless, watching the waves. She didn't know that her grandfather saw and understood that Olive had discovered how to see good in even the smallest things, and then, too, she was learning how to let go, and let things be as they are.

She sat on the ground below the old cottonwood tree, patted the pages of Lottie's story into place and began to read the last of it. As she began to read, Olive felt as though she could hear her grandfather's voice deep inside the waves as they tossed droplets of water at her, dancing to the liquid laughter of the lake.

The Legend of the Blue Water Stones

Forevermore

Under the midday sun, a warm summer breeze began to dance about, taking on its own shape. It danced upwards, then down, then to and fro, disturbing nothing in its path. The little breeze stretched wide and long and zipped though narrow spaces, making a whistling sound. The whistling sound caused it to dance more. And the more it danced, the more joyful it became. It was not long before it began to playfully lift leaves and tiny twigs off the ground, tossing them into the air. The little breeze traveled from North Point along the shoreline, rustling leaves and stirring sand particles as it went. When it reached Stony Point, it was strong enough to scatter leaves.

It touched down by Whisperleaf and whirled happily upwards, lifting Deweye into the warm summer air. Up he went, overjoyed, and calling to his friends, "Farewell! Farewell!"

> *"It is I, Dew-Eye!*
> *Light and Swift*
> *I spin, then fly*
> *And dream to drift*
> *In the summer sky*
> *The breeze will lift*
> *Me up so high*
> *What a an honor*
> *To be Dew-Eye!"*

He began to twirl and dance with immeasurable happiness as he was whisked away far above Whisperleaf. It was all he had ever hoped for, it was his true path. All of Stony Point wished him a farewell as they watched him vanish into the blue sky.

Even Solosong made a rare appearance as Deweye flew away, belting out his favorite song as he went. Solosong sat in Whisperleaf's highest leafless branch looking down at all the lake creatures. It was a fine day. Skeedad had returned with Daysee and Deweye and a stone named OldSkogmo, who had been gone since Whisperleaf was a young tree, many crow lifetimes.

Whisperleaf, stooped and brittle, called to her friend, "OldSkogmo? Is it really you? You have finally returned?"

At that very moment, the old stone awoke. He tipped forward a little as his eyes began to open wide. He blinked, looked around, then smiled a huge contented smile. It was as though his long slumber had only been for an instant. His smile faded slightly when he saw that his friend, Whisperleaf, was very old indeed and yet very grand in her wisdom.

"OldSkogmo, you see, I am quite old now and so very weary. I don't have much time left…" she said, as though reading his thoughts.

"Good friend Whisperleaf." It was all he could say. They were the first words he had uttered in many seasons. Skeedad and Daysee were elated to hear their friend finally speak.

All the lake creatures wanted to hear the tale of their journey. Skeedad and Daysee wanted to hear about OldSkogmo's travels and his days with Brookdweller and Mr. Littlejohn.

Everyone knew that OldSkogmo told things best of all. They began to gather around, hoping he would begin. The sun was already low in the sky. Solosong was quietly perched in Whisperleaf's uppermost branches.

The blue heron, the whitetail deer with her fawn and the geese had all returned in the final light of day.

And so, as the yellow sun dipped to the horizon, causing the sky to burst into orange and pink, then fade to dusky lavenders, OldSkogmo thought of all the tales he would tell. He watched reflectively as the moon appeared in the early night sky. He thought of Brookdweller, of Aunt Olive as a young girl, of the obstinate but good hearted Mr. Littlejohn,.

He thought of the times before the Drawer of Forgotten Things and of the days within it. He considered the persistence and kindliness of Lottie and goodScout, both unusual for their species. Then there was the unforgettable Deweye. Never had there been such a creature. And then, he thought of cheerful Daysee and of noble Skeedad. There were tales to tell about the world within a human dwelling and tales to tell about still other journeys. All the stories would be told. All the stories would keep for another time.

A hush fell over the lake creatures as OldSkogmo began to speak. He looked again to the giant yellow moon rising in the darkening sky, his eyes twinkling as though he were about to reveal a long awaited secret.

"Ahem! Listen carefully, dear friends. Many moon times ago, long, long before Solosong, long before Meadowsteps, Silverdew and even our friend, Blue Heron, it was our good fortune to know a young cottonwood tree who would come to be with us here on Stony Point beside the blue water forevermore.

Although young, she was quiet and kind and asked many wise questions. She grew strong and tall and over many winters bent closer and closer to the earth, sharing her wisdom with all who listened. In time, we began to call her Whisperleaf."

And so it was. As twilight blanketed the day, OldSkogmo told the legend of Whisperleaf's life as she stood nearby, old and bent, but filled with pride and joy because she knew she would live in their hearts forevermore. And it would always be good.

The End

- 10 -

THE PERMANENCE OF WAVES

The day they were to leave for Kansas, Olive was unable to find her grandfather. Lottie and Scout were gone, too. She'd been so preoccupied with packing the car that she'd lost track of time and had not noticed them slip away. Slip away? Did they slip away, or had they wandered off? Had they told her they were going somewhere together, or had they just gone?

She looked out the window. Her grandfather's car was still there, so was his bicycle. Someone was walking toward the cabin. She couldn't quite make out who it was, just that the person was carrying some sort of object. She watched as a young woman came into view. The woman looked so carefree, dressed in jeans and a warm sweater jacket for the brisk spring morning. For a moment, Olive thought she recognized the woman. Surely she wasn't coming to tell her some news about her grandfather's whereabouts. Olive felt a sense of dread come over her.

She flung open the door before the woman had a chance to knock, causing her to jump back and drop a large basket of muffins on the ground. The woman scrambled to pick up the rolling muffins while Olive stood watching her.

"Sorry…" Olive said.

"Goodness, you startled me! Olive?" The woman pushed her jacket hood back to reveal a disorganized mop of white blond hair. Her eyes were as blue as the sky and her face was rosy from the crisp morning air.

"Sorry. Yes, I'm Olive. Have we met?"

"Well, it's been a long, long time. But yes, I'm Matilda Peterson. Our family's cabin is down by the creek." She pointed vaguely toward the woods. "You and I used to play together during summer vacations when we were kids." Seeing Olive's hesitation, she added, "It was really an awfully long time ago."

"Oh yes. Matilda," Olive said uncertainly. "I'm so sorry. It's good to see you again. Come in, come in." She stood back, opening the door wide, embarrassed at her own lack of hospitality.

"Oh well, thanks but I can't stay. I was just bringing your grandfather some muffins I baked this morning. They're blueberry, his favorite." Matilda continued standing on the front step and handed the basket of remaining muffins to Olive.

"You know, I can't find him this morning," Olive said suddenly looking toward the driveway, oblivious to the fact that Matilda had awkwardly shifted the spilled muffins to her other hand.

"What do you mean?" Matilda raised her eyebrows slightly.

"It's just that we are driving back to Kansas today, and I've been packing the car. I didn't notice him leave or remember either of them say anything about going anywhere. His dog is gone, too."

"Scout?" Matilda gasped. The missing dog seemed to alarm her. She put her free hand to her mouth.

"I really don't even know how long they've been gone. I'd wanted Grandpa to come back with us to Kansas. He's getting up there in years, you know, and I've been concerned about him being up here all alone. He's so frail. I guess I've just been so preoccupied with that. I don't know." She looked toward the driveway again.

Matilda followed Olive's gaze, then laughed. "Frail? I never thought of him as frail. We can barely keep up with your grandpa. He's an amazing man."

"I'm really pretty worried," Olive said, ignoring Matilda's comment.

"I'm sure he's fine. Don't you think they just went for a little walk or something?" Matilda answered.

"I don't know. I guess. I'd probably better go look for them though." Olive wondered why Matilda seemed so unconcerned. Sure her grandfather was an amazing man, but he was an old man nonetheless. He could fall or get lost or have a sudden heart attack or any number of mishaps. It was up to her to protect him, wasn't it? She just wanted Matilda to leave.

"Thanks so much for the muffins. Here, let me throw those away." She reached for the stray muffins that Matilda was still holding.

"Okay. Look, if you can't find him let us know. Our phone number is right next to his phone. He wrote it on the wall, actually. I'm sure he's just fine. Maybe if you wait a little while, they'll be back." Matilda turned to go and added, "Nice seeing you again. Next time you're here, come down for coffee. You're always welcome."

Olive barely heard this. She felt as though she was colliding with a world where it was no big deal when people wandered off without a word to anyone and where it was okay for old people to live alone with no one around for miles. A world where freshly baked blueberry muffins arrived at your doorstep with great nonchalance marked by references to the past and vague invitations for the future. She waved at Matilda, then went back inside to find her hiking boots. She had to unpack her suitcase to find them. As she was crisscrossing the shoelaces, she wondered if she might be overreacting. Maybe Matilda was right. She should wait a bit. They might return at any moment.

She stepped outside just as a noisy flock of Canadian geese soared high above her in the light blue sky. She watched as a few of the geese broke the V formation. Their flight looked suddenly jumbled and disorganized. And yet they continued on, in their forward direction, honking loudly.

Olive inhaled deeply. At that moment she felt there was nothing more exhilarating in the world than a crisp cool Minnesota morning in the heart of Otter Tail County. She felt energized. What could it hurt to go look for her grandfather and niece? And besides, what if they were in trouble? If he had only told her where they were going, she wouldn't have to wonder.

She followed the long winding driveway to Stony Point Trail and then stopped. Which way should she go? She tried to think like her grandfather, then smiled to herself realizing how impossible that would be. Someday, somehow, she decided right then and there, she would spend more time up here in these north woods. She would learn everything her grandfather knew, all the things he had tried to teach her over the years. She would ask him to tell her his stories again. He was such a wonderful storyteller.

Years ago he'd shown her a trail through the deep woods that he'd marked with pieces of painted rope tied around tree trunks. She'd been fascinated with the trail's twists and turns through underbrush and ravines, how tall the trees were, how dark it became the deeper into the woods they went. Then the sunlight would peek through as they reached a partial clearing. She would feel like heaven itself had opened to them. Never once had she felt uneasy. Her grandfather knew the way. He was the smartest, most capable man she had ever known. A momentary pang of grief filled her heart as she thought of him helpless and lost in these same woods.

Olive cut across Stony Point Trail into the woods, careful to step high over a sagging barbed wire fence. She noticed a tree with a piece of red painted rope tied around it. Confident that this was the old trail, she set out to find the next tree marker. It wasn't exactly as she'd remembered it. There were trees down and decaying from a tornado that had ripped through the area a year before. One of the trees lay on the ground like a fallen soldier with its red rope marker still tied to it. Downed trees obscured the tiny path, and she had to look hard into the woods for the next trail markers.

She stood staring for a moment feeling almost blinded by the monotony of tree trunks and under-brush. When she finally spotted a marker, she noticed another one several feet ahead of it and still another as the path took a turn into a sort of low grassy area. She stepped over a huge fallen moss-covered tree, startling a black squirrel who scampered out to scold her. Three white-tailed deer bounded ahead of her, disturbed by her presence. She watched as they disappeared into the deep woods, amazed at how quietly they escaped her view.

Her grandfather had often told her that time stood still in the woods. Not the time of day or time of year, but the idea of time itself. Olive had paid little attention to his words then. But now, alone on this unfamiliar trail, she understood fully what he'd meant, and it amazed her. She meandered along picking up her pace when she spotted another marker.

At first it seemed curious that the painted rope color had been changed to green, but she decided that he must have run out of the red paint. The green was harder to see and the trail became nearly nonexistent. Just when she thought she'd lost the path, it would reappear and there would be another green rope marker to identify it. It was odd though. Olive didn't quite remember the trail leading into a more sparse birch woods, nor did she remember it cutting through tall reeds and cattails along the back water of the lake. At one point, she came so close to the water that she frightened a flock of mallards that rushed upwards, squawking angrily.

Then it dawned on her that something was terribly wrong. The trail markers were further and further apart. Some weren't even painted. It disturbed her to think that perhaps her grandfather had wandered out here aimlessly marking a trail that led nowhere. She'd been more worried about his aging physical condition than his mental capacities. He'd always been sharp as a tack, an articulate thinker. Was it possible that his mind was becoming clouded and dulled with age?

Olive continued onward in the direction that seemed most likely to her. She glanced up at the sky, trying to determine time of day and direction. How long had she been following this trail? There were over a hundred acres of woods all around her. Could Lottie and her grandfather be just up ahead? Deep inside she felt a whirl of panic begin to rise. She started to turn back, but everything looked the same. Refusing to give in to the panic, she stopped. Okay, so I'm a little turned around. This shouldn't be difficult. I can do this, she thought.

Suddenly a memory of her mother swimming ahead of her in the lake took hold of her. The lake water shimmered in the bright sunlight. Her mother was wearing her favorite blue one-piece bathing suit, which made her creamy white skin appear even more blanched. Her white bathing cap with the little yellow flowers made her look like an egghead. She swam easily, kicking her legs, swimming further and further away. She rolled onto her back and bobbed up and down in the waves, motioning for Olive to follow. Wait, wait for me, Mother. But the memory faded, and Olive was alone again in the woods. Her mother had always loved the water more than the woods.

Olive closed her eyes and thought about the waves caused by her mother's swimming rolling into the shore against the blue water stones. A wind was beginning to stir the treetops and a chill had returned to the air. As if she could will herself to find her way back, she visualized the blue water stones along the beach. She decided to walk into the wind. Even as she did this, the dull sound of a barking dog sounded far in the distance.

"Scout? Scout! Scout! Scout!" she yelled. Somehow her shouting seemed ridiculously futile. She stopped abruptly. She began to walk briskly toward the direction of the barking dog, listening so intently that she lost her footing. She tripped and fell forward, landing face down in a bramble of spiky underbrush. Olive wanted to cry. How could this be possible? She couldn't be lost. She didn't bother to get up for a long time.

When she did get up, she discovered that her jeans were ripped and her left shin was lacerated and bleeding.

She groaned. With her shirt sleeve, she mopped at the blood oozing from her wound. As she did this, she realized that the barking dog sounded even closer than before. When Olive stood up, a pain shot through her leg. She gained her footing by leaning heavily on a large cottonwood tree. She listened. Was it her imagination or was the dog coming closer?

"Scout! Scout!" she shouted. Just then it occurred to her that maybe the barking dog wasn't Scout. Maybe it was some rangy vicious wild dog. Or a wolf. Was that possible? Wolves? In Otter Tail County? She laughed out loud, but then stopped, shuddering slightly when she remembered the coyotes she'd heard howling in the night. She put her head in her hands. I've got to get a grip here, she thought.

There was a faint rustling in the nearby underbrush. Olive took shallow breaths, covering her eyes with her hands.

"Olive?"

"Grandpa?" Olive jerked her head up to see Scout bounding over to her, her grandfather and Lottie close behind. They were coming from the opposite direction, which made Olive feel disoriented. Had she found them or had they found her?

"Aunt Olive, what are you doing way out here?" Lottie asked, clearly puzzled. "Oh my gosh, what happened to your leg?"

But her grandfather had already begun to apply a makeshift bandage to Olive's leg. "It looks superficial. We need to get you back so we can clean this up." He took her by the arm and led her forward, testing her ability to bear weight.

"Wait. Wait. This is nuts. You shouldn't be helping me. I came out here looking for you three." She looked from her grandfather to Lottie to Scout. They looked fine. It was obvious they were not lost and that they had not been lost at all.

"Looking for us? Well for heaven's sake, why," her grandfather asked, truly astonished.

"I couldn't find you anywhere, Grandpa. When I was finished packing the car, I looked for you. I was pretty worried, because, you know..." Olive hesitated. It all seemed foolish now. She was the one who was lost and injured, and they were the ones helping her find the way back.

"Well, I'm sure sorry, Olive," her grandfather said.

"But I followed your old trail markers. The rope color changed from red to green and then there were some that weren't painted any color at all. It was confusing." Olive brushed bramble twigs from her hair.

"Green," her grandfather said thoughtfully, stroking his chin. "The green rope marks an old game trail. The red markers are the hiking path. You must've taken a wrong turn. You wandered way off course."

Olive moaned. Of course. He had marked two trails. She shook her head, not knowing what to say.

"Grandpa was showing me the big birdhouses he made for the herons. They're gigantic," Lottie said.

"We were just across the road in the woods...not far at all. I'm surprised you didn't see us," her grandfather added. He took his cap off and placed it on Olive's head the way he had when she was a little girl.

Olive patted the cap into place and laughed. Here was her grandfather coming to her rescue. Why had she been so worried about him? He was as solid and constant as he had ever been.

For the rest of her days, Olive would remember this moment. And she would always remember driving away with Lottie, smiling and waving good-bye to her hero, her great grandpa, standing beside the cabin, old and bent but wise and capable, with Scout by his side. She would look past him at the waves rolling steadily into shore, splashing over his blue water stones, memorizing it all.

The Permanence of Waves
was written for all ages.

The drawings are submitted
in remembrance of the
author's childhood at Stony Point.

Drawings by C.J. Clark

Drawings by C.J. Clark

TO ORDER THIS BOOK

The Permanence of Waves

If not available at your favorite bookstore,
order from

LANGMARC PUBLISHING

P.O. Box 90488 • Austin, Texas 78709-0488
1-800-864-1648
www.langmarc.com • e-mail: langmarc@booksails.com

- -

Please send payment with order:

_____ copies of *The Permanence of Waves*
at $14.95   _____
Sales tax (Texas residents only) 8.25%   _____
Shipping $2.50 1 book; $1 additional books _____
Amount of check enclosed: _____

Or Credit card: _____
Expires: _____   small 3-4 digit no: _____

Your name: _____

Address: _____

_____

Phone number: _____
e-mail: _____

FOR PRINTS OR BOOKS YOU MAY CONTACT

C. J. CLARK

cjclark144@att.net
1420 Cypress Creek Road, Ste. 200-144
Cedar Park, Texas 78613

LaVergne, TN USA
25 August 2010
194607LV00002B/9/A